Nanotechnology for Producing Novel Cosmetics in Japan

Edited by

Kunio Shimada

Copyright © K. Shimada, 2009
All rights reserved. No part of this book may be reproduced in any form, by Photostat, microfilm, retrieval system, or any other means, without the written permission of authors.
Original Japanese language edition is published 2007.

ISBN978-4-7813-0113-6 C3047 ¥30000E
Printed in Japan

CMC Publishing Co., Ltd.
1-13-1 Uchikanda, Chyoda-ku, Tokyo   101-0047 JAPAN

Preface

Nanometer is one billionth of one meter. Figuratively speaking, one nanometer is "the size of one apple if the radius of the earth is one meter" or "the size of one grain of sand in Tokyo Domed Baseball Stadium if the stadium is one meter". Economists predict that the market of nanotechnology products will expand to $273 billions by 2010. Nanotechnology is expected to particularly accelerate biotechnology and IT developments in Europe and Japan, respectively. However, nanotechnology is already widely used in the field of cosmetics, in which new raw materials are introduced quickly. The cosmetics market is small. In Japan, the annual sale is only $15 billion. Despite of the small size, the cosmetics market clearly reflects the stability of the nation and the economic strength of people. Therefore, it was cosmetics that used nanotechnology ingredients first in Japan.

Let me overview changes in the period. In the 1960s, when I was born, the economy grew rapidly in Japan. Olympic games were held in Tokyo, and the Shinkansen (super express train) line was completed. Today, eight years after the start of the 21st century, much seems to have changed. Probably, much of what is remaining from the 20th century will disappear in ten years. Generally speaking, it needs fifteen years for "centuries" to change, like the warmth and light of the day remaining after the sunset for a while. Two centuries ago, in 1815, Victor Marie Hugo wrote in "Les Miserables" that Waterloo was not just a historic battlefield but it changed the history in Europe. The Industrial Revolution started in 19th century after this last war of Napoleon Bonaparte. I think we are today in such as a turning point. I have worked in the cosmetic field for long. I decided to edit this book because we should review nanotechnology in our technological history.

When Dr. Daisaku Ikeda (SGI President), who is my life-long teacher, talked with Dr. Linus Carl Pauling, who is the father of modern chemistry, Dr. Pauling said "About ten million different substances have been discovered or created by chemists and other scientists and many of their properties have been described. Their structures are now so well understood that chemists and physicists can sometimes predict the kinds of substances that must be produced in order to achieve certain desirable properties. The modern world has been shaped by discoveries made by physicists and chemists during the last century or two. As long as politicians and other people in power refrain from hindering them, those scientists can continue contributing to the well-being of all peoples. (quotation form A Lifelong Quest for Peace: A Dialogue)". I believe that the mankind has just started enjoying the profits and gifts of cosmetics nanotechnology. It would be a great pleasure if this book would provide information and knowledge to young and other scientists working in the field of cosmetics.

Lastly, I want to appreciate Mr. Hirokazu Tomoda, an editor of CMC Publishing Co., Ltd., and all parties concerned for the help in editing this book.

16 March, 2009
Kunio Shimada, Ph.D.

On publishing an English version

This book is a translation of "*Keshohin Kaihatsu To Nanotekunoroji* (Production of Cosmetics and Nanotechnology)" published in Japan in 2007. The original book has been favorably received by those engaged in related researches and other fields since it appeared on the market. There have been cries for sharing the useful information contained in this book with oversea researchers, and we decided to publish an English version. There has been no other book in which world-leading researchers and scholars on cosmetics write on their R&D technologies, and we expect that this book would be a help for researchers of associated fields.

The sentence of this book is translation of the Dr. Kunio Shimada. There may be difference in the nuance with the meaning of original, but please understand it for the property of translation.

CMC Publishing Co., Ltd.

Editor (translator)

Profile
Kunio Shimada, Ph. D. is the General Manager of the Life Science Products Division of NOF CORPORATION.

He has been at the present post since 2000, after acting as the group leader of evaluation studies on 2-methacryloyloxyethyl phosphorylcholine polymers (product name: Lipidure®) at Tsukuba Laboratory of NOF CORPORATION. He is a member of the permanent committee on the Society of Cosmetic Chemist of Japan (SCCJ) and won the Editor Award of the Japan Oil Chemists' Society (JOCS) in 2000. He entered NOF CORPORATION in 1997 after working for a cosmetics company. He graduated the Graduate School of Engineering of the Oita University. He was born in 1955 in Tokyo, Japan

Books
Production of Cosmetics and Nanotechnology" (CMC Publishing Co., Ltd.)
"Choosing Additives and Formulations and Responding to the Latest Regulations (in Japanese)" (Technical Information Institute Co., Ltd.)
"Gel Troubleshooting (in Japanese)" (Johokiko Co., Ltd.)

List of Contributors

Toshiyuki Suzuki	Kao Corporation
Hiroshi Fukui	Shiseido Co., Ltd.
Takahiro Suzuki	Nihon L'Oreal K.K.
Kazuyuki Takagi	Mizuho Industrial Co., Ltd.
Akihito Shundo	PRIMIX Corporation
Mitsutoshi Nakajima	University of Tsukuba
Isao Kobayashi	National Food Research Institute
Tohru Okamoto	Shiseido Co., Ltd.
Sadanori Ban	Nippon Menard Cosmetic Co., Ltd.
Yuji Sakai	POLA Chemical Industries, Inc.
Toyokazu Yokoyama	Hosokawa Micron Corporation
Hiroyuki Tsujimoto	Hosokawa Powder Technology Research Institute
Takatsugu Yoshioka	Iwase Cosfa Co., Ltd.
Keiko Iwasaki	Iwase Cosfa Co., Ltd.
Eiichiro Misaki	Kao Corporation
Kunio Shimada	NOF Corporation
Asako Mizoguchi	NOF Corporation
Hiroshi Oda	NOF Corporation
Yuka Shimoyama	Showa Pharmaceutical University
Makiko Fujii	Showa Pharmaceutical University
Yoshiteru Watanabe	Showa Pharmaceutical University
Setsuko Yamane	Tokyo Medical and Dental University
Kazunari Akiyoshi	Tokyo Medical and Dental University
Keiji Terao	CycloChem Co., Ltd.
Shinobu Ito	I.T.O.Co., Ltd.
Tomohiro Imura	National Institute of Advanced Industrial Science and Technology
Tokuma Fukuoka	National Institute of Advanced Industrial Science and Technology
Tomotake Morita	National Institute of Advanced Industrial Science and Technology
Masaru Kitakawa	National Institute of Advanced Industrial Science and Technology
Atsushi Sogabe	National Institute of Advanced Industrial Science and Technology
Dai Kitamoto	National Institute of Advanced Industrial Science and Technology
Nobuhiko Miwa	Prefectural University of Hiroshima

Yoko Yamaguchi	NANOEGG Research Laboratories Inc.
Rie Igarashi	NANOEGG Research Laboratories Inc.
Hidetoshi Kondo	Dow Corning Toray Co., Ltd.
Laurent Martin	BASF Japan Ltd.,
Toshiro Sone	Yakult Honsha Co., Ltd.
Takayuki Tsuboi	Ichimaru Pharcos Co., Ltd.
Kohei Watabe	Q.P.Corporation
Sadaki Takata	Shiseido Co., Ltd.
Kazuyoshi Tsubata	Nippon Menard Cosmetic Co., Ltd.
Hiroyuki Asano	Nippon Menard Cosmetic Co., Ltd.
Naoyuki Enomoto	JGC Catalysts and Chemicals Ltd.
Yoshihiro Yano	NOF Corporation

Contents

PART1 Introduction

Chaper1 Emulsification technologies for Skin Care Cosmetics
Toshiyuki Suzuki

1.1	Introduction	1
1.2	Basics of emulsification and recent technologies	1
1.2.1	Types and condition of emulsification	1
1.2.2	Production of emulsions and different emulsification conditions	2
1.2.3	Understanding emulsion and microemulsion using phase diagrams	2
1.2.4	Liquid crystals in emulsification and application of the D phase	3
1.2.4.1	Liquid crystal emulsification	3
1.2.4.2	D phase emulsification	4
1.2.4.3	Preparing nanoemulsion using microemulsion	4
1.3	Emulsification technology in recent skin care cosmetics	5
1.3.1	Multi-lamellar emulsions and gels prepared using self-organization lipids	5
1.3.2	Preparing makeup cleansing using bicontinuous D phase	5
1.3.3	High moisture emulsification using inverted hexagonal liquid crystals	5

Chaper2 Powder Technology in Makeup Cosmetics
Hiroshi Fukui

2.1	Introduction	8
2.2	Constituents of makeup cosmetics	8
2.3	Physical properties of powders	10
2.3.1	Extensibility	10
2.3.2	Adhesion	10
2.3.3	Absorbency	10
2.4	Optical properties of powders	11
2.4.1	Color correction	11
2.4.2	Responding to changes in photoenvironment	13
2.4.3	Concealing unevenness on the skin	13
2.4.4	Adding three-dimensional impression	13
2.4.5	Lasting of makeup effects	13

2.4.6	Blocking ultraviolet rays	14
2.5	Surface modification of powder	15
2.5.1	Coating with metal oxides	15
2.5.2	Surface treatment with oils, fats, metalic soaps, and fatty acids	16
2.5.3	Surface treatment with amino acid system compounds	16
2.5.4	Surface treatment with siloxane coating	16
2.5.5	Surface treatment with fluorine-based polymers	17
2.5.6	Surface treatment with biomaterials	17
2.6	Conclusions	18

Chaper3 Composite Powder
Takahiro Suzuki

3.1	Introduction	19
3.2	Shapes and physical properties of extender pigments	19
3.2.1	Aspect ratio and a powder shape index	19
3.2.2	Luster and scatter reflection	20
3.2.3	Bulk density and sensory evaluation	20
3.2.4	Adhesion of powder	22
3.2.5	Coefficient of kinetic friction of powders	23
3.3	Shapes and physical properties of coloring pigments	24
3.4	Compositing powders	25
3.4.1	Objectives of compositing powders	25
3.4.2	Coating the surface of flaky powder	25
3.4.3	Coating the surface of fine pigment particles	26
3.4.4	Composite powder of extender pigment and fine pigment particles	27
3.4.5	Light interfering pearl pigment	29
3.4.6	Compositing organic pigments	30
3.5	Conclusions	31

PART2 Application of Nanotechnology to Cosmetics [Equipment]
Chaper1 Nanoemulsion Manufacturing Equipment
Kazuyuki Takagi

1.1	Introduction	33
1.2	Use of emulsification technologies	34
1.2.1	Effective mechanical power for emulsification: Shearing	34
1.2.2	Emulsification by prescription and mechanical emulsification	36
1.2.3	Equipment for emulsification	36
1.2.3.1	Super-high-speed shear mixer	36
1.2.3.2	High pressure homogenizer	37
1.2.4	Emulsion polymerization	38
1.2.5	Preparation of nanoemulsion	39
1.2.5.1	Prescription example and a preparation method of nanoemulsion	39

1.2.5.2	Fat emulsion	40
1.2.5.3	Liposome	40
1.3	Supercritical method	41
1.4	Other nanoparticle manufacturing methods and processing examples	41
1.4.1	Manufacturing methods of carbon nanotubes and fullerene	41
1.4.2	Dispersion of fine particles of titanium dioxide	41
1.4.3	Novel methods for preparing medical supplies using nanotechnology	42
1.4.4	Use of polymer micelles	42
1.4.5	Bionanofiber (cellulose nano fiber)	42
1.5	Future of nanotechnology	42

Chaper2 Latest Mixing Technologies for Cosmetics Production
Akihito Shundo

2.1	Introduction	44
2.2	Conventional mixing technologies	44
2.3	Recent trends of cosmetics and demands for mixers	45
2.4	Thin-film spin system high-speed mixer "T.K. FILMICS®"	46
2.5	Principles and effects of T.K. FILMICS®	46
2.6	Examples of applications	48
2.6.1	Emulsification of liquid paraffin: sharp particle size distribution and particle size control	48
2.6.2	Solubilization of lotion	48
2.6.3	Liposome	49
2.6.4	Fine particles in cream containing polymer thickener	50
2.7	Conclusions	51

Chaper3 Development of Microchannel Emulsification Technology
Mitsutoshi Nakajima and Isao Kobayashi

3.1	Introduction	52
3.2	Manufacture of monodisperse emulsion using microchannel array	53
3.3	CFD simulation and analysis of droplet production	54
3.4	Development of silicon asymmetric straight-through microchannels and non-silicon straight-through microchannels	55
3.5	Application of microchannel emulsification	56
3.6	Conclusions	56

[Emulsification]
Chaper4 Fast, Simple and Easy Production of O/W Fine Emulsion
Tohru Okamoto

4.1	Introduction	59
4.2	Preparation of fine emulsion using techniques of surface chemistry	59
4.2.1	Inverse phase emulsification method	60
4.2.2	HLB temperature emulsification method	60
4.2.3	D phase emulsification method	61
4.3	Preparation of nanoemulsion using the condensation method	61

4.3.1	Preparation of nanoemulsion using the solubilization region	61
4.3.2	Nano dispersion wax	63
4.3.3	Nanoemulsion preparation using polyoxyethylene·polyoxypropylene random copolymer dimethyl ether	64
4.3.4	Preparation of nanoemulsion from uniform solutions	65
4.4	Nanoemulsion preparation using high pressure homogenizer	66
4.4.1	Preparation of nanoemulsion using water soluble solvents	67
4.4.2	Nanoemulsion of surfactant-amphiphile-oil-water system (milky lotion and cream)	68
4.5	Conclusion	70

Chaper5 Technologies for Stabilizing Emulsions
Sadanori Ban

5.1	Introduction	72
5.2	Types of emulsions	72
5.3	Preparation of emulsions	73
5.3.1	Emulsification using surface chemistry	73
5.3.2	Mechanical emulsification	74
5.4	Destruction and stability of emulsions	75
5.4.1	Destruction processes of emulsions	75
5.4.1.1	Creaming	75
5.4.1.2	Flocculation	75
5.4.1.3	Coalescence	76
5.4.1.4	Ostwald ripening	76
5.4.2	Stabilization of emulsions	76
5.4.3	Evaluation methods of emulsions	77
5.5	Thermodynamics of emulsions	77
5.6	Conclusions	79

Chaper6 Control Technologies for Emulsion Films
Yuji Sakai

6.1	Introduction	81
6.2	Compatibility between moisturizing and occlusive functions	81
6.2.1	Emulsion film to increase both moisturizing and occlusive functions	81
6.2.2	Dispersions of water into the lipophilic parts	82
6.2.3	Structure reinforcement of hydrophilic parts	83
6.2.4	Evaluation of moisturizing and occlusive functions	86
6.2.5	Observation of the conditions of the emulsion films	87
6.2.6	Continuous using evaluation	87
6.3	Improvement of sensory evaluation	90
6.3.1	Emulsion film structures of enhanced occlusive functions and improved sensory evaluation	90
6.3.2	Investigation of surfactants	90
6.3.3	Stability of emulsion prepared using propylene glycol alginate	91
6.3.4	Conditions of emulsion films and sensory evaluation	92

[Powders and Fine Particles]

Chaper7 Application of Composite Nanoparticles to Cosmetics
Toyokazu Yokoyama and Hiroyuki Yokoyama

7.1	Introduction	95
7.2	Preparation of inorganic oxide nanoparticles and applications	96
7.2.1	Preparation of inorganic oxide nanoparticles using the vapor-phase method	96
7.2.2	Characteristics of inorganic composite nanoparticles	96
7.2.3	Dry composition of inorganic oxide nanoparticles by mechanical techniques	98
7.3	Preparation and application of composite nanoparticles of biocompatible polymers	100
7.3.1	Preparation of PLGA nanoparticles by spherical crystallization	100
7.3.2	Application to whitening cosmetics	100
7.3.3	Application to scalp care	102
7.4	Conclusions	104

Chaper8 Dispersion of Fine Particle Titanium Dioxide and Fine Particle Zinc Oxide
Takatsugu Yoshioka and Keiko Iwasaki

8.1	Introduction	105
8.2	The present condition of sunscreen products	105
8.3	Particle sizes and shapes of ultraviolet scattering agents	107
8.4	Dispersion of particle titanium dioxide and particle zinc oxide	108
8.5	Assignment and future development	111

Chaper9 Optical Analysis of the Skin and Development of the Multilayered Powder
Eiichiro Misaki

9.1	Trends of cosmetics development and backgrounds	113
9.2	Optical analysis of the skin and targets of development	114
9.2.1	Multiple-viewpoint image analyzer	114
9.2.2	Optical characteristic of beautiful and normal skins	115
9.3	Designing and developing optics materials (multilayered powder)	115
9.3.1	Thickness distribution of substrate films	116
9.3.2	Refractive index with light absorption considered	117
9.3.3	Orientation of board-shaped particles	117
9.3.4	All interference layers	118
9.4	Development and inspection of base makeup cosmetics	118
9.5	Conclusions	119

PART3 Trends in developments
[Emulsification]
Chaper1 Polymer Micelles
Kunio Shimada

1.1	Introduction	121
1.2	Synthetic polymer micelles	121

	1.3	Ingredients for cosmetics	122
	1.4	Encapsulating active elements and maintaining effect	123
	1.5	Conclusions	126

Chaper2 Lipid Capsules
Asako Mizoguchi, Hiroshi Oda, Yuka Shimoyama, Makiko Fujii and Yoshiteru Watanabe

2.1		Introduction	128
2.2		Penetration of liposomes into the skin	129
	2.2.1	Penetration of water-soluble materials (*in vitro*)	129
	2.2.2	Examining the water holding capacity and water absorption capacity (*in vivo*)	130
2.3		Conclusions	131

Chaper3 Nanogel Engineering of Cholesterol-bearing Pullulan
Setsuko Yamane and Kazunari Akiyoshi

3.1		Introduction	132
3.2		Formation of nanogel of cholesterol-bearing pullulan	132
3.3		Formation of CHP nanogel-based hydrogel	133
	3.3.1	Hydrogel by association of CHP nanogels	133
	3.3.2	Nanogel cross-linking hydrogel	134
3.4		Functions of CHP nanogel	134
	3.4.1	Interaction of nanogels with hydrophobic molecules	134
	3.4.2	Interaction of nanogels with cyclodextrin	134
	3.4.3	Interaction with proteins and molecular chaperone function	134
	3.4.4	Nanogel-calcium phosphate hybrid nanoparticles: nanogel template mineralization	135
3.5		Interaction between CHP nanogel and colloids	136
	3.5.1	Interaction with surfactants	136
	3.5.2	Interaction between hydrophobized polymers	136
	3.5.3	Interaction with liposomes	137
	3.5.4	Stabilization of O/W emulsions	137

Chaper4 Technology for Using Cyclodextrins in the Field of Cosmetics
Keiji Terao

4.1		Introduction	139
4.2		What is cyclodextrin?	139
4.3		Purposes of using cyclodextrins in cosmetics	140
4.4		Stabilization effect	141
	4.4.1	Vegetable oil containing unsaturated fatty acid triglyceride	141
	4.4.2	Vitamin A (Retinol)	141
	4.4.3	Phthalimide caproic acid peroxide (PIOC)	142
	4.4.4	Linoleic acid (Vitamin F)	142
4.5		Reduction effect	143
	4.5.1	Reducing unpleasant smell (deodorizing effect)	143
	4.5.1.1	Iodine	143

4.5.2	Reducing irritation	144
4.5.2.1	Salicylic acid	144
4.6	Controlling release	144
4.6.1	Menthol	144
4.6.2	Tea tree oil	145
4.7	Improving bioavailability (Vitamin E (tocopherol) and coenzyme Q10, as examples)	145
4.8	Conclusion	145

Chaper5 Possibility of Antioxidation Vitamin Derivative Capsules
Shinobu Ito

5.1	Introduction	148
5.2	Derivative of vitamin C with surface activity	149
5.3	Derivative of water-soluble vitamin E with an ability to form transparent gel: TPNa	150
5.4	Ascorbyl-2-phosphate (AP)	150
5.5	High concentration of the ascorbic acid in the cell	151
5.6	Promotion of collagen synthesis and inhibition of tyrosinase activity by APP	151
5.7	Production of micro and nano capsules (ITO-Nano DDS) using APP	152
5.7.1	Multilayered liquid crystal structure of lipid-containing APP capsules of the self-emulsification type (ITO-Nano DDS Capsules)	152
5.7.2	Electrophoresis of APP capsules	152
5.8	Synergy with fullerene	153
5.9	Conclusions	154

Chaper6 Biosurfactants as Cosmetic Ingredients
Tomohiro Imura, Tokuma Fukuoka, Tomotake Morita, Masaru Kitakawa, Atsushi Sogabe and Dai Kitamoto

6.1	Introduction	155
6.2	What is biosurfactant	155
6.3	Structure of biosurfactants	156
6.4	Examples of biosurfactant use	157
6.5	Microbe production of biosurfactants	157
6.6	Surface chemical characteristics of biosurfactants	159
6.7	Biochemical characteristics of biosurfactants	160
6.8	Application of biosurfactants for skin care cosmetics	161
6.9	Conclusion	163

Chaper7 Application of PLGA Nanoparticles to Skin Care Cosmetics
Hiroyuki Tsujimoto and Nobuhiko Miwa

7.1	Introduction	165
7.2	Safety and application of PLGA nanoparticles to skin care cosmetics	165
7.2.1	Safety of PLGA nanoparticles	165
7.2.2	Stability of PLGA nanoparticles	166
7.3	Functions of PLGA nanoparticles and application to skin care technology	166
7.3.1	Skin penetration of PLGA nanoparticles	167

	7.3.1.1	Evaluation of penetrability of PLGA nanoparticles and dermis deliverabilty of drugs enclosed in the nanoparticles (using excised human skin specimens and modified Bronaugh diffusion chamber) ··· 167
	7.3.1.2	Evaluation of skin penetrability of drugs enclosed in PLGA nanoparticles (using rats and a horizontal diffusion cell) ··· 168
	7.3.1.3	Evaluation of the skin penetrability of solution components (using a three-dimensional artificial skin and a Franz diffusion cell) ··· 169
7.3.2		Example of using composite nanoparticles in cosmetics ··· 170
7.3.3		Skin care and scalp care technologies of PLGA nanoparticles ··· 170
	7.3.3.1	Skin care technology ··· 170
	7.3.3.2	Example of functional cosmetics containing PLGA nanoparticles (for sensitive skin) ··· 171
	7.3.3.3	Scalp care technology ··· 172
7.4		Conclusions ··· 173

Chaper8 NANOEGG® and NANOCUBE®
Yoko Yamaguchi and Rie Igarashi

8.1	Introduction ··· 175
8.2	New DDS technology—birth of NANOEGG—) ··· 176
8.3	Skin regeneration effect of NANOEGG® (reducing stains and wrinkles) ··· 178
8.4	Skin regeneration by bio-mimetic technology NANOCUBE® ··· 179
8.5	Mechanism of skin regeneration by NANOCUBE® ··· 181
8.6	Effects of NANOCUBE® on human ··· 182
8.7	Conclusions and the future prospects ··· 183

Chaper9 Emulsification and Nanotechnology of Silicone and Application to Cosmetics
Hidetoshi Kondo

9.1		Introduction ··· 184
9.2		Emulsification of the silicone ··· 185
	9.2.1	Emulsion polymerization method (silicone fluid in water) ··· 185
	9.2.2	Suspension polymerization methods (silicone fluid in water) ··· 186
	9.2.3	Mechanical emulsification methods (silicone fluid in water) ··· 186
9.3		Polyether-modified silicones ··· 187
9.4		Nanodispersion of silicone fluids ··· 187
9.5		Silicone vesicles ··· 188
9.6		W/Si nanoemulsions ··· 189
9.7		Conclusions ··· 190

Chaper10 Bio-Drug Delivery System
Laurent Martin

10.1		Introduction ··· 192
10.2		Limits of current technologies ··· 193
10.3		Overview of trigger release mechanisms ··· 193
	10.3.1	Release by enzymatic digestion ··· 193
	10.3.2	Enzymatic release: an exact approach ··· 193

10.4	Micro- and macrosized particles for enzymatically activated technologies	194
10.4.1	Marine collagen	194
10.4.2	Plant proteins	195
10.4.3	Polysaccharide-based encapsulation	195
10.4.4	Nanoencapsulation	195
10.5	Properties and performance of micro- and nanospheres and capsules	196
10.5.1	Enzymatic digestion *in vitro*	196
10.5.2	Penetration vs storage	197
10.5.3	Pharmacokinetic	200
10.5.4	Membrane selection	201
10.6	Perspectives and conclusions	202

Chaper11 Lamellar Vesicles of Monoglyceride
Toshlro Sone

11.1	Introduction	203
11.2	Screening and particle diameters of monoglycerides	204
11.3	Differences in effectiveness to the skin by particle diameter	205
11.3.1	Barrier function	205
11.3.2	Skin surface morphology	206
11.3.3	Transglutaminase I (TGase-1)	206
11.4	Stabilization	208
11.4.1	Phase transition temperature	208
11.4.2	Protective colloid action	209
11.5	Conclusions	210

Chaper12 Microcapsule Drug: Active-ingredient Delivery System
Takayuki Tsuboi

12.1	Active-ingredients Delivery System (ADS)	212
12.2	ADS - Microcapsules	214
12.3	Possibilities of ADS	217

Chaper13 Hyaluronic Acid of Super Low Molecular Weight
Kohei Watabe

13.1	Introduction	219
13.2	What is hyaluronic acid?	219
13.3	Properties of hyaluronic acid	220
13.4	Use of hyaluronic acid in cosmetics	220
13.5	Development of hyaluronic acid of a super low molecular weight	220
13.6	Properties of the super low molecular weight hyaluronic acid	220
13.6.1	Low viscosity	220
13.6.2	Superior solubility	221
13.6.3	High stability of viscosity (to heat and pH)	221
13.6.4	Compatibility	221
13.7	Checking the performances of super low molecular weight hyaluronic acid	222

13.7.1	Penetration into the skin	222
13.7.2	Moisturizing effect	222
13.7.3	Differences in moisturizing effect by the molecular weight of hyaluronic acid	223
13.7.4	Penetrability into hair	224
13.8	Future prospects	225
13.9	Conclusions	225

[Powders and Fine Particles]

Chaper14   Pearl Pigments Coated with Barium Sulphate
Sadaki Takata

14.1	Introduction	227
14.2	Evolution of hybrid powders by surface modification	227
14.3	Development of functional powder by controlling the shape of powder surface using barium sulphate	229
14.3.1	Development of hybrid powder with a new optical property of obfuscating sagging on a face	229
14.3.2	Development of hybrid powder of ideal reflection characteristics by numerical computation	231
14.4	Conclusions	235

Chaper15   Shape-controlled Composite Powder of Titanium Dioxide and Sericite
Kazuyoshi Tsubata and Hiroyuki Asano

15.1	Introduction	236
15.2	Sericite of Furikusa, Aichi Prefecture	237
15.3	Preparation of composite powder and physical property evaluation	238
15.3.1	A preparation of composite powder	238
15.3.2	Transmittance of ultraviolet and visible light	239
15.3.3	Light reflecting properties	239
15.3.4	Spreading characteristic of the powders	240
15.3.5	Adhesion on the skin	241
15.4	Application to powder foundation	241
15.4.1	Measurement of UV blocking performance	241
15.5	Conclusions	242

Chaper16   Light Diffusive Inorganic Powder
Naoyuki Enomoto

16.1	Introduction	244
16.2	Light diffusing spherical powder	244
16.2.1	Characteristics and effects of HOLLOWY N-15	244
16.2.2	Optical properties	245
16.2.3	Effects of mixing in cosmetics	246
16.3	Platelet-shaped light diffusive powder	246
16.3.1	Characteristics and effects of Cover Leaf® AR-80	246
16.3.2	Optical properties	247
16.3.3	Effects of mixing in cosmetics	248
16.4	Conclusions	248

PART4 Trends in Patents
Chaper1 Trends of Nanotechnology-related Patents
Yoshihiro Yano

1.1	Introduction	251
1.2	Intellectual property rights and patent rights	251
1.3	Scope of the Patent Act	252
1.4	Procedure of obtaining patent rights	252
1.5	Trends of nanotechnology patents	254
1.6	Trends of patents of nanotechnologies in medicine and cosmetics	258
1.7	Conclusions	259

PART 1　Introduction

Chapter 1
Emulsification Technologies for Skin Care Cosmetics

Toshiyuki Suzuki

1.1 Introduction

There are two general kinds of skin care cosmetics: (1) skin protection cosmetics for keeping the skin beautiful and healthy, and (2) skin cleanliness cosmetics for keeping the skin clean. Emulsions, such as creams and milky lotions, are widely used as a basis of skin care cosmetics of both kinds, which should contain both a moderate amount of oil and various amounts of water-soluble ingredients that are active to the skin. Emulsification technologies, which are for preparing emulsions and keeping them stable, have also been cleverly used to quickly wash out dirt and color cosmetics that finished their duty from the skin. Moisturizing cosmetics, which keeps the moisture of the skin, and UV care cosmetics for protecting the skin from the damage of ultraviolet rays also contain emulsions for forming the lipid structure that resembles the stratum corneum and a strong water-repelling film on the skin. Developed applications have also been made, such as preparing emulsions in minute droplets of nanometer size for greatly altering the appearance and sensory evaluation.

1.2 Basics of emulsification and recent technologies

1.2.1. Types and condition of emulsification

Emulsion is a system consisting of two or more mutually insoluble liquids, such as water and oil, in which one exists dispersed in forms of droplets in the other. The diameter of emulsion particles is usually around 0.1–100 μm (100 to 10^5 nm)[1]. Emulsions are classified into O/W type and W/O type by whether the continuous phase (an outside phase) is water or oil. The former is hydrophilic, and the latter forms a water-repelling film when applied on the skin. There are also multi-layer (multiple) emulsions of O/W/O type, in which O/W emulsion is dispersed in oil, and W/O/W type, in which W/O emulsion is dispersed in water (figure 1).

As the size of emulsion particles is reduced to 200 nm or smaller, which is shorter than the wavelength of the light, the appearance of the emulsion changes

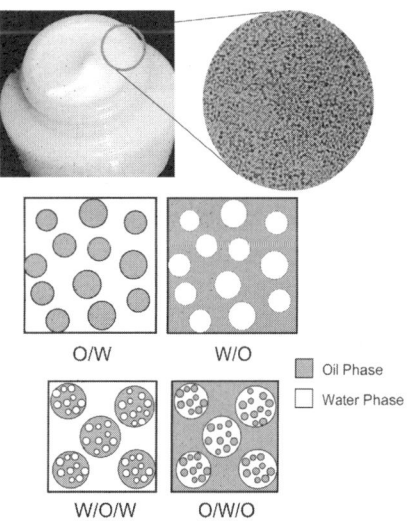

Figure 1　States and models of emulsions

from semitransparent to transparent. Emulsions with minute particle diameters of the 20–200nm level are called nanoemulsions (nano size emulsions)[2]. Like nano size emulsions, "microemulsions" consist of particles of minute diameters. Originally, the name was given to denote emulsions that have particles of very small diameters. Recent studies have shown that microemulsion is a system that is thermodynamically stable, is in a condition that can be called swelling micelles, and is clearly different from ordinary emulsions[3,4].

1. 2. 2 Production of emulsions and different emulsification conditions

Emulsion consists of oil, water and surfactants. Figure 2 shows changes in composition during two different emulsification processes. An emulsification method A in the figure is called the phase inversion emulsification method, in which O/W emulsion is produced by dissolving and dispersing surfactants in the oil phase and then adding the water phase. B is called the agent-in-water method, which involves adding surfactants in the water phase and then adding the oil phase. Both produce emulsions of the same composition (starmark in the figure), but the phase inversion emulsification method generally results in smaller particles than the other. This is because during phase inversion, infinite associations of liquid crystals and surfactants (the D phase) are formed, sharply reducing the interfacial tension between oil and water and producing smaller sizes even under the same mixing conditions[5]. Therefore, the phase diagram of three ingredients is used to analyze how conditions of the system change by the dissolving conditions and emulsification process used.

1. 2. 3 Understanding emulsion and microemulsion using phase diagrams

Figure 3 shows a phase diagram[6] of polyoxyethylene(9.7)-nonylphenylether (7 wt%), water, and cyclohexane. "I" in the figure is 1-phase domain, "II" in the figure is 2-phase domain, and "III" in the figure is 3-phase domain. Nonionic surfactants of the ethylene oxide (EO) type undergo property changes from hydrophilic to hydrophobic along with rises in temperature. Curve (a) is a solubility limit curve of oil into a micelle water solution. In the domain on the right of the curve, the amount of oil exceeds the limit, and oil and the micelle water solution coexist as two separate phases. Curve (2) is a solubility limit curve of water into an inverse micelle oil solution. In the domain on the left of curve, the amount of water exceeds the limit, and water separates. Curve (3) is a cloud point curve, at which surfactants coagulate and

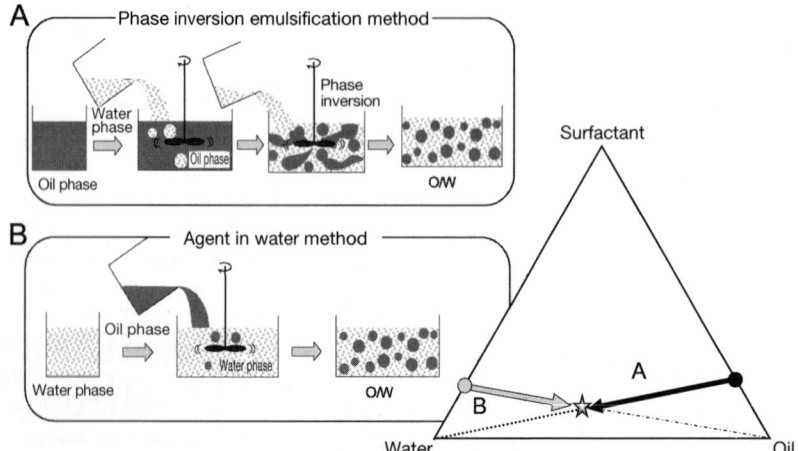

Figure 2 Emulsification processes and a diagram of composition change

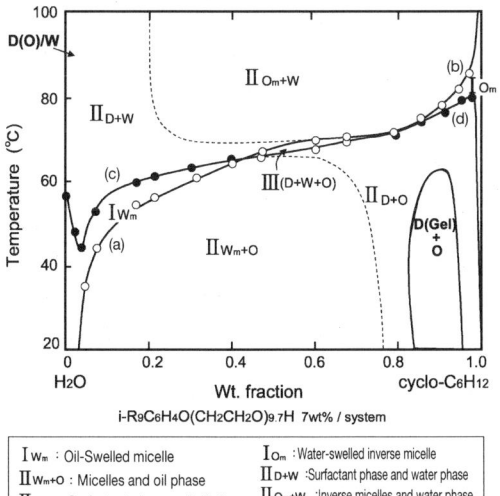
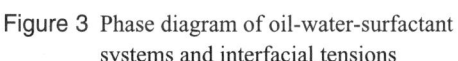
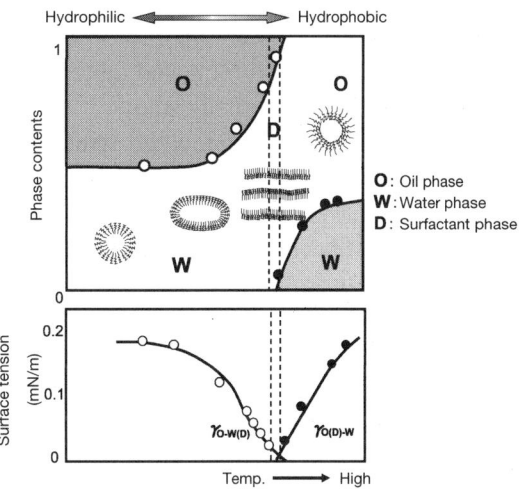

Figure 3 Phase diagram of oil-water-surfactant systems and interfacial tensions

Figure 4 Interfacial tensions and dissolved conditions of oil-water-surfactant systems

separate and the mixture becomes cloudy. Curve (d) is a cloud point curve of a nonaqueous system. Below this temperature, surfactants containing water separate from the oil phase.

The domains between the solubility limit curves and the cloud point curves are the micelle and inverse micelle domains of Phase I, in which oil and water, respectively, are dissolved. As the temperature approaches the cloud point, solubility sharply increases, and the micelles and inverse micelles become swelled with large amounts of oil and water, respectively. This state is called microemulsion (the former is O/W microemulsion, the latter is W/O microemulsion). The name of emulsion is given, but it is 1-phase system that is stable and apparently different from ordinary emulsions in thermodynamic property. Beyond the 1-phase domains, there is a 3-phase domain consisting of oil, water and surfactants (the D phase). The D phase is infinite associations of molecules formed under a balanced condition of hydrophobic and hydrophilic properties of surfactants. As shown in figure 4, the surface tension of oil and water remarkably drops by the formation of the D phase. The mean curvature of surfactants becomes zero at the D phase, and both oil and water phases become bicontinuous[7,8].

1. 2. 4 Liquid crystals in emulsification and application of the D phase

When infinite associations of molecules such as the D phase and liquid crystals are formed during emulsification (figure 5), minute-particle emulsions are formed. Efficient emulsification methods have been reported, such as "liquid crystal emulsification", "the D phase emulsification" and "producing nanoemulsion from microemulsion", which involve preparing infinite associations first. Lamella liquid crystals and D phases are used to produce O/W emulsions, and inverted hexagonal liquid crystals are used to prepare W/O emulsions[5]. Production of O/W and W/O emulsions using cubic liquid crystals has also been reported[9,10].

1. 2. 4. 1 Liquid crystal emulsification

Liquid crystal emulsification[11] is a technology that disperses and maintains the dispersed phase (oil in an O/W emulsion) into liquid crystals of surfactants and produces minute emulsification particles (figure 6). The use of the liquid crystal phase for emulsification orients surfactant molecules efficiently on the oil-water interface, decreases the interfacial tension, and strengthens the interfacial film. Surfactants of two chain type, which are easy to form lamella liquid crystals, are an appropriate emulsifier. Multivalent alcohols such as glycerin strengthen liquid crystal films and increase the amount of oil maintained in liquid crystals.

1.2.4.2 D phase emulsification

The D phase emulsification method, which was developed by Dr. Sagitani[12], involves producing O/W emulsion by adjusting the HLB of non-ionic surfactants using multivalent alcohol to produce the surfactants phase (the D phase), adding an oil phase while stirring, mixing the mixture until it becomes transparent gelatinous emulsion, and diluting with water. Divalent alcohols such as nonionic surfactants of EO type and 1,3-butanediol are used to produce the D phase.

1.2.4.3 Preparing nanoemulsion using microemulsion

Using a phenomenon that the solubility of nonionic surfactants increases remarkably at around the HLB temperature, nanoemulsion is produced by preparing a microemulsion, which is a soluble system, and cooling until it becomes a semitransparent minute-particle emulsion

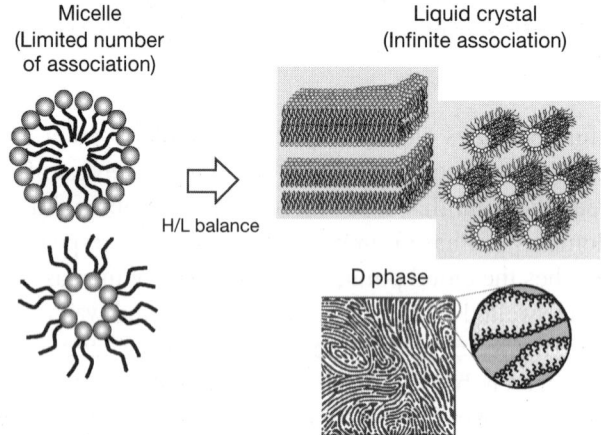

Figure 5 Infinite associations of amphipathic molecules

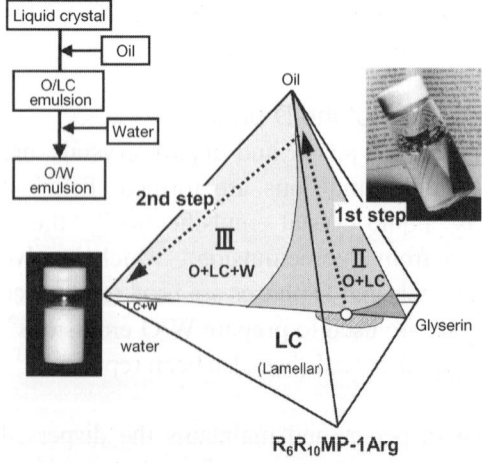

Figure 6 Process of liquid crystal emulsification

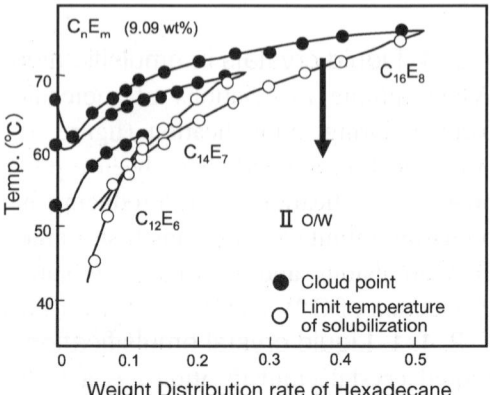

Figure 7 Preparation of nanosize emulsions using microemulsion

(figure 7)[13]. Although the formative domain of a microemulsion depends on the HLB in the system, at a similar HLB, larger surfactants show higher solubility of both hydrophobic and hydrophilic molecules and are thus more appropriate for this purpose.

1. 3 Emulsification technology in recent skin care cosmetics

1. 3. 1 Multi-lamellar emulsions and gels prepared using self-organization lipids

The healthy skin, which is also beautiful, contains stratified aggregation of intercellular lipids, such as ceramide, which is amphipathic, contributing to keeping the moisture of the skin and functioning as a barrier. Moisturizing cosmetics, which can be called as a standard of skin care cosmetics, also contain amphipathic lipids that resemble ceramide. When an emulsion is prepared using polar lipids (amphipathic lipids) that self organize as a main constituent, the emulsion looks ordinary but has a multilamellar structure of emulsified particles (figure 8)[14]. High pressure processing of such a multi-lamellar emulsion has been reported to make the nano-size emulsified particles to arrange regularly and produce transparent colloidal crystals (figure 9)[15].

1. 3. 2 Preparing makeup cleansing using bicontinuous D phase

Cosmetic products prepared using a bicontinuous D phase, which has high solubility performances, have been developed and applied to cosmetics. Water proof oil cleansings are a typical example. Because D phases with a bicontinuous structure have a high performance of dissolving both oil and water, oily dirt elutes quickly even under the presence of water. In ordinary nonionic surfactants with ethylene oxide (EO) added, the domain where the D phase is formed and hydrophilicity and hydrophobicity are in balance is easily affected by temperature. A stabilization method has been devised, which involves using surfactants that have hydroxyl (-OH) bases (figure 10)[16].

1. 3. 3 High moisture emulsification using inverted hexagonal liquid crystals

Inverted hexagonal liquid crystals have strong hydrophobicity and maintain a certain amount of water by encasing in the crystals. However, it separates into two layers of liquid crystals and water at high water contents. Unlike ordinary W/O emulsions, stable emulsions of high inner phase ratios of water contents exceeding 95 % are formed when a small amount of non polar oil is used together with surfactants that are easy to form inverted hexagonal liquid crystals such

	Lipids emulsion
Artificial ceramide	10.0 wt%
Stearic acid	6.0
Cholesterol	3.0
Cholesteryl ester isostearate	1.0
Squalane	10.0
Monoarginine-hexyldecylphosphate	0.5
Glycerin	3.0
Pure water	66.5
	100.0

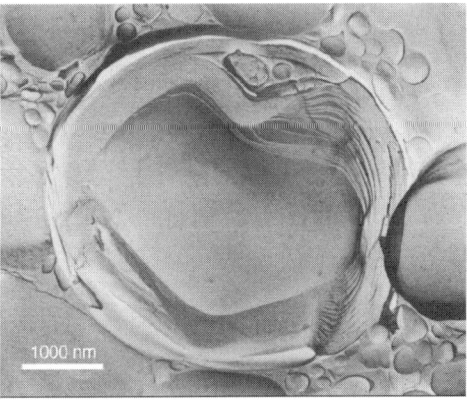

Figure 8 Formulation of lipids emulsion and electron microscope (cryo-cem) image

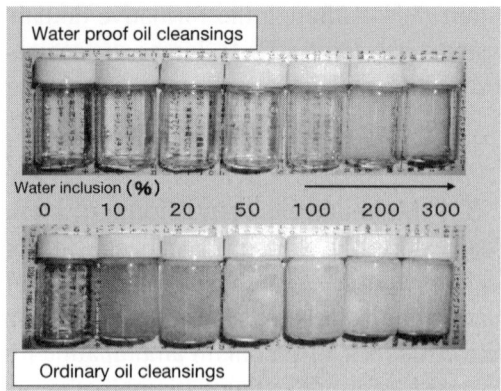

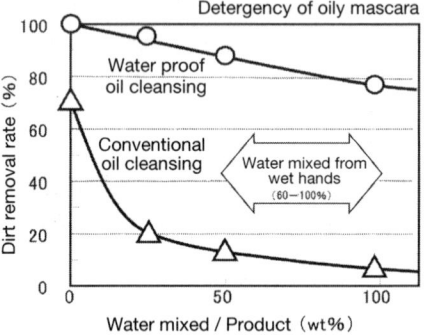

Figure 9 Transmission electron microscope (TEM) images of gel-type nano emulsions

Figure 10 Capability of water proof oil cleansings

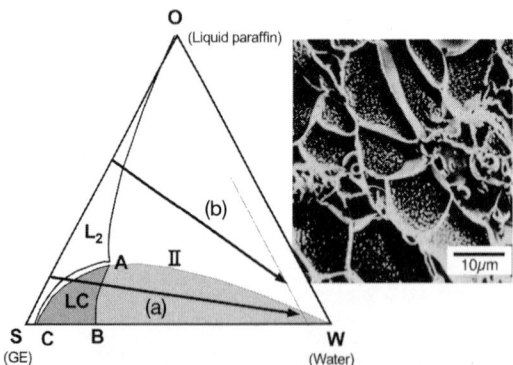

Figure 11 High hydrous emulsion phase diagram of W/O liquid crystal emulsions and cryo-cem image of high water-content emulsion

as long-chain alkyl glyceryl ether (GE) (figure 11)[5,11]. Emulsification using inverted hexagonal liquid crystals is characterized by the formation of stable emulsion when the amount of oil is small. This is a result of liquid crystals maintained in the continuous phase. When there is little water as in figure 11 (a), the continuous phase is liquid crystals, but the continuous phase becomes oil at high oil concentrations, destabilizing the emulsion. W/O emulsions of high water contents prepared by liquid crystal emulsification are widely used as the base of UV care cosmetics because they give a refreshing feeling while showing high water repellency.

References

1. W. Clayton, "Theory of Emulsion" 4th, Blackston Co., New York (1943)
2. C. Solans *et al.*, *Langmuir*, **20**, 6594-6598 (2004)
3. K. Shinoda *ed.*, "Solvent Properties of Surfactant Solutions", Marcel Dekker, New York (1967)
4. K. Shinoda, H. Saijyo, (in Japanese) *J. Jpn. Oil Chem. Soc.*, **35**, 308 (1986)
5. T. Suzuki, "New High Stabilization Technique of the Emulsion (in Japanese)", pp 35-44, 64-79, Technical Information Institute Co., Ltd. (2004)
6. K. Shinoda, H. Saito, *J. Colloid Interface Sci.*, **26**, 70 (1968)
7. K. Shinoda, "Solution and Solubility (in Japanese)", pp1-8, 141-181, Maruzen Co., Ltd. (1991)
8. B. Lindman, H. Wennerstrom, H. Kunieda, *Topics in Current Chemistry*, **87**, 1-83 (1980)
9. C. Rodriguez, K. Shigeta, H. Kunieda, *J. Colloid Interface Sci.*, **223**, 197 (2000)
10. H. Kunieda, K. Aramaki, T. Nishimura, M. Ishitobi, (in Japanese) *J. Jpn. Oil Chem. Soc.*, **49**, 617 (2000)
11. T. Suzuki, H. Takei, S. Yamazaki, *J. Colloid Interface Sci.*, **129**, 491 (1989)
12. H. Sagitani, T. Hattori, K. Nabeta, M. Nagai, (in Japanese) *Nippon Kagaku Kaishi*, **1399** (1983)
13. T. Tomomasa, M. Kawauchi, H. Namazima, (in Japanese) *J. Jpn. Oil Chem. Soc.*, **37**, 1012 (1988)
14. T. Suzuki, G. Imokawa, A. Kawamata, (in Japanese) *Nippon Kagaku Kaishi*, **1107** (1993)
15. H. Iwai *et al.*, The proceeding of 223rd ACS National Meeting, COLL73 (2002)
16. H. Tsunoda, (in Japanese) *J. Soc. Cosmet. Chem. Jpn.*, **39**(1), 3(2005)
17. Y. Suzuki, H. Tou, (in Japanese) *J. Jpn. Oil Chem. Soc.*, **36**, 588 (1987)

Chapter 2
Powder Technology in Makeup Cosmetics

Hiroshi Fukui

2.1 Introduction

It is an important matter for all people to be beautiful. Makeup cosmetics, which are used to actualize this dream, have a long history. In ancient times, people applied natural pigments on the body and face for protection and ritual purposes. A palette for makeup, which dates back to about 10,000 B.C., was found during an excavation in Egypt, suggesting that ancient Egyptians used old forms of almost all modern-day cosmetics. Ancient Egyptian women emphasized their eyes by drawing thick eye lines with antimony kohl, black oxide manganese, burnt almond, carbon black and iron oxide and applying bright eye shadow which grounded malachite, turquoise, copper green rust and charcoal. They grew the eyebrows long and dyed them black to match with the eyes, and the eyelashes were either dyed black or removed. They colored the lips crimson or rosy with rouge and polished and colored the finger and toe nails with orange alcanna. Both men and women applied yellow ocher on the body, which is an origin of today's foundations, and only men used orange. In addition, blood vessels at the temples and the chest were colored blue to give a clear contrast to the base skin color. Rouge is believed to have been introduced in Japan in the seventh century. At that time, rouge was made from dyer's saffron and was expensive, and it was in the Edo era when rouge spread to public. Rouge in lipsticks appeared in the 1910s. Foundation was introduced in Japan in the 1950s. Since then, various types of foundation have been developed, including emulsions and the oil type, enabling natural and beautiful makeups. Today, foundations are the main stream of base makeup, and face powder is applied on top of the foundation to absorb oil. Powder technology is indispensable in producing today's functional makeup cosmetics and is described in this chapter.

2.2 Constituents of makeup cosmetics

Makeup cosmetics consist of powders (pigments) scattered in various kinds of bases. There are many kinds of cosmetics, which vary in function, efficacy and use. Their agent types and constituents are shown in table 1[1]. There are simple bases, which are prepared by mixing water, oil and moisturizing agents, and emulsions, which also contain surfactants. The hydrophilicity and lipophilicity of powders are determined by into which phase the powder is dispersed. Some bases also contain antioxidants and perfumes, and the interactions between powders and the compounds should also be taken into account when designing products. In makeup cosmetics, coating and coloration performances are especially important in terms of makeup effects, and these functions mainly come from their powder constituents. Powders used in

Chapter 2 Powder Technology in Makeup Cosmetics

Table 1 Makeup cosmetics and constituents [1]

Part of application		Face — Face powder and foundation					Cheeks — Rouge				Lips — Lipstick			Eyes — Eye shadow			Eyes — Eyebrow			Eyes — Eyeliner						Eyes — Mascara				Nails — Nail enamel	
Category	Type	Oily	Emulsion	Compact	Powder	Sheet*	Kneaded	Liquid	Sheet*	Compact	Kneaded	Stick (emulsion)	Stick (oil)	Emulsion	Oily	Compact	Pencil	Oily	Compact	Pencil	Polymer emulsion	Emulsion	Volatile oily	Oily	Solid	Polymer emulsion	Emulsion	Volatile oily	Oily	Emulsion	Solvent
Raw materials — Base	Oils / Fats	○	○	○			○	○		○	○	○	○	○	○	○	○	○	○	○		○	○	○	○		○	○	○		
	Wax	○	○	○			○	○		○	○	○	○	○	○	○	○	○	○	○					○		○	○	○		
	Fatty acids		○									○																			
	Higher alcohols		○									○																			
	Fatty acid esters	○	○	○		○	○	○	○	○	○	○	○	○	○	○	○	○	○	○		○	○	○	○	○	○	○	○	○	○
	Hydrocarbons	○	○	○		○	○	○	○	○	○	○	○	○	○	○	○	○	○	○		○	○	○	○		○	○	○	○	○
	Surfactants		○		○									○								○				○	○			○	○
	Metallic soaps	○		○																											
	Plasticizers																													○	○
	High molecular compounds	○	○	○		○	○	○	○	○				○		○			○		○	○	○			○	○			○	○
	Inorganic thickener		○					○	○					○										○			○	○			
	Volatile oil (solvent)							○	○															○							
	Polyol		○			○		○				○	○										○								
	Inorganic powders	○	○	○	○	○	○	○	○	○	○	○	○	○	○	○	○	○	○	○	○	○	○	○	○	○	○	○	○	○	○
	Purified water		○			○		○				○		○								○					○			○	○
Coloring material	Organic coloring material	○	○	○	○	○	○	○	○	○	○	○	○	○	○	○	○	○	○												
	Inorganic natural coloring material	○	○	○	○	○	○	○	○	○	○	○	○	○	○	○	○	○	○												
	Pearly pigment		○	○	○		○	○		○		○	○	○	○	○															

*Paper face powder

Table 2 Powders used in makeup cosmetics [2]

Classification		Raw materials
Extender pigment		Talc, Kaolin, Mica, Sericite, Calcium carbonate, Magnesium carbonate, Silicic acid anhydride, Barium sulphate, *etc.*
Color pigment	Organic	(Synthetic) food, medical supplies, Tar pigment for cosmetics (Natural), Carotene, Carthamin, Carmine, Chlorophyll, *etc.*
	Inorganic	Red iron oxide, Yellow iron oxide, Black iron oxide, Ultramarine, Prussian blue, Chromium oxide, Carbon black
White color		Titanium oxide, Zinc oxide
Pearly pigment		Pearly essence, Bismuth oxychloride, Titanium dioxide-coated mica, Iron oxide coated titanium dioxide-coated mica, *etc.*
Others	Metallic soap	Mg stearate, Ca stearate, Al stearate, Zn myristate acid, *etc.*
	Synthetic polymer powder	Nylon powder, Polyethylene powder, Polymethyl methacrylate, *etc.*
	Natural product	Wool powder, Cellulose powder, Silk powder, Starch powder, *etc.*
	Metal powder	Aluminum powder, *etc.*

makeup cosmetics are shown in table 2[2]. Color and white pigments are used to control the color and the coating performance of products. Extender pigments are diluents of color pigments, control the color of products, and adjust the extensibility and adhesion of the products on the skin, absorbency of sweat and the sebum, and luster. The physical properties of the powders are described below.

2.3 Physical properties of powders

2.3.1 Extensibility

Extensibility is the ability to spread on the skin and gives a smooth touch. Talc and mica have good extensibility. Especially, Fe-substituted synthetic mica is also "soft" as well as "spreads well" and is "smooth" probably attributable to the strains generated by different elements coordinated within the crystal lattice[3]. Not only plate-shaped powders but also spherical powders of about 5 to 15 μm are used to improve the extensibility of products. Spherical powders can be inorganic, such as silica and alumina, and polymers, such as nylon and polyethylene. Dynamic friction coefficient is used as an index to show extensibility, and smaller values show higher extensibility.

2.3.2 Adhesion

Adhesion or adhesiveness on the skin determines the easiness of adhering the powder on the skin and is also related to makeup finish and lasting of the makeup. Mineral residue was widely used in the past. Today, adhesion is improved by modifying the surfaces of powders. To prevent powders on the skin from mixing with sweat and sebum and smearing, powders are processed so as to be water and oil repellent.

2.3.3 Absorbency

Absorbency is the ability to absorb sweat and sebum and is demanded to erase the greasy luster of the skin and prevent makeup from smearing. Persons who excrete large amounts of sebum are prone to smearing of makeup, particularly on the T zone, i.e. near the nose and the forehead, where excretion of sebum is large. Kaolin and calcium carbonate have high sweat and sebum absorbency. Porous powders are also used. Porous silica beads and porous cellulose powders are spherical and can also improve extensibility.

2.4 Optical properties of powders

One of the roles of powders in makeup cosmetics is to control visible light, including color correction and concealment. The concealment power of powder is related to its refractive index and particle size. Powder of a large refractive index scatters much light and has a high concealment effect. Foundations and face powder consist mainly of extender pigments (talc and mica) that do not have concealment effect but are soft to use; and titanium dioxide and zinc oxide, which have high refractive indices, are added to give concealment. Powders control visible light to erase bruises, stains and freckles by correcting colors, conceal unevenness of the skin such as pimple traces and seat pores, and add three-dimensional impression. Powders also block ultraviolet rays. These are described below.

2.4.1 Color correction

The color of the skin is determined by pigment information inside the skin such as hemoglobin and melanin. In spectral reflectance curves of beautiful skin, reflections are generally high at around 400 to 450 nm, which makes the skin to look transparent, and at around 650 to 700 nm, which gives a look of good circulation. Correcting the colors of these wavelengths can actualize beautiful skin. Blue and green pigments were once used to correct the colors, but the pigments caused drops in brightness and dark colors due to light absorption by the pigments and subtractive color mixture.

A correction method was developed that can correct colors by not causing drops in brightness. The method involves use of the additive color mixture effect by the transmittance and reflection of interference light of titanium dioxide mica. Interference colors are the colors of rainbow seen on soap bubbles, shells and oil films. As shown in figure 1, reflected light B and reflected light C, which reflected after passing a thin film, of incident beam interfere with each other and intensify colors of certain wavelengths. Therefore, the wavelengths of interference light change by the thickness of the film, the refractive index and the angle of the incident beam. Electron micrographs of mica are shown in figure 2. Titanium dioxide coats mica in layers, and various interference colors can be obtained by changing the thickness. However, ordinary titanium dioxide mica gives only slight interference colors because the light is mixed with diffused light, making the powder to look white. To highlight the interference colors, making the surface black is effective, like soap bubbles are clearly visible in front of a black back. Based on this concept, a pearl pigment that gives vivid interference colors was developed by producing black low-dimensional titanium oxide by reducing some layers of titanium dioxide into titanium oxide[4].

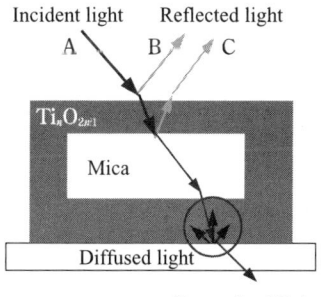

Figure 1 Optical properties of titanium dioxide mica

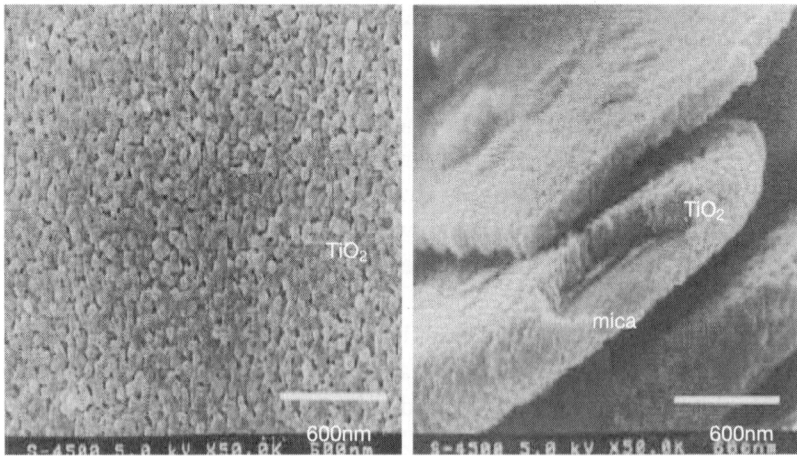

Figure 2 Electron micrograph of titanium dioxide-coated mica

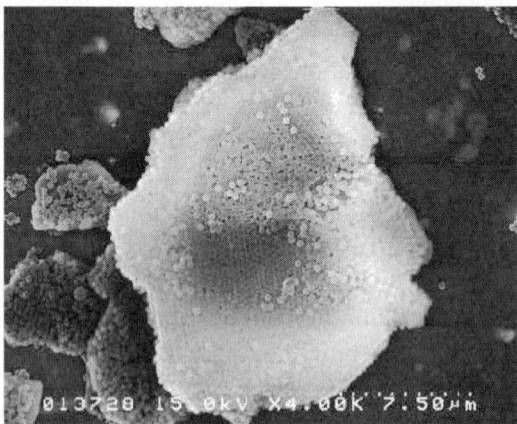

Figure 3 Electron micrograph of plate particle coated with spherical PMMS

This pearl pigment, which gives various colors depending on film thickness and oxidation state of titanium oxide, is highly resistant against light, heat and chemicals. Because only mica and titanium oxide are used to produce an infinite number of colors by interference of light, this pearl pigment is called "Infinite Color". It is used not only in cosmetics but also in paints. Although this titanium dioxide mica can cover bruises naturally, it produces shiny dots deteriorating makeup effects. A composite powder was developed (figure 3) by coating the surface the titanium dioxide mica with spherical powder of polymethyl methacrylate (PMMA) to control the shine. Because the shine of titanium dioxide mica is controlled by diffused reflection on the surfaces of PMMA powders while the refractory indices in the red domain are corrected using the red interference colors, the resultant composite powder can actualize translucent and smooth makeup finish.

Fluorescence can also be used to correct skin color. Reduced zinc oxide is an example. Fluorescence emitted by zinc oxide when irradiated by ultraviolet rays, which are included in solar radiation and fluorescent lamps, increases the reflectance at around 500 nm, the blue to green domain, and gives a translucent look[5]. There has also been a practice in which the reflectance at long wavelengths was increased using the fluorescence of dyer's saffron pigment (carthamin).

2.4.2 Responding to changes in photoenvironment
Faces wearing makeup look whitish under the sun and yellowish and dark under fluorescent lamps. Photochromic powder has been developed to solve this problem, in which iron is doped on titanium dioxide[6]. The powder turns light brown when exposed to ultraviolet rays and prevents the whitish look under the sun. Color rendering pearl pigment that changes its color by UV exposure has also been developed, in which photochromic characteristics are further developed and the interference color effect and brightening effect of titanium dioxide mica are enhanced[7]. For example, coating pearl pigments with red interference colors with iron-doped titanium dioxide, which has photochromic characteristics, gives pigments that are white when not exposed to UV but develop vivid red interference color when exposed to UV because the outer layer turns dark. In other words, the powder turns from white to red by UV exposure. This color rendering pearl pigment emits red interference colors under fluorescent lamps, which create a relatively yellowish photoenvironment, and prevent the skin from looking yellowish. Under the sun, the total reflectance drops, preventing the face from looking whitish. Thus, the makeup looks natural under both photoenvironments.

2.4.3 Concealing unevenness on the skin
Unevenness on the skin such as wrinkles and sweat pores can be concealed using the defocusing effect by light scattering of spherical powders, etc. Plate-shaped powders are coated with spherical powders to reduce the powdery texture and appearance, increasing the adhesion and improving the defocusing effect. The red interference pearl pigment coated with nanoparticles of PMMA described above is one of such applications and combines the color correction effect by interference light and the concealing effect by light scattering.

2.4.4 Adding three-dimensional impression
To add three-dimensional impression to a face, it is effective to light the front of the face and darken the sides as if wearing black stockings. Powder of high angle dependence can be prepared by forming a layer of low-dimensional titanium oxide on the surface of a titanium dioxide mica-base pearl pigment. The powder makes the front to look bright and the sides to look dark, enhancing a three-dimensional impression. Powders that are designed to not only darken the sides but also scatter light from the front and sides in different ways have also been prepared by coating plate-shaped powders with various phases of particles (figure 4). For example, "reflex board powder"[8] is a composite powder in which titanium dioxide mica powder that has red interference colors is coated with plates of barium sulphate. When applied on a face, the composite powder gives a natural color on the forehead and cheeks where the light is irradiated from the top because the plate-shaped crystals diffuse part of the light and the red interference colors of titanium dioxide mica develop a color similar to the skin. On the other hand, on shaded parts such as sags, where light is irradiated aslant, the plate-shaped crystals diffuse almost the entire light, making the shaded parts inconspicuous. In a sensory evaluation, the powder made the face to look facing slightly upward because sags were concealed, and a computer simulation analysis of appearances calculated the ages younger than their actual ages.

2.4.5 Lasting of makeup effects
It is important to keep beautiful makeups long. Therefore, cosmetics are required to not dissolve

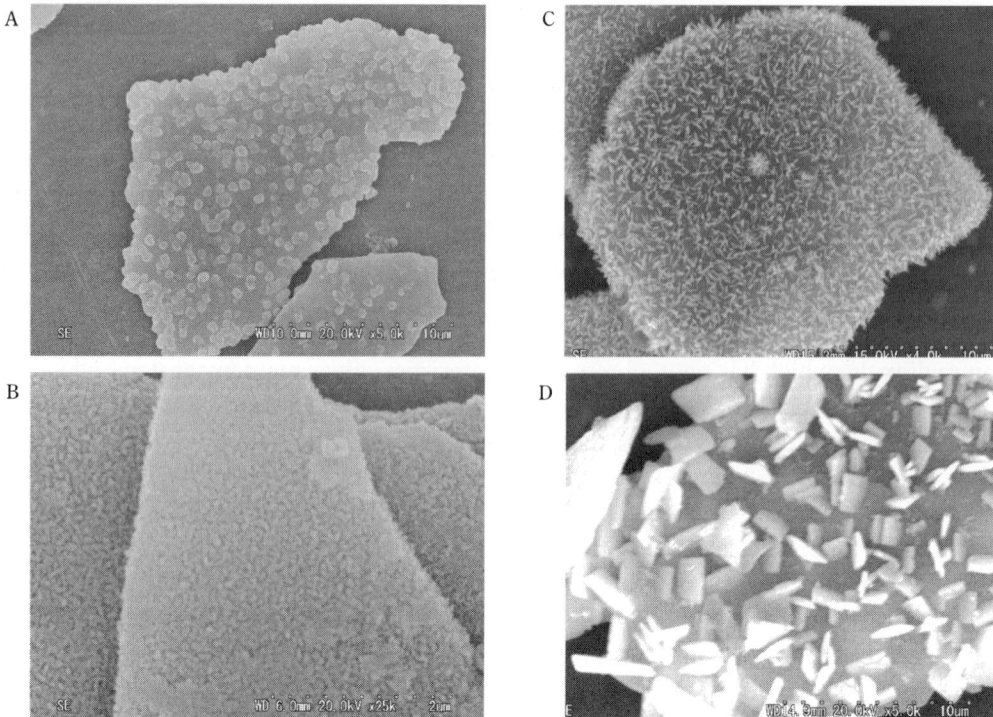

Figure 4 Plate-particle coated with shape-controlled particles
A: Grain-shape, B: Coral-shape, C: Fiber-shape, D: Plate-shape

in sebum and sweat secreted from the skin, leading to the development of silicone treatment and fluorine treatment. Besides these methods, a composite powder has been developed in which plate-shaped powder is coated with shape-controlled powder that solidifies sebum to also improve optical properties. This is called "clear quality powder" and is prepared by allowing fiber-shaped crystals of zinc oxide to grow on the surface of a pearl pigment as if covered by nanometer size feathers and is a functional optical powder (figure 4 C). The fibers of zinc oxide give a superior diffusive reflection, and the resultant makeup is translucent and natural. Zinc oxide also solidifies sebum, enabling the makeup to last long.

2. 4. 6 Blocking ultraviolet rays

Ultraviolet rays are known to cause photoaging and other adverse effects on the skin. There are two kinds of ultraviolet rays that reach the ground surface: UVA (320 to 400 nm) and UVB (280 to 320 nm). Cosmetics are demanded to block both UV ranges and not look unnaturally whitish when applied on the skin. Fine particles that are both translucent and block UV are being actively developed. There are two mechanisms for blocking UV: absorbing and scattering. Fine particles of titanium oxide smaller than 100 nm and zinc oxide have these two functions. For example, titanium oxide absorbs UV because it has a band gap of about 3 eV, which is equivalent to the energy of light of about 410 nm in wavelength, and electrons of titanium dioxide are excited by absorbing light of wavelengths shorter than 410 nm. After absorbing UV, electric charge separation occurs, and photocatalysis takes place. When used in cosmetics, the photocatalystic activities are usually controlled to not cause decomposition of other components and adverse effects on the skin, but there are attempts to actively use the photocatalystic activities. In all cases, stability and safety are most important. The light scattering characteristics of fine particles are determined by the sizes of the particles. When particles are much larger than the wavelength of the light, the dispersion is in proportion to the cross sectional area of the particle in the geometrical domain, and the light intercepting area is larger in smaller particles. When the particle size in the Mie domain similar to the wavelength of the light, light scattering becomes greatest at particle size of about 1/2 of the wavelength λ. When the particle size is much smaller than the wavelength of the light and is in the Rayleigh domain, the light scattering coefficient decreases in proportion to the sixth power of the particle size, and transparency increases. The particle size dependency of UV absorption and scattering of titanium oxide and zinc oxide in the Mie domain has been computationally calculated, and the particle sizes optimum for blocking each wavelength have been determined[9].

2.5 Surface modification of powder

The purposes of surface modification of powder in cosmetics are: (1) improving the disadvantages of the powder, and (2) giving a new function to the powder. (1) may involve improving dispersibility, enhancing resistance against light, heat and solvents, and restraining catalytic activities on other constituents. (2) includes adding hydrophobicity and/or hydrophilicity and other properties demanded to cosmetics. Surface modification frequently results in products of an entirely novel form, and surface modification of powder is an indispensable technology in development of cosmetics.

2.5.1 Surface treatment with metal oxides

To block the catalytic activities of powder, a method that involves coating with silica-alumina has been proposed. Coats of metal oxides have also been used for purposes other than blocking catalytic activities. Coating ultramarine with metal oxides to improve its resistance against acid and heat has been investigated. Ultramarine has sulfur in its structure, which is easily oxidized into hydrogen sulfide by heat, mechanical crushing and acid, causing the color to fade. In cosmetics, generation of hydrogen sulfide itself is also a problem because it smells and blackens silver containers. Coating with silica-zinc oxide or zinc oxide showed a remarkable effect in controlling oxidization. Producing cosmetics that can block both ultraviolet and

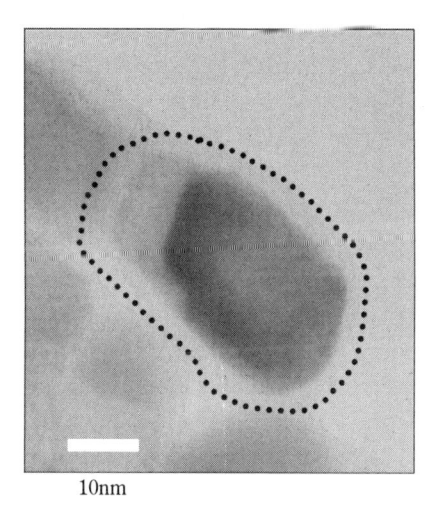

Figure 5 Conceptual diagram of functional nanocoating

infrared rays have also been proposed, which involves either producing fine particles from metal alkoxide or metal-acetylacetonate using vapor deposition and coating the surface again by vapor deposition, or coating flaky substrates with titanium oxide and zirconium oxide. Skin care powder in which zinc oxide is coated with silica has also been developed[10]. This innovative composite powder selectively attracts and inactivates "urokinase", which is an enzyme that causes rough skin. An electron micrograph of this powder is shown in figure 5. A zinc oxide particle of about 30 nm is coated with a very thin layer of silica, which is about 2 to 3 nm thick. The powder has a much higher skin care effect than zinc oxide or silica alone. This is likely because the coat of silica adds negative zeta electric potential to zinc oxide, which inactivates urokinase, and the negative electric potential attracts urokinase.

2. 5. 2 Surface treatment with oils, fats, metallic soaps, and fatty acids

This method adds lipophilicity to powders by simply adhering metallic soaps residues and fatty acids on powder surfaces or by causing chemical reactions, such as esterification or etherification, using functional groups such as hydroxyl groups on the surfaces. For example, spherical silica can be processed by esterification with higher fatty acid esters and higher alcohols and alkyl etherification with alkyl silyl. There is also a treatment method that uses lipophilic polymers, such as polyethylene and polyamide, and hydrophilic polymers, such as vinyl polymer and cellulose derivatives. Coloring an acrylic system polymer by adding a dye and treatment has also been reported. There is also a method that involves hydrolyzing fatty acid esters on the surface of titanium dioxide, esterifying higher alcohol residues with hydroxyl groups on the surface of titanium dioxide, producing mineral residues with the remaining fatty acid residues and metal ions on the surface, and making both lipophilic. This method can be performed at a relatively low temperature because the oxidization point of titanium dioxide is used. There is also a method that uses emulsions, which involves preparing an emulsion from hydrocarbons, higher fatty acids, higher alcohols or ester oils by adding surfactants, dispersing powder, and making the surface of the powder lipophilic by destroying the emulsion by adding ethanol, etc. The method is effective for improving the adhesion and extensibility on the skin.

2. 5. 3 Surface treatment with amino acid based compounds

The method was developed to improve the lipophilicity of pigments and their adhesion on the skin. There are methods that use 1) N-acyl amino acid and its amides, 2) aluminum salt and resin acid, such as fatty acid and rosin, and 3) basic amino acid and fatty acid. A representative method uses N-lauroyl-L-lysine[11]. N-lauroyl-L-lysine is simply mixed as fine particles with pigments, or used after dissolving in acid or alkaline water solvent, neutralized, and deposited. The resultant pigments have been reported to have an improved touch.

2. 5. 4 Surface treatment with siloxane coating

Most siloxane coating processes use dimethylsiloxane and methylhydrosiloxane. Particularly the latter compound has Si—H groups, which are highly reactive. The groups crosslink with each other, polymerize, and form a network of polysiloxane on the surface of the powder. There are patented methods of producing highly hydrophilic talcum powder by mixing siloxane and talc and baking together and by dissolving siloxane in organic solvents, adding zinc octylate as a catalyst for bridging, and baking. Attempts have also been made to perform these reactions by mechanochemical reactions after coating powders with metal oxides. Powders are treated with both siloxane and other oils for cosmetics to produce two way foundations, and fine particles of mica are used to produce makeup cosmetics of high transparency. There have been

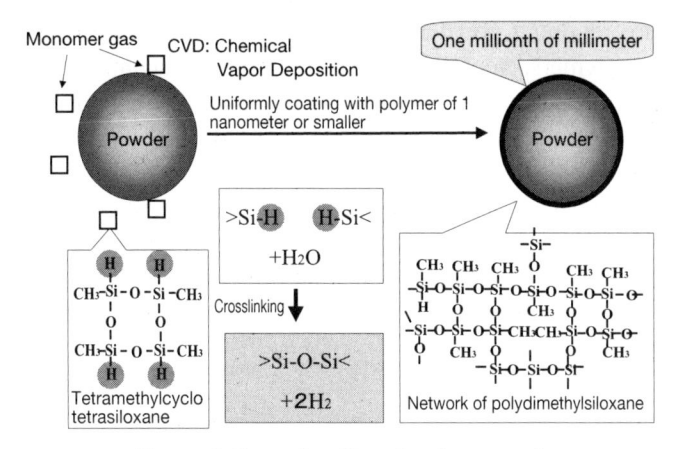

Figure 6 Example of functional nanocoating

many reports on combined treatment, including those on coating titanium dioxide with silica or alumina in forms hydrates and further treatment with siloxane and those on combining siloxane-treated powders with silicone and surfactants. There are also silane coupling agents that have alkyl groups, siloxane that has a hydrosilyl group, and siloxane that has a hydrolytic group at one end. Compounds with branched siloxane chains and acrylic siloxane copolymers have also been used for surface treatment[12].

Functional nanocoating is an example of using chemical vapor deposition (CVD), which was developed for semiconductors, in powder surface modification for cosmetics[13]. This technique is also used to produce column fillers for high performance liquid chromatography[14], stable dried flowers, and other products. Functional nanocoating involves forming an ultra thin network film of cyclic siloxane on the surface of powder and introducing unsaturated compounds into the residual Si—H groups by hydrosilylation. The dispersibility of products can be improved by the groups introduced, and the technique is used to produce lipsticks of bright colors and two-way foundations. Functional nanocoating can fix medicines, sterilizers, ultraviolet ray absorbents, pigments, enzymes and hormones, and can produce safe functional powders that are not absorbed percutaneously[15].

2. 5. 5 Surface treatment with fluorine-based polymers

Sebum causes makeups to smear and the face to look shinny and feel sticky when seeps above the makeup. Powders that are coated with fluorine compounds are resistant to both water and oil, and thus can be used to produce cosmetics that highly resist sebum. There are patented techniques of coating powder surfaces with polymers that have perfluoroalkyl groups, their phosphates and phosphate esters, silane, and siloxane; and combining siloxane and fluorine compounds has also being investigated.

A method for adhering perfluoroalkyl phosphonate on the surface of pigments has been reported, which involves making an aqueous colloid of perfluoroalkyl phosphonate diethanolamine salt acidic under the presence of the pigment[16]. Siloxane with perfluoroalkyl groups introduced into its siloxane skeleton as an oil repellant and siloxane into which alkoxy silanes are introduced as reaction groups are also used for coating powders[17].

2. 5. 6 Surface treatment with biomaterials

Surface treatment with biomaterials has also been practiced because cosmetics are applied on the skin. Moisture retaining powders have been developed by coating with or enclosing various hydrophilic polymers, such as hyaluronic acid, collagen, chitin, chitosan, salts of deoxyribonucleic acid, chondroitin sulfate, and agar. Coating with hydrogenated lecithin gives a smooth texture, high skin adhesion, and moisture retaining characteristics. There are also powders coated with silk fibroin, porous fibroin with perfume adsorbed, and proteins stained red with gold colloid. Phosphorylcholine, which is the main constituent of the cell membrane, has emulsification, solubilization, thickening and moisturizing effects. Surface treatment of the compound so as to have reaction sites suppresses protein adsorption and improved biocompatibility[18]. Because cosmetics are applied on the skin, treatment with biomaterials also gives a good impression to users.

2.6 Conclusions

Because the powder technologies mainly used in makeup cosmetics were described, probably too much weight was given on optical properties and surface treatment. In the entire field of cosmetics, powder technologies that are deeply connected to living bodies have been developed, such as those that involve active use of reduction action of nanoparticles and photocatalysts. However, there are many unclarified points in nanoparticles in terms of effects and safety, and future detailed studies on connections with living bodies are awaited.

References

1. M. Tanaka, S. Kumagai, *Chemistry and Chemical Industry* (in Japanese), **40**, 115 (1987)
2. T. Mistui, *ed.*, "New cosmetics studies (in Japanese)", NANZANDO Co., Ltd. (1993)
3. M. Abiko, *Fragrance Journal*, **34** 6, 65 (2006)
4. T. Kimura, *Hyomen*, **34**, 56 (1996)
5. K. Ohno, *Fragrance Journal*, **22**, 11 (1994)
6. K. Ohno, *J. Soc. Cosmet. Chem. Jpn.*, **27**, 314 (1993)
7. K. Ogawa, Preprints of the 43th Scientific Meeting of the SCCJ (in Japanese), (1998)
8. K. Yagi, K. Ogawa, T. Kanemaru, K. Joichi, N. Kunizawa, R. Takano, *IFSCC Magazine*, **9** 2, 109 (2006)
9. P. Stamatakis, B. R. Palmer, G. C. Salzman, C. F. Bohren, T. B Allen, *J. Coatings Tech.*, **62**, 95 (1990)
10. E. Kawai, Y. Kohno, K. Ogawa, K. Sakuma, N. Yoshikawa, D. Aso, *IFSCC Magazine*, **5**, 269 (2002)
11. K. Esumi, S. Yoshida, K. Meguro, *Bull. Chem. Soc. Japan*, **56**, 2569 (1983)
12. M. Kamei, *Fragrance Journal*, **30** 6, 81(2002)
13. H. Fukui, T. Ogawa, M. Nakano, M. Yamaguchi, Y. Kanda, "Controlled Interphases in Composite Materials", *ed.* by H. Ishida., p. 469, Elsevier Science, Publishing Co. Inc., New York (1990)
14. Y. Ohtu, H. Fukui, T. Kanda, K. Nakamura, O. Nakata, Y. Fujiyama, *Chromatographia*, **24**, 380 (1998)
15. Patent No. 1635593
16. I. Tanaka, *Journal of the Japan Society of Colour Material*, **72**, 2, 67 (2006)
17. K. Hosomi, *Fragrance Journal*, **32**, 2, 46 (2004)
18. K. Miyazawa, A. Hirayama, K. Sakuma, N. Sumita, K. Maeno, K. Takei, *Engineering Materials*, **54**, 8, 24 (2006)

Chapter 3
Composite Powder

Takahiro Suzuki

3. 1 Introduction

Makeup cosmetics contain various kinds of powder that differ in shape and physical properties. They are classified by shape into flakes, spheres, fibers and micro particles, and different kinds are used for each purpose. Powder plays the main role in covering the entire surface of the skin, adjusting its color, covering stains, freckles, and wrinkles, which are local defects, and improving the appearance. Composing different kinds of powder and covering one with another are effective for making use of different functions and to mutually cover disadvantages.

3. 2 Shapes and physical properties of extender pigments

3. 2. 1 Aspect ratio and a powder shape index

The majority of extender pigments applied on the entire face is flat and flaky. This is because the flaky shape is appropriate for spreading smoothly and fixing firmly on the skin surface. Mica and talc powder used in foundations have particle diameters of about 5–20 μm and a thickness of about 0.2–0.5 μm. Aspect ratio and power shape index are used to show the shapes of powder (figure 1). Aspect ratio is the mean particle diameter divided by thickness. The particles of talc prepared by dry grinding are relatively thick, thus have small aspect ratios of about 10–30, and look relatively mat. On the other hand, mica flakes prepared by wet grinding have aspect ratios of 40–80 and look lusty and transparent. The mean particle diameter of powder can be determined by sedimentation speed measurement, laser beam particle diameter measurement, zeta potential particle size analysis and ultrasonic attenuation particle diameter measurement. The effects by the thickness and grain size distribution of the particles vary by measuring method, and the shapes of the particles must be taken into account for determining mean particle diameter. On the other hand, thickness is usually determined by measuring the thickness of particles of almost the mean particle diameter on scanning electron microscopic images and calculating the mean. The thickness can also be roughly estimated by calculating the aspect ratio from the relationship between Laser beam particle diameter measurement and sedimentation speed. The optical characteristics, such as luster, transparency and scatter reflection, of flaky powder depend mainly on the aspect ratio[1].

As thickness increases, the powder spreads less, adheres less to the skin and each other, and increases in bulk and filling densities. Powder foundation that contains a large amount of thick powder is hard itself and is hard when used. Because these properties depend on the aspect ratio and the thickness of the powder, the powder shape index, which is calculated by dividing the aspect ratio by thickness as shown below, is a more useful index.

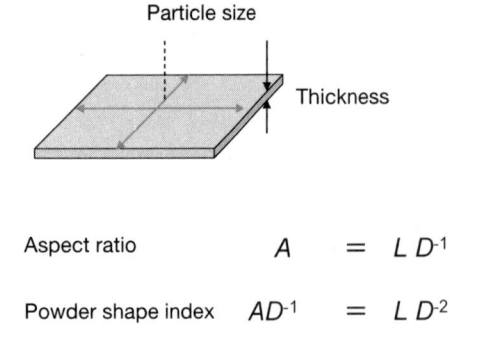

Figure 1 Aspect ratio and powder shape index of flaky powder

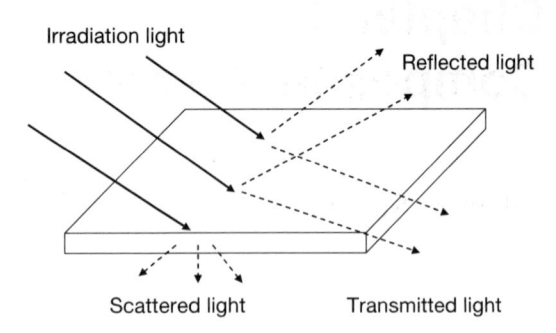

Figure 2 Luster and scatter reflection of a flaky powder

Powder shape index = aspect ratio ÷ thickness
= particle diameter ÷ (thickness)^2

In other words, when powers that have the same aspect ratio but different particle sizes are compared, the smaller powder has a larger powder index because the smaller powder is thinner. Powder with a larger powder shape index adheres better, is easier to form a card house structure, and thus has a lower bulk density. The physical properties of foundation cakes, such as the hardness, adhesiveness to the skin, and oil absorbency, depend mainly on the bulk density, and thus powder shape index is useful[1].

3.2.2 Luster and scatter reflection

Of the optical properties of powder, scatter reflection is determined mainly by aspect ratio and particle size distribution. As shown in figure 2, irradiated light either penetrates through or regularly reflects at the surface of flaky powder and little scatters. On the other hand, light scatters on the side surfaces of thick and flat particles. Because the side surfaces are rough and/or uneven, light that penetrated through particles and light that reflect on the side surfaces are scattered to every directions. Side surfaces account for a large percentage of the total surface area in thick and large particles, small-diameter particles and powder of non-uniform grain size, resulting in increased scattering. This is why talc, which has a small aspect ratio, is not transparent and looks white and mat. Talc is usually prepared using the dry grinding method at low cost. There are talc products prepared using the wet grinding method to increase the aspect ratio and washed to remove fine fragments to make the grain size uniform. Such products are lusty and transparent.

3.2.3 Bulk density and sensory evaluation

Powder foundation that is soft on the skin is favored. The soft feeling is provided by low bulk density. Because talc has a large bulk density and becomes firm easily, foundation cakes that contain a large amount of talk are hard, and the hardness is felt when scraping off the powder from the press-molded cakes with a sponge and when putting the powder on the skin. On the other hand, mica has a small bulk density, and cakes that contain a large amount of mica entrain a large amount of air and are soft. Powder can be scooped onto a sponge just by rubbing it softly, and can be put on the entire face just by touching the skin with the sponge softly. Because the powder of a low filling density spreads on the skin together with the entrained air, it is soft and light to use. The thickness of talc is 2–5 folds larger than that of mica. Mica, which has a high aspect ratio and large powder shape index, has a small bulk density and is thus soft and light to use. Similarly, porous powder gives the feeling of softness because it has a small bulky density[2].

The more uniform the particle size, the more the bulk density can be reduced. On the contrary, bulks of large grain size distribution contain large amount of particles of small aspect ratios and result in increased filling density and hard products, which are not soft to use. The bulk density of powder used in cosmetics is determined by filling a scaled test tube with a specified weight of the powder, tapping the test tube for several hundreds of times, and measuring the volume and weight when the filling density maximizes and stabilizes. A relationship between bulk density and powder shape index is shown in figure 3. The figure shows that the bulk density is lower for larger powder shape index. As shown in figure 4, powder of large powder shape index entrains a large amount of air, is easy to form a card house structure, and results in low bulk density. Similarly, at the same aspect ratio, the smaller the particle size results in large powder shape index, smaller bulk density and thus softer to use. When powder particles are thick, large load is exerted on the contact unit area of the card structure, resulting in the bulk powder to tighten and the resultant cake to be firm. On the other hand, thin powder particles easily forms the card house structure with a large amount air entrained because the load exerted per unit contact area is small. As a result, the larger the powder shape index, the smaller the bulk density. On the contrary, powder that contains much fragments and fine

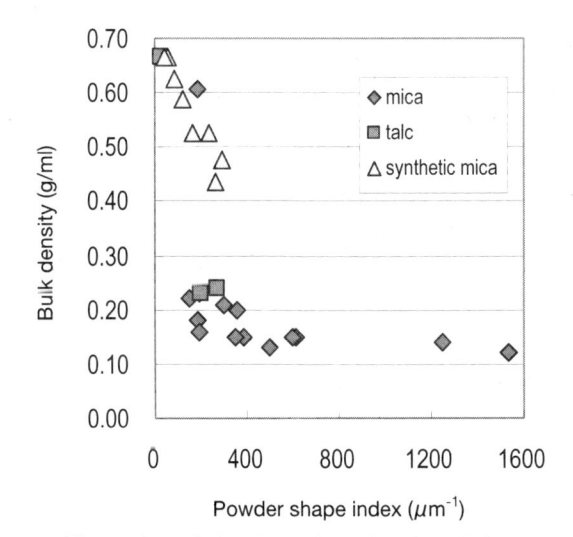

Figure 3 Bulk density and powder shape index

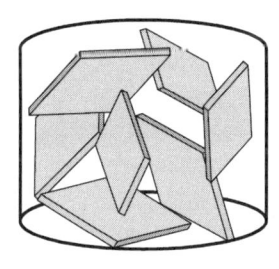

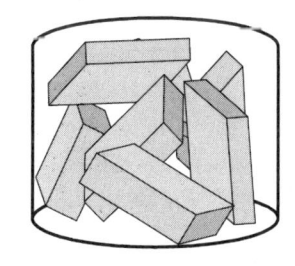

Powder of high aspect ratio   Powder of low aspect ratio

Figure 4 Difference in bulk density by aspect ratio and filling characteristics

particles is prone to tightening, and thus the resultant product becomes hard. Refined powder from which fragments were removed has low bulk density and results in products that entrain much air. Powder that has a uniform grain size and a large powder shape index gives a feeling of soft to use.

The majority of commercially available powder foundations have mean particle diameters of several to 20 μm when measured using the dry measuring method, and the smaller the mean grain size the softer to use. Because grain size values measured using the dry method are also affected by the cohesiveness of the powder, the values are different from the actual grain size distribution of the powder used. Foundation cakes contain flaky powder, fine particles and spherical particles of color pigments, surface coating agents, and 10 to 20% of oil ingredient to enhance the molding bond. The oil improves the molding nature of the powders and reduces the dispersibility of the powders. Therefore, bulks that have small mean particle diameters even measured by the dry method are little affected by oil and low in cohesion among particles and thus disperse well on the skin and are soft to use. In this way, the feeling of using actual products is affected also by the oil components in the bulk.

3. 2. 4 Adhesion of powder

A comparison of adhesion between various powders used in cosmetics, such as mica, talc and synthetic mica, is shown in figure 5. The adhesion of powder under the dry condition depends mainly on the powder shape index. In other words, the force of adhesion per unit area mainly depends on the van der Waals forces when the powder has flat and chemically stable surfaces. The force of adhesion between powder particles is larger for thinner particles with larger powder shape index as shown in figure 6. Such a dependency is more notable in powders of more uniform grain size distribution. Adhesion between particles is small in thick particles because the load per unit contact area is large.

Under a condition in which static electric force can be ignored, the bonds between powder particles via liquid, such as oil, are affected by the capillary attraction due to differences in pressure between the inside and outside of the liquid film (F_p) and the surface tension of the liquid film (F_s) as well as the van der Waals forces (F_w)[3–5] as shown in the following equation:

Adhesive force between powder particles = $F_w + F_p + F_s$

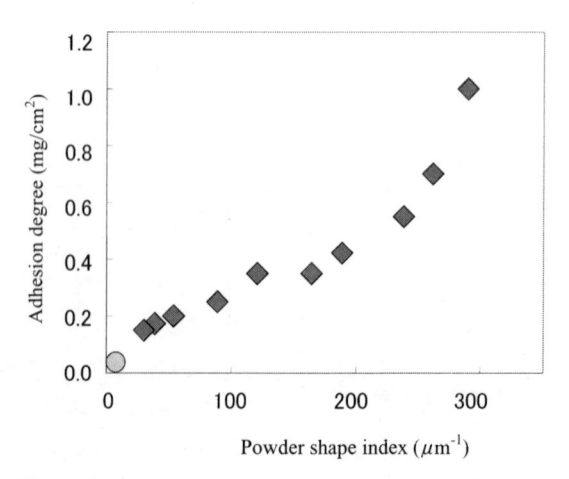

Figure 5 Powder shape index and adhesion of cosmetics

Strong adhesion ⬅➡ Weak adhesion

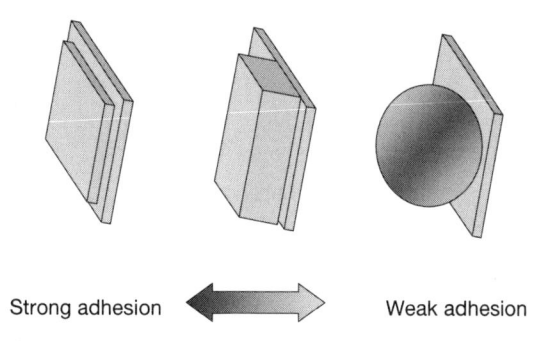

Figure 6 Particle shape and adhesive force between particles

Oil of strong polarity is suitable as a molding agent because it enhances bonds. When applied on the skin, the polar oil that coats the particles facilitates the particles to adhere onto the skin. The oil also makes the surface of the particles hydrophobic and increases the resistance against sweat and water, but polarity too strong results in adsorption of water and thus should be adequate. Polar oil is also prone to mixing with free fatty acids on the skin adsorbing sebum, making the skin shinny, formation of an unpleasant makeup film, and accelerated running of the makeup[6].

3. 2. 5 Coefficient of kinetic friction of powders

The smoothness and extendibility of various kinds of powder can be evaluated by measuring the coefficient of kinetic friction. Powder foundation that has a small coefficient of kinetic friction spreads smooth and uniformly on the skin, is soft to use, and results in good makeup. The coefficients of kinetic friction of various kinds of powder used in foundations are shown in table 1. The coefficient of kinetic friction of flaky boron nitride is as small as 0.2 or smaller, and thus it is superior to talc and mica in extendibility and sensory evaluation. The coefficients of kinetic friction of main flaky fillers used in foundation are 0.1–0.5, but those of fine particle pigments, such as iron oxide and titanium oxide, are as high as 0.6–1.0. There is a trend of the smaller the particle, the larger the coefficient of kinetic friction. Microparticles that are added as color agents are a big cause for deteriorating extendibility. The extendibility of a bulk that contains oil is affected also by the adhesiveness of the oil. Viscous oil, which is cohesive, raises the coefficient of kinetic friction of the bulk, and low viscous oil with lubricant characteristics, such as wax, decreases the coefficient of kinetic friction.

Because spherical and semi-spherical powders, which have smooth curved surfaces,

Table 1 Coefficients of kinetic friction of various powders used in cosmetics

Pigment	Coefficient of kinetic friction
Talc	0.2–0.3
Mica	0.3–0.5
Sericite	0.3–0.5
Synthetic mica	0.3–0.5
Boron nitride	0.15–0.3
Barium sulphate	0.3–0.5
Iron oxide	0.5–0.7
Titanium oxide	0.6–1.0

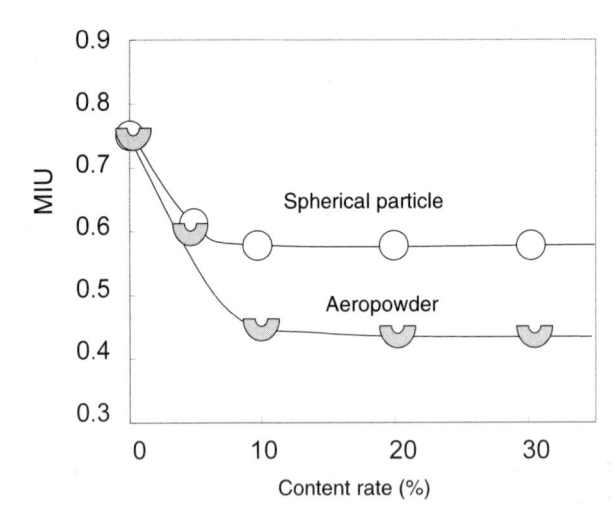

Figure 7 Changes in extendibility of the bulk powder by addition of spherical particles

reduce friction areas and serve as lubricant improving the extendibility of other powders, the coefficient of kinetic friction of the whole bulk is reduced by adding this kind of powder. Changes in the coefficient of kinetic friction (MIU) caused by addition of spherical or bowl-shaped aeropowder to mica are shown in figure 7[2]. A 10% addition of the malleable powder sharply reduced the coefficient of kinetic friction.

3. 3 Shapes and physical properties of coloring pigments

Makeup cosmetics contain various kinds of color pigment to adjust the color of the skin. Major inorganic color pigments contained in foundation are red, yellow and black iron oxide and white titanium oxide. Inorganic pigments are superior to organic pigments in concealment and stability. Organic pigments can provide bright colors but are prone to fading by sebum, sweat, and photo oxidation and are not suitable for light-color cosmetic products, such as foundation, but are mostly used in lipsticks, nail enamels, and other point makeup products.

Fine particles of iron oxide and titanium oxide are 200 nm to several hundreds of nanometers. Fine particles of titanium oxide used to block UV are smaller and colorless because light of visible wavelengths penetrates through the particles. These ultra fine particles of pigments give a totally different feeling from the flaky extender pigments when applied on the skin[7].

As the particles are small, their end surfaces are big braking the movements of the powder particles, resulting in reduced fluency[8]. Particles of iron oxide and titanium oxide have indeterminate forms to enhance concealment and light scattering characteristics, reducing the rolling characteristics and extendibility. Their coefficients of kinetic friction are as large as 0.6–1.0 and are much larger than those of mica and talc (0.2–0.5). Because their dispersibility is low, they are spread on the skin together with flaky particles in the bulk.

The dispersibility of the pigment particles affects the chroma and uniformity. When color pigment is applied on the skin in clumps, the coloring effect drops, and it is necessary to apply a large quantity of powder. This results in thick and unnatural makeup. When the clumps get wet with sebum, chroma rises, causing changes in the color of the makeup film and darkening of the skin. Because such a change is easy to occur in sweat pores and the sulcus cutis, it may result in highlighting sweat pores and wrinkles. On the other hand, good dispersion of primary pigment particles results in good chroma even with a little amount of foundation, effective makeup and good sensory evaluation. Uniform distribution of primary pigment particles enables the effect of adjusting the color of the skin to be well manifested and improves durability and the total performance of the product.

3.4 Compositing powders

3.4.1 Objectives of compositing powders

Extender pigments and color pigments, which are mixed in foundations, have makeup functions that depended on their physical properties. Each has advantages and disadvantages related to purpose and usage. Pigment particles are composited to improve optical performances and comfort of use. However, it is not easy to composite powders, which mutually differ in physical properties and shape. Because the dispersibility of fine pigment particles, such as iron oxide and titanium oxide, is low, relatively strong force is needed to mix materials. Therefore, during the mixing process, flaky powder is prone to crushing, which leads to loss of transparency, luster, and softness. The coloring effect is lost when the dispersibility of a color pigment is low, which also causes a problem of the makeup film being prone to changing color when getting wet with sweat or sebum. Because these fine particles have large coefficients of kinetic friction and low lubricity, their inclusion greatly affects the sensory evaluation of the resultant foundation. The differences in the extendibility of foundation attributable to its fine particle component is sensed when the foundation is applied on the skin. To resolve the problem of low dispersibility and improve the feeling when used, composite particles are prepared by uniformly dispersing pigment particles on the surfaces of flaky powders.

Flaky powders that have large powder shape indices and aspect ratios are lustrous and may cause glaring and shiny makeup. The light scattering characteristics of the powders are lost when they get wet with sweat or sebum, causing the desired makeup effects such as concealment to be lost. The light scattering and light absorbing characteristics are improved by covering the surfaces of the flaky powders with micro particles, which reduces glaring and color changes. Compositing powders that differ in refractive index is also effective for enjoying the coloring effects of interfering lights and their combined effects.

Major objectives and uses of compositing powders are shown in table 2. There are methods that involve forming chemical bonds, binding organic molecules such as oil, and forming a cover of inorganic oxides and pigments on substrate powders.

3.4.2 Coating the surface of flaky powder

The surface of powder can be coated either with oil or fine powders. Oil coating, which involves making the powder water-repellent, prevents makeups from running and smearing by sweat

Table 2 Coating and composition of the cosmetic powder

Substrate powder	Coating agent	Application
Flaky powder Fine pigment particles Composite powder	Hydrophobic oil agent Modified Organic silicone	Water-repellency, Extendibility, Dispersibility, Adhesion, Press molding
	Perfluoro agent	Sebum-resistance, Sweat-resistance
	Inorganic film	Dispersibility, Stability, Sensory evaluation
Flaky powder	Inorganic pigment Organic pigment	Uniform dispersibility, Sensory evaluation, Optical effect, Coloring effect,
Fine pigment particles	Inorganic pigment Organic pigment	Uniform dispersibility, Optical effect, Coloring effect

and sebum and maintains good appearance for a long time.

In powder coating, the surface of the substrate powder is coated with a single or two or more layers of fine particles. Covering the surface of flaky powder with fine particles of iron oxide and/or titanium oxide is effective for improving the dispersibility and sensory evaluation, adjusting chroma, and using colors created by interference light of specific wavelengths.

The feeling of use also changes by composition. Different coating and compositing techniques are used depending on purpose. The characteristics of general methods are described below.

To make powder water-repellent, the surface of the powder is coated with hydrophobic oil. The oil can be fatty acids, higher alcohols, esters, lipids, hydrocarbon oils and silicone, which are used depending on the purpose. Attaching long chains of alkyl groups from the powder surface toward the outside makes the powder smooth and easy to spread on the skin. Esters are superior in adhesion to and extendibility on the skin and water-repellency. Lipids and the fatty acids are effective in keeping the products in shape and improve the feeling of use. However, fatty acids, lipids, and esters have structures similar to the components in sebum; and thus they cause the oil absorption capacity and sebum resistance of the bulk powder to drop. This is because a film of the coating agent fills up the spaces among powders, mixes with sebum and promotes the formation of a continuous oily film. Fixing stearic acid on the surface of synthetic mica of an aspect ratio of 50 at 1% in weight ratio, the oil absorption capacity of the mica is halved. This is because stearic acid, which covers the surface, is a fatty acid that belongs to the same group with that of a main constituent of sebum and assists the formation of a continuous sebaceous film. Thus it is rather disadvantageous in terms of preventing shiny and smearing makeup. This is one of the reasons for using silicone to form a water-repellent coat. Silicone has a comparatively low compatibility with the sebum and causes the oil absorption capacity to drop only slightly, and thus the resultant powder is sebum resistant.

A problem of silicone is the touch. Methyl hydrogen polysiloxane (abbreviated as methicone), which is widely used in cosmetics, cannot much reduce the squeaky feeling of powders, and thus methicone-coated powders are rather dry and dusty. The dryness is mainly attributable to the siloxane frame structure. To reduce the dryness, alkyl silicone oil, in which alkyl chains consisting of 12–18 carbons are bound to side chains, and poly siloxane, in which side chains and branched chains are substituted by organic chains such as esters and acryl, are used[9]. These organic chains give more smooth touch than methicone and are effective in controlling the dry feeling. Lipophilicity rises as the percentages of organic chains connected to poly siloxane increases, and the sebum resistance drops. The balance between poly siloxane and organic chains is an important factor for improving the feeling of use and making long-lasting makeup films.

Perfluoro oil agent is used to increase both water repellency and oil repellency. Perfluoro coating increases oil repellency but decreases the affinity to the skin. Because affinity to the skin is related to lipophilicity, a method has been investigated in which perfluoro oil is combined with another organic agents to give sufficient water repellency and oil repellency by reducing the amount of perfluoro oil used[1].

3. 4. 3 Coating the surface of fine pigment particles

Fine particles of iron oxide and the titanium oxide have particle sizes of 200 nm to several hundred nanometers, poor lubricity, low dispersibility and low extendibility on the skin. Coating the surface of these fine particles prevents cohesion between the particles, improves dispersibility in oil agents and chroma, forms films to prevent quality changes caused by chemical reactions with other components, and prevents the makeup from turning dark and changing color when it gets wet with sebum.

The surface of these fine particles is coated as in flaky powders. Changes in the color of makeup films are caused by drops in light scattering effects and increases in transmissivity by wetting of the pigment particles by sweat and sebum. Red and yellow iron oxide fine particles used in foundations should maintain stable chroma and colors over a long time and are combined in ratios that satisfy the tastes of users. Therefore, perfluoro is widely used to coat the surface of pigments to provide oil repellency as well as water repellency to prevent the effects of sebum. Perfluoroalkyl phosphate, which is used to coat the surface of cosmetic powders, can increase water and oil repellency, but is adhesive and may adhere primary particles. Thus, coating should be performed by controlling the dispersion state of the particles. In addition, special attention is needed on the combination with the oil agent contained in the product because increased oil repellency decreases dispersibility in the oil phase. Various hybrid surface coating agents have been developed to improve the adaptability with oil agents, but all have advantages and disadvantages[10].

When dispersion fine particles of pigments in an emulsion such as liquid foundation, drops in repulsion between the droplets may occurs when mixing and dispersion of the particles are insufficient and when insufficiently coated particles remain at the oil-water interface and coalesce. This destabilizes the emulsion, which becomes prone to separating into oil and water phases. Coating the surface of fine particles thoroughly improves the dispersion state in oil and stabilizes of the emulsion.

The surface of composite powder of several hundreds nanometers coated with particles of color pigments is inferior to the substrate flaky powder in smoothness and extendibility. Therefore, the surface of the composite powder is also coated to improve the feeling of use. When the substrate or pigments are wet with sweat or sebum, they lose the light scattering characteristics, causing changes in chroma. Coating the outermost layer prevent the powder from getting wet[11].

Surface coats of composite powder also serve a bonding agent between the substrate and micro particles. For the coating agent to form crosslinks between particles, the molecules must have at least two binding groups and at least a certain length, and thus methicone, acrylic acid polymers and modified organic silicones into which alkyl chains are bound to improve the texture are used.

In the process of mixing materials for producing powder foundations, the dispersibility of both the particles and the oil agent to add as a molding agent is important. Surface coating agents also increase the diffusibility of the oil agent and make the products light to use because the amount of oil can be reduced. Because the particles are uniformly dispersed when lightly stirred, the coating prevents particles from breaking during the mixing process of flaky powders of high aspect ratios and facilitates scaling up.

3. 4. 4 Composite powder of extender pigment and fine pigment particles

Composite powder of extender pigments, such as mica and talc, and fine pigment particles, such as iron oxide and titanium oxide, increases the efficiency of mixing raw materials, improves the feeling of applying the product on the skin, and improves coloring characteristics and the durability by uniformly dispersing the fine pigment particles. Both extender pigments and fine pigment particles, such as titanium oxide, are prone to aggregation alone and low in dispersibility. When composite powder is produced by coating the surface of extender pigment with fine pigment particles, the cohesion among powder particles drops, and dispersibility is sharply improved. Fine pigment particles on the surface also scatter light, which is otherwise reflected straightly by the flaky substrate, and improve luster and chroma.

Flaky powder of mica and talc spread smoothly on the skin and covers it uniformly. On the other hand, smaller pigment particles of iron oxide and titanium oxide, which are 200 nm to several hundreds of nanometers, are low in dispersibility and extendibility. When the particles accumulate and aggregate in sweat pores and sulcus cutis, they cause unnatural finish and reduced durability. To improve the problem, the fine particles are uniformly dispersed and fixed on the surface of the flaky substrate in advance, which facilitates the particles to be uniformly applied on the skin together with the flaky powder.

The thickness of flaky powder that has an aspect ratio of 50 and a mean particle diameter of 10 μm is 0.2 μm. Spreading fine pigment particles of 0.4 μm in diameter in one layer on this substrate, the thickness of the composite powder becomes 1 μm, and the apparent aspect ratio becomes 10. The powder shape index, which is calculated by dividing aspect ratio by thickness, decreases from 250 in the flaky powder to 10 of the composite powder. Therefore, the bulk density increases, and the compound powder becomes heavy powder. The surface coated with the fine pigment particles is not smooth, and the composite powder is less adhesive than the flaky substrate also because the adhesive area among powder particles per unit area and unit space is reduced.

The number of fine particles that can be fixed on the surface of this flaky powder in a single layer is several hundreds the maximum. Because the coat layer formed on both sides of the substrate becomes 4-fold thicker than the substrate, the weight coating rate of the fine particles per flaky powder is 4-folds larger the maximum when the substrate and the coat have are the same in specific gravity. Generally, the powder foundation consists of 50% to 70% flaky extender pigment powder and 10% to 20% fine pigment particles of iron oxide and titanium oxide. In other words, ordinary powder foundations have 4-times larger amount of flaky powder than fine particle pigments. Therefore, replacing several percentages of the flaky powder by the aforementioned composite powder (flaky powder : fine particle = 1:4) is sufficient for giving the necessary amount of fine pigment particles.

Examples of calculated powder constitution ratios of makeup films produced when pressed foundation is applied on the skin are shown in table 3. The makeup film on the skin is estimated to contain 2–4 layers of flaky powder on average and 0.3–0.5 layers of pigment particles when it is assumed that the density of the makeup film is 0.3 mg/cm^2, 67% is flaky powder of mica and talc of aspect ratios of 30–50 and particle diameter of 10 μm, and 17% is pigment particles of iron oxide and titanium oxide of aspect ratios of 1–2 and particle diameter of 0.4 μm. The amount of powder foundation applied on the skin is usually 0.1–0.5 mg/cm^2 [6] although the amount may vary by the part of the face. Thus, fine pigment particles are discontinuously dispersed covering the entire face. There are unevenness on the skin, such as sweat pores,

Table 3 Calculation powder constitution ratio of powder foundation applied on the skin

	Particle size	Thickness	Specific gravity	Ratio of combination	Quantity of application	Powder thickness	Number of powder layers
	μm	μm	g/cm^3	%	mg/cm^2	μm	
Flaky powders	10	0.15–0.4	3	67	0.2	0.7	2–4
Fine pigment particles	0.5	0.2–0.4	4.5	16	0.05	0.1	0.3–0.5
Oil agents, Others				17	0.05		
Total				100	0.3	0.8	2–5

wrinkles and the texture of the skin itself. Needle-shaped and powder of indeterminate forms are prone to accumulating in such pores and wrinkles. Flat powder with large aspect ratios is estimated to form a uniform film of uniform color on the skin as calculations showed that flatter the powder the more layers can be formed. However, flaky pigments are low in light scattering and masking effects, thus cannot cover stains and spots. This may lead to increased use of color pigments, which causes color changes and unpleasant feeling at use.

There are tiny spaces among powder particles in a makeup film on the skin. The density of such a film is slightly larger than the bulk density of the bulk powder determined by the tap method. The bulk densities of products on the market are about 0.5–1 g/cm^3. When the bulk density of a makeup film on the skin is assumed to be 1 g/cm^3, the apparent thickness of the makeup film is about 3 μm when 0.3 g/cm^2 is applied. Therefore, the thickness of the composite powder to be mixed in the base makeup is better be smaller than several micrometers. Spherical powders, porous powders, and fibrous powders, which are larger than or thicker than 5 μm project out from the makeup film, disturbing the uniformity, and may cause the makeup to run. The large particles are low in adhesive power and easily fall off from the crista cutis of skin but accumulate in sweat pores, wrinkles and ditches, scattering light there and erasing the shades. Talc and other powders that are thick and have small aspect ratios are widely used on loosened skin that lost brightness due to aging, *etc.* Spherical particles, which easily accumulate in depressed parts, are used to reduce the thick and unnatural look of talc and enhance the covering effect.

3. 4. 5 Light interfering pearl pigment

Pearl pigment prepared by laminating titanium oxide, which has a large refractive index, on transparent flaky substrate of mica and other substances of small refractive indices reflects light, provides a bright look and increases chroma. Because the wavelength of the light interfered can be controlled by adjusting the thickness of the lamina, the color can be adjusted depending on purpose. The thickness of the lamina reaches 200 nm to 1 μm, and on a substrate of a particle diameter of several μm, whiteness increases and chroma drops because light is highly scattered at the end surfaces. When large substrates of several tens of micrometers are used, strong luster and chroma are provided, thus the pearl pigments are used in point makeup cosmetics that need luster and color, such as eye shadows and lipsticks.

As a person ages, the amount of substances that absorbs light such as melanin increases in the skin, and thus the amount of light reflected and radiated on the skin surface decreases, causing the skin to look dark and blotchy. To improve the appearance of the skin, the reflection

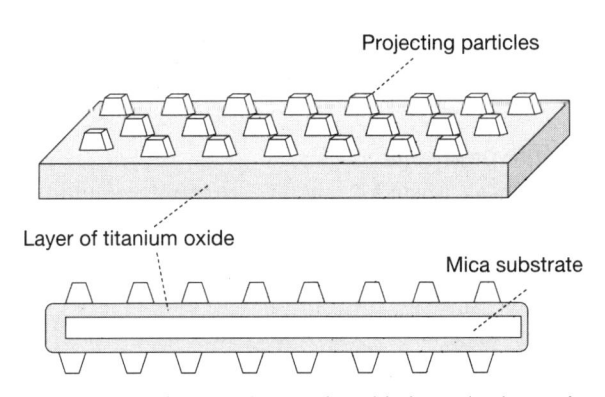

Figure 8 Multilayered composite powder with the projection to the outer layer

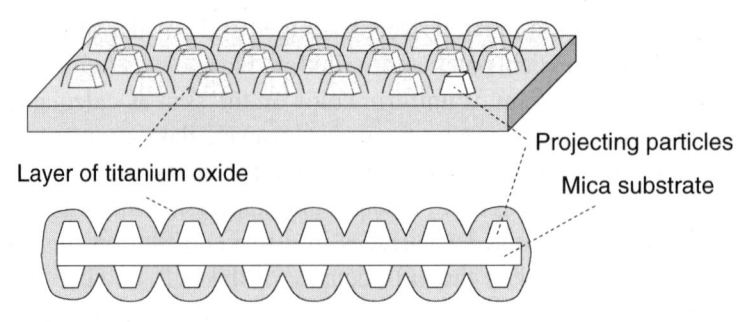

Figure 9 Multilayered composite powder with projections board

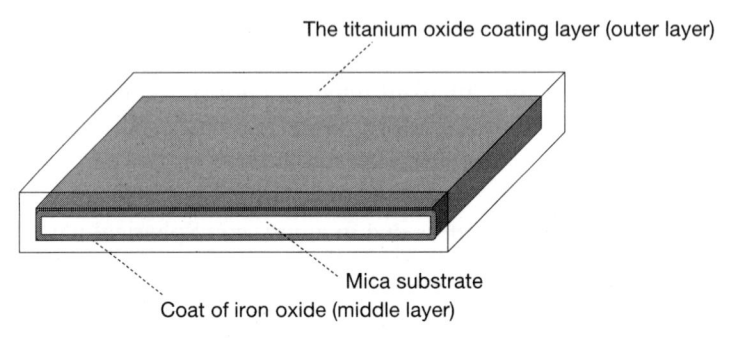

Figure 10 Multilayered composite powder with double coats

of light should be increased. Composite powder shown in figure 8 is used for this purpose, in which projecting particles are attached on the surface of pearl pigment to enhance scattering of light while keeping the light volume high to suppress glary look[12]. A composite pearl pigment has also been developed (figure 9), in which barium sulphate, which has a refractive index higher than that of mica, is attached scattered on a mica substrate and the entire powder is coated with titanium oxide[13]. There is also a multilayered pearl pigment (figure 10) prepared by inserting a thin layer of iron oxide between mica and titanium oxide and is used to provide high chroma[14]. It is also necessary to increase the dispersion of light inside the skin to make the skin look transparent[15]. There is also a composite powder consisting of mica with projections of titanium oxides coated with silicone polymers[16].

Light interfering pearl pigment, which highly reflects light, is prone to shining and losing transparency over time by sebum secretion. To suppress the effects of sebum and increase light scattering, composite powder in which nanofiber of zinc oxide are scattered has been developed[17]. The powder has been reported to keep transparency over long time due to the sebum solidification function and the diffusion reflection characteristics of zinc oxide.

3. 4. 6 Compositing organic pigments

Point makeup cosmetics, such as lipsticks and eye shadows, contain organic pigments to improve coloring. Organic pigments give high chroma but are inferior to inorganic pigments in concealment and chemical structural stability and are easy to fade. When micro particles of organic pigments are added, the pigments may dissolve in its oil component and into sweat and or sebum by the activity of surfactants. The color may fade by the photocatalystic reactions of titanium oxide and UV rays. Therefore, organic pigments are inappropriate for base makeup, which is to be applied on the entire face, but are used mainly in lipsticks, nail enamels and eye shadows.

To effectively use the coloring performances of organic pigments, they are stabilized by compositing with inorganic pigments, surface processing and crosslinking. Covering the surface of titanium oxide particles, which highly reflect and scatter light, with organic pigments result in bright chroma and clear colors even when prepared into lipsticks. Composite pigments prepared by coating organic pigments with silica or polymers are resistant against fading, are stable and can be used in foundation makeup.

3. 5 Conclusions

The recent major stream of developing ingredients for makeup cosmetics is to develop powders of high efficacies to efficiently improve the overall performance of the entire bulk. Because the selling price of products cannot be much increased, the majority of raw materials are common ones, and a small amount of novel high-performance ingredients is added to improve the effects of products. Compositing powders is attracting attention to develop such products. To develop and use composite powders, one must know the disadvantages and makeup performances of each powder and must understand the methods for improving the disadvantages while improving the demanded functions and effects. Recent development of technologies has enabled the effects of product properties on the look of the skin and the feeling at use to be understood in detail. Powder materials are now developed and improved rationally based on designing of micro structures. Development and application of composite powders will continue backed by detailed analysis of makeup effects and progress of evaluation technologies.

References

1. T. Suzuki, *Fragrance Journal.*, **34** (4), 45 (2002)
2. T. Suzuki *et al.*, Latest technologies for creating functions and developing materials of cosmetics and applicational technologies (in Japanese), p. 371, Gijutsu Kyoiku Publishing Ltd. (2007)
3. T. Kani *et al.*, *Powder Technol.*, **176**, 99 (2007)
4. T. Kani *et al.*, *Powder Technol.*, **176**, 108 (2007)
5. A. Fukunishi *et al.*, *J. Soc. Powder Technol. Jpn.*, **41**,162 (2004)
6. N. Arimura *et al.*, *J. Soc. Cosmet. Chem. Jpn*, **22** (3), 149-154 (1988)
7. H. Tsujimoto *et al.*, *Cosmetic Stage*, **1** (4), 46 (2007)
8. T. Uchiyama *et al.*, *Kagaku Kogaku Ronbunshu*, **22** (3), 655 (1996)
9. M. Kamei, *Fragrance Journal.*, **34** (6), 81 (2002)
10. T. Tanaka, *Fragrance Journal.*, **35** (3), 21 (2007)
11. S. Toui *et al.*, *Fragrance Journal.*, **35** (3), 16 (2007)
12. K. Yagi, *Fragrance Journal.*, **35** (3), 10 (2007)
13. T. Noguchi, *Cosmetic Stage*, **1** (1), 72 (2006)
14. E. Misaki *et al.*, Proceedings of 24th IFSCC, 45 (2006)
15. A. Matsubara, *Fragrance Journal.*, **35** (1), 61(2007)
16. M. Igarashi, *Cosmetic Stage*, **1** (2), 62 (2006)
17. K. Ogawa *et al.*, *J.Soc. Cosmet. Chem. Jpn.*, **39** (3), 209 (2005)

PART 2 Application of Nanotechnology to cosmetics [Equipment]

Chapter 1
Nanoemulsion Manufacturing Equipment

Kazuyuki Takagi

1.1 Introduction

Nanotechnology is attracting attention as a new technology because nanoparticles have a critical particle diameter at which the physical properties change greatly. The physical properties of nanoparticles are unique and, in many cases, useful. Here, particles of about 100 nm are regarded to be nanoparticles.

There are two general methods for manufacturing nanoparticles.
(1) Breakdown method (Top down method)
This method involves crushing big particles using mechanical power to produce nanoparticles. Possible mechanical powers are: compression and pressing, which act directly on particles, impacts (figure 1) and shearing (figure 2). Some emulsification methods, such as forced emulsification, also belong to this category.
(2) Buildup method (Bottom up method)
The method involves preparing a solution and growing crystals to produce nanoparticles. Some emulsification methods, such as emulsion polymerization and drying emulsions, also belong to this category.

Supercritical methods can also be classified into these two. The long-used method that uses the low surface tension of a supercritical fluid and its expansive force while returning to ordinary pressure is a breakdown method. A recently developed method that involves growing crystals using the dissolubility of a supercritical fluid is a buildup method. Both need production devices. Mechanical power plays the main role in breakdown methods, and prescription is the key in buildup methods.

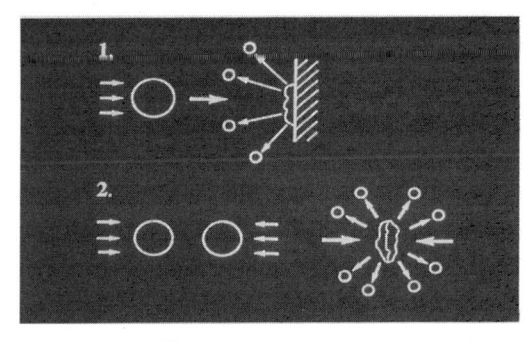

Figure 1 Impact force

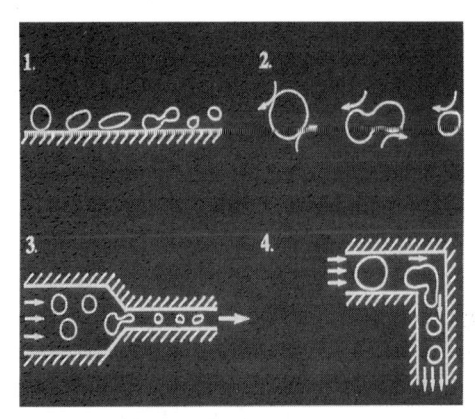

Figure 2 Shearing force

Figure 3 Cavitation

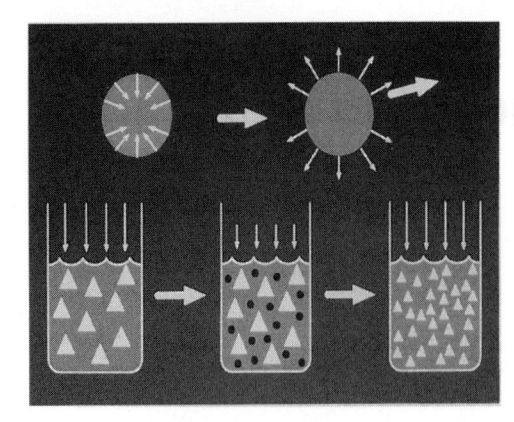

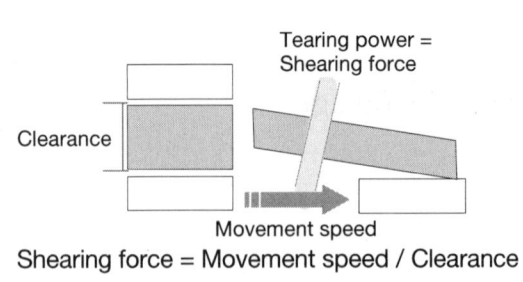

Figure 4 Shearing force

Generally, particle sizes of up to 100 nm can be produced using breakdown methods. It is believed difficult to produce particles smaller than 10 nm with methods other than buildup methods.

Effective mechanical power for producing fine particles differs by purpose. For example, for crushing primary particles in a dispersion process, compression and pressing mentioned in buildup methods are effective. Shearing, impacts and cavitation (figure 3) are effective in emulsification processes.

1. 2 Use of emulsification technologies

There are both breakdown and buildup methods for manufacturing nanoemulsions using emulsification technologies.

1. 2. 1 Effective mechanical power for emulsification: Shearing

Of mechanical powers effective for emulsification, shearing is most effective. Strong shearing forces are highly demanded. Shearing force is, for example, the power exerted when a specimen inserted between two plates is torn off by moving one of the plates (figure 4). The larger the power is needed to tear off the specimen as the distance between the two plates are narrower and the faster the plate is moved. Many emulsions show thixotropy, in which the speed of the moving plate does not increase linearly but suddenly at a certain point. Thus, the two plates must be close together because the shearing force would not be exerted on the specimen when the clearance between the plates is big.

Particle size distribution is important for the stability of emulsions. Emulsions become unstable when there are differences in particle sizes because small particles are taken in by big particles. Most of shear mixers used for preparing emulsions have narrow clearances, such as homogenizers (figure 5) and UltraMixer (figure 6). This can be explained in terms of etymology.

Emulsification and homogenization comes from homogenize, thus a device suitable for emulsification is a homogenizer, which has a narrow clearance. On the other hand, a mixer suitable for producing dispersions is a dispersion mixer (figure 7).

In the case of emulsion polymerization, shearing force too large results in insufficient emulsification, and shearing force too weak causes increases in cohesion. Large-scale impeller are widely used to keep shearing force uniform throughout the tank.

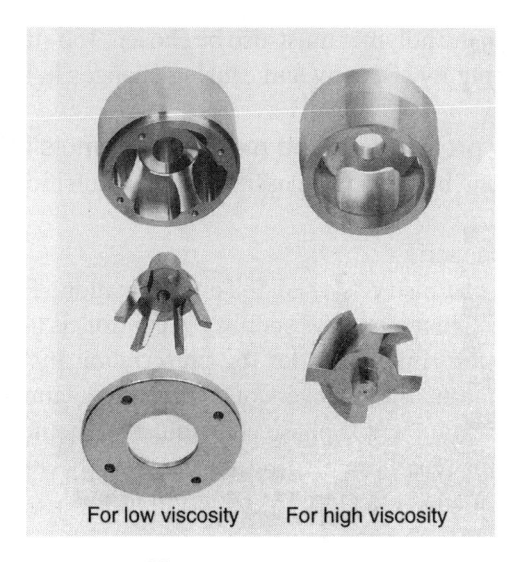

Chapter 1 Nanoemulsion Manufacturing Equipment

For low viscosity   For high viscosity

Figure 5 Homogenizer

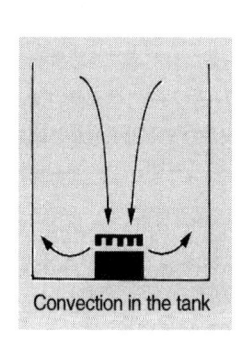

Convection in the tank

Figure 6 Ultramixer

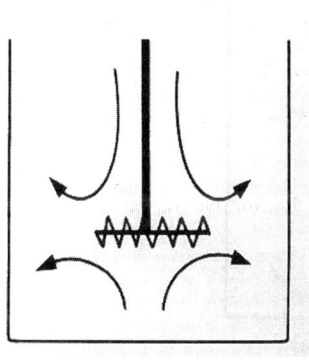

Figure 7 Dispersion mixer

Equipment appropriate to the polymer must also be chosen. Too strong shearing force cuts the polymer into pieces, reducing its viscosity and causing changes in its rheological properties.

1.2.2 Emulsification by prescription and mechanical emulsification
Emulsification methods can be broadly classified into emulsification by prescription and mechanical emulsification.
(1) Emulsification by prescription
In this method, surface chemistry is used for emulsification. The surface tension can be controlled by selecting emulsifiers, and the stability is controlled using differences in specific gravity and electric repulsion. Emulsification by prescription includes soap emulsification, phase inversion emulsification, phase inversion temperature emulsification, liquid crystal emulsification, gel emulsification and D phase emulsification methods. In late years, choosing mixing blades suitable for each emulsification method becomes important[1]. Polymer emulsification is included in emulsification by prescription.
(2) Mechanical emulsification
This method uses mechanical power, and the use of a super-high-speed shear mixer or high pressure homogenizer is necessary to produce nanoparticles.

1.2.3 Equipment for emulsification
A number of equipment is used in an emulsification process. Here, equipment that can prepare nanoemulsion is briefly described.
1.2.3.1 Super-high-speed shear mixer
Super-high-speed shear mixers are widely used for preparing emulsions mainly of 500 nm to 5 μm although other sizes can be prepared depending on prescription.
(1) Homogenizer (figure 5)
Emulsification is performed by the strong shearing and impact forces generated between the stator and turbine blades, which rotate fast. Rotating the turbine blades at a high speed produces cavitation (difference in pressure) between the upper and lower parts of the blades and pumps up the fluid from the bottom. The clearance between the stator and blades is about 0.5 mm, and the fluid receives strong shearing force when it passes through the narrow space.

Composite mixing blades (figure 8) that are combined with an anchor mixer are also widely used. The mixers have been widely used in diverse types of industry in late years because they can prepare small particles of a uniform size and keep the system stable even with a small amount of surfactants using mechanical power. There are low and high viscosity types. The low viscosity type can be used for a viscosity of up to about 8 Pas, but the discharge decreases at a viscosity higher than about 4 Pas. Thus the high viscosity type is usually used for viscosity higher than 4 Pas.
(2) Ultramixer (figure 6)
An ultramixer suctions a product, which flows down by its own weight, into the mixer by the suction force produced by cavitation and dragging force of the rotary hood, and expels it from the side. A strong shearing force of three to five folds larger than that of homogenizer is applied on the produce because the upper rotary hood has triple slits and the lower fixed part has double slides. The mixer can apply a high shearing force to highly viscous products. It can be used up to a high viscosity of about 100 Pas (100 000 c.p.s.).

These years, the demand is increasing for mixing machines that can exert stronger shearing forces than homogenizers on highly viscous products. Sometimes, homogenizers and dispersion mixers are used by improving them into super-high-speed rotary specifications (10 000 rpm

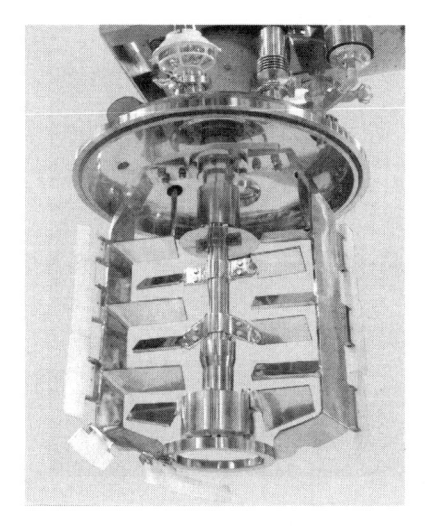

Figure 8 Composite mixing blade
(Homogenizer and scraper mixer)

Figure 9 Composite mixing blade
(Ultramixer, anchor mixer, and dispersion mixer)

the maximum in a production machine), but the use of ultramixers is increasing. Installation of the upper part of the ultramixer is also possible.

(3) Dispersion mixer (figure 7)

There are no narrow slits, but the tips of the blades are sharpened to effectively crush powder aggregates. It is suitable for suspensions in which powder is dispersed in a liquid and is used to efficiently disperse highly viscous polymers. Dispersion mixers are widely used for emulsification, but attention is necessary because resultant clearances are prone to be not uniform when carried out in a scale-up system. Installing a baffle plate to disturb eddies facilitates convection to occur and improves mixing. Composite mixing blades are also widely used (figure 9).

1. 2. 3. 2 High pressure homogenizer

In a high pressure homogenizer, pressure determines the dispersion and particle diameter. Because it uses stronger mechanical power than homogenizers, it may cut the chains of polymers. Because the opportunities of the collision increase by Brownian movement when emulsification particles become small, the composition must be carefully examined to produce stable emulsions.

(1) Valve-type homogenizer (figure 10)

A valve-type homogenizer consists of a high pressure pump and homo valves. Shearing force is applied by passing a product through narrow slits under a pressure of 10 to 100 MPa. Impact force is them applied by colliding against the impact ring, and almost simultaneously cavitation

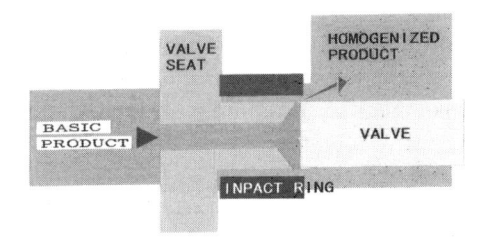

Figure 10 Valve-type high pressure homogenizer

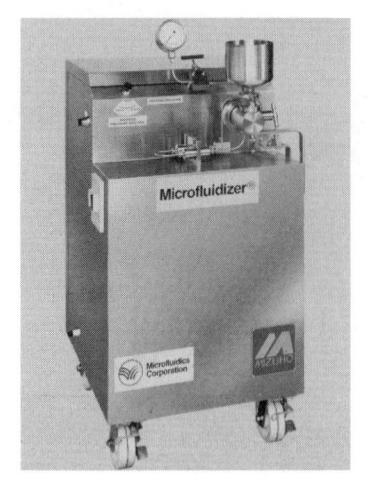

Figure 11 Microfruidizer

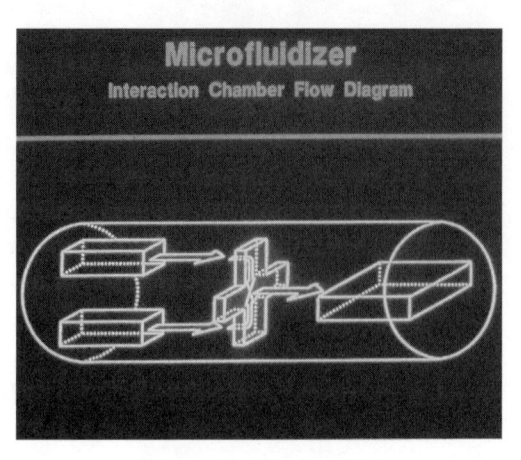

Figure 12 Chamber flow

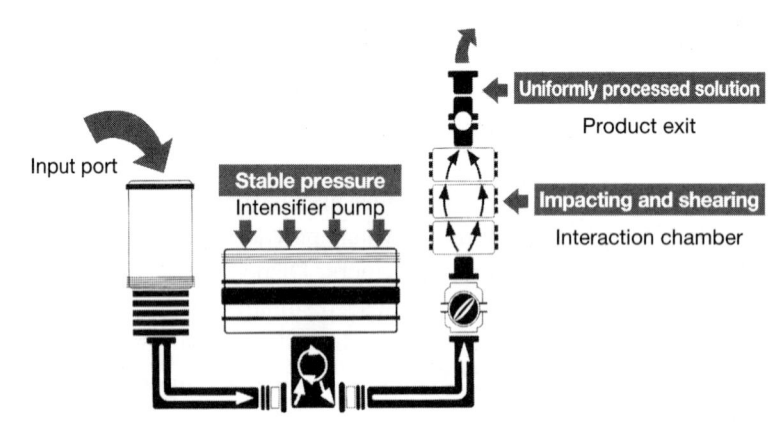

Figure 13 System flow

power is applied by reducing the pressure to an ordinary level. There is also a two-staged valve method, and application of pressure in two stages sometimes gives good effects on the stability and formation of fine particles. The second stage uses low pressure.

In food, valve-type homogenizers are used to reduce the size of fats in milk to improve its stability and for producing ice cream and yogurt.

(2) Microfluidizer: Flow fixed type high pressure homogenizer (figure 11–13)
In a microfluidizer, the flow of materials is divided into two in the chamber, a super high pressure of 275 MPa is applied to pass the materials through narrow tubes, and strong shearing force is applied. The flows are joined, and impact force is applied by making the materials to collide with other. Because the process is performed under a super high pressure, it is effective for products that do not disperse in other equipment and for preparing fine emulsions of 500 nm or smaller. It is also used for producing surfactant-free cosmetic creams and cell homogenization.

1. 2. 4 Emulsion polymerization

This liquid phase method for manufacturing microparticle polymers can produce polymer particles of 50 nm to 1 μm. Prescription design involves choosing four ingredients of monomers, polymerization initiators, surfactants and water, and preparation method. Controlling the particle diameter and reducing encrustation and adhesion are the keys. Strong shearing force is not needed. Shearing force too strong may destroy emulsification particles causing shear adhesion, but force too weak may cause insufficient emulsification and resultant molecule adhesion. Large scale impeller and anchor blades are appropriate as they can exert relatively low shearing force uniformly over the entire system. There are other methods for manufacturing polymers, such as suspension polymerization and soap-free emulsion polymerization, but the widely used method for manufacturing nanoparticles is the emulsion polymerization method.

1.2.5 Preparation of nanoemulsion
1.2.5.1 Prescription example and a preparation method of nanoemulsion
Prescription
Water 71%
Liquid paraffin 25%
Surfactant 4%
(TWEEN/SPAN HLB=10)
Processing conditions
Pre emulsification: 70°C in a homogenizer at 5000 rpm for 30 min.
High pressure emulsification: in a microfluidizer at 172 MPa for 1 pass.

The result of processing is shown in figure 14. Five μm in the homogenizer became 100 nm or less after it was processed in the microfluidizer.

This data shows that the amount of surfactant can be reduced by using strong mechanical power. In other words, this show that the amount of surfactant can be reduced to 0.18% by using the microfluidizer from that in the homogenizer to produce the particles of a diameter of about 5 μm.

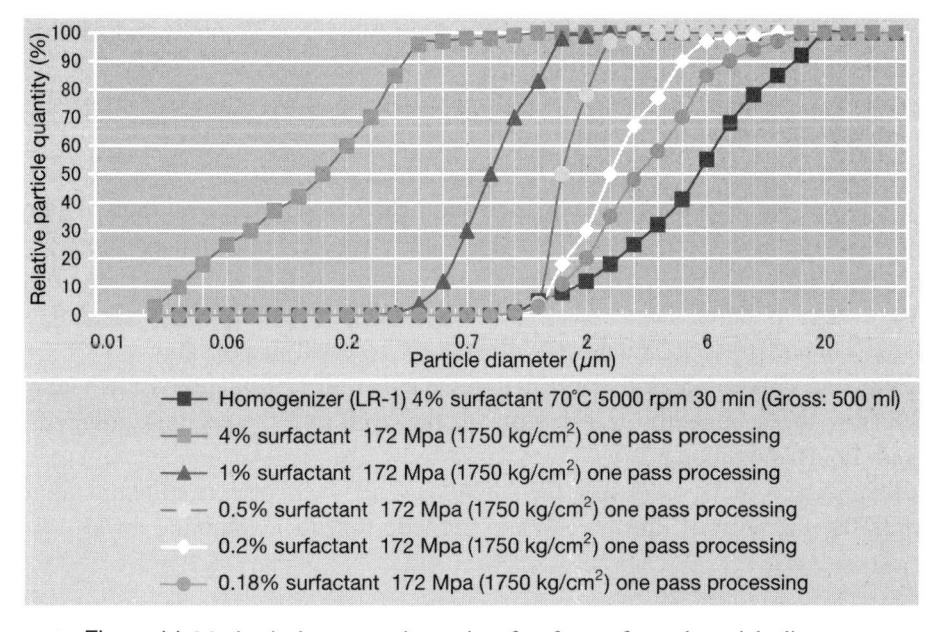

Figure 14 Mechanical power and quantity of surfactant for each particle diameter

Nanoemulsions have been reported to penetrate fast into the epidermis remain long[2].

1.2.5.2 Fat emulsion

In the medical supplies industry, there is an injection drug called fat emulsion, which consists of emulsified soybean oil prepared by using lecithin as a surfactant.

>Prescription: Soybean oil: 10.0%, lecithin: 1.2%, water: 88.8%
>Process: 1) 70°C in a homogenizer at 3000 rpm for 5 min
>   2) in a microfluidizer at 172 MPa
>Results: 1) 14.226 μm
>   2) One pass: 0.473 μm
>   Two passes: 0.416 μm
>   Three passes: 0.108 μm

Emulsions of a particle size of about 200 nm to 400 nm have been conventionally produced using valve-type high pressure homogenizers. Particles of 100 nm can be produced using a microfluidizer, which applies high pressure. Prescriptions have been improved, and lipid nanospheres of 50 nm are produced today.

Particles smaller than 100 nm can pass through the liver.

1.2.5.3 Liposome

There is a drug called liposome, which is prepared using the bimolecular membrane formation property of lipids. Because the biomembrane consists of a bimolecular membrane structure, this is called bio-mimetic film.

(1) Usefulness of liposome preparations

Liposomes are affinitive to the skin, moisturize the skin, improve the skin barrier functions (improving crow's-feet), improve the penetration and absorbency into the intestinal mucous membrane (penetration of polymers: increased blood concentration without changing the structure of the polymers), release the contents in a controlled manner, antioxidize, and act as lipid nanotubes.

(2) Manufacturing methods of liposomes[3]

Freeze-drying empty liposome method: Liposome preparations of about 100 nm in diameter are prepared by freeze drying sterile empty liposomes filled in vial bottles and adding a water solution of a drug.

Mechanochemical method: Lipid powder is dispersed in a water solution of a drug using a high pressure homogenizer to apply shearing force.

Spray drying method: A volatile organic solvent in which lipid is dissolved and saccharides are dispersed as water-soluble cores is spray dried and mixed with a water solution of a drug.

Lipid dissolution method: Lipid is dispersed in a volatile organic solvent, the dispersion is dried by nitrogen bubbling, etc. and the dried dispersion is mixed with a water solution of a drug.

Polyhydric alcohol method: Lipids dissolved or swelled in a polyhydric alcohol, such as propylene glycol and glycerin, are mixed with a water solution of a drug.

Warming method: This involves instantly hydrating and swelling lipid powder at a temperature above both the phase transition temperature of the lipid and the melting point of electric charged oily lipids and mixing.

In the Japanese cosmetics industry, the safety of three years at normal temperatures must be shown to use the term of liposome or bio mimetic film for explaining or advertising the product

1.3 Supercritical method

The supercritical method involves creating a state in which there is no gas-liquid interface by raising the temperature applying high pressure. There are breakdown and buildup methods.

(1) Breakdown methods using supercritical technology

They are relatively old. There is a method that involves passing a super critical fluid through pores of porous powder or spaces between secondary aggregates, by using the very low surface tension property of a super critical fluid, and returning the pressure, at which the super critical fluid swells and form microparticles.

Another method uses pressure differences and involves dispersing particles in a super critical fluid and spraying through fine holes to crush into microparticles using the impact and shearing forces.

(2) Buildup method using supercritical technology[4]

They are new and use the dissolution characteristics of super critical fluids to make material that is dissolved in a super critical fluid to deposit by quickly expanding it. It uses a super critical fluid as a solvent. There are the SAS method (Supercritical Anti-Solvent), which uses a super critical fluid as a poor solvent, and RESS (Rapid Expansion of Supercritical Solution).

Using super critical CO_2: Because the critical temperature of CO_2 is low, super critical CO_2 can be used for materials that are easy to denature by heat. The critical temperature of CO_2 is 31°C, and the critical pressure is 7.4 MPa.

Using super critical ethanol: Super critical ethanol is suitable for fine particles of organic metals. The critical temperature of ethanol is 240°C, and the critical pressure is 8.1 MPa.

Using super critical water: The critical temperature of water is 374°C, and the critical pressure is 22.1 MPa. There are methods for preparing nanoparticles within super critical water in which the hydrothermal synthesis reaction is used. Preheating, mixing and the mixing ratio to the super critical water affect the particle size.

There are patented surfactant-free emulsification methods which use supercritical technologies.

1.4 Other nanoparticle manufacturing methods and processing examples

1.4.1 Manufacturing methods of carbon nanotubes and fullerene

Physicochemical methods are described below:

Chemical vapor deposition (CVD): It involves heating and reacting methane, hydrogen gas and a metal catalyst at 100°C in a vacuum chamber that can heat up to 1000 to 1100°C. The productivity is high.

Arc-discharged method: It involves vaporizing carbon by high-temperature plasma and depositing carbon on an electrode afterwards.

Laser ablation method: It involves producing high-temperature carbon vapor from graphite field at 900 to 1400°C.

1.4.2 Dispersion of fine particles of titanium dioxide

Titanium dioxide particles of 300 nm have been used in cosmetics to increase the effect of blocking ultraviolet rays. Today, fine particles of 30 nm are widely used. However, fine particles are pone to aggregation and require advanced dispersion technologies and technologies for preventing aggregation.

Dispersion by prescription: Adjusting the electric charges on the surface, addition of dispersants, and uniform dispersion in a gel

Dispersion suing mechanical power: Use of high pressure homogenizers and ultramixers

Titanium dioxide as small as 30 nm absorbs ultraviolet rays rather than reflect them. This is a good example of nanotechnology.

1. 4. 3 Novel methods for preparing medical supplies using nanotechnology[5]
Fat soluble compounds and insoluble compounds are converted into nanoparticles to increase the surface area and solubilize. Submicron particles and suspending technologies are also used to control destabilization caused by the second cohesion between particles. For example, the insoluble drug delivery method (SkyePharma PLC), involves high pressure shearing of insoluble main ingredients and a surfactant in a microfluidizer.

1. 4. 4 Use of polymer micelles[6]
A method for preparing nanoparticles of the polymer micelle type, which does not need strong mechanical power, was developed by Professor Kazunori Kataoka (School of Engineering, the University of Tokyo). He reported the construction of a nano structure device of the polymer micelle type and its development as a target-bound nanocarrier.

1. 4. 5 Bionanofiber (cellulose nano fiber)
The Swiss Federal Laboratories for Material Testing and Research and The Swedish Pulp and Paper Research Institute have developed biodegradable materials by producing nanofibers from straw and wood pulp.

1. 5 Future of nanotechnology[7]
(1) Handling of fine particles after preparation is important.
Because it is difficult to keep nanoparticles stable, various methods are being investigated.
1) Functional particles
Core resin particles are coated with titanium oxide of 15 nm in particle size.
2) Investigation of suspension systems
Wet milling is investigated because it is easier to prevent aggregation of suspension systems than that of dry powder.
(2) Many nanotechnologies used today were found accidentally[8].
1) Discovery of the carbon nanotube (1991)
2) Discovery of the nanothermometer
When carbon nanotubes that contained gallium were observed under an electron microscope, gallium was found to stretch and contract within the nanotubes when heated by an electron beam.
   Nanotechnologies are predicted to make dramatic progresses propelled by the development of advanced measurement and evaluation methods.
(3) Collaboration with the other fields[9]
1) Nano biotechnology developed from nanotechnology and biotechnology.
2) Nano mixing
(4) Improved nanolevel evaluation by progresses of measuring and analytical instruments
Use of deep probes that are made of carbon nanotubes in atomic force microscopes (AFM) has enabled nanolevel and micron level forms to be correctly evaluated.
   However, evaluation of nanoparticles is increasingly difficult. For example, appropriate sampling conditions and dispersing conditions must be determined for measuring grain distribution.

(5) Safety of nanoparticles

The majority of nanoparticles were created by people and do not exist in nature. The safety of nanoparticles to living bodies is discussed. As a measure, they can be handled within an isolator.

When the surface area increases, the risk of dust explosion increases. A possible measure is to handle them under nitrogen to prevent them from reaching explosive conditions.

References

1. K. Takagi, New emulsion technology searching for a harmonization of formulation and mechanical emulsification method, Fragrance Journal, Special Edition, 19, 131 (2005)
2. H. Imai, Y. Sakai, *Shikizai Kyokaishi*, **78** (1), 28(2005)
3. H. Kikuchi, Liposomes based on nanotechnology (in Japanese), *PHARM TECH JAPAN*, **19** (1), 99 (2003)
4. S. Yoda, Preparation of nanoparticles using supercritical fluid (in Japanese), *Shikizai Kyokaishi*, **76** (4), 142 (2003)
5. T. Morita, From micro DDS particles to nanoparticles and from nanoparticles to solubilization (in Japanese), *PHARM TECH JAPAN*, **20** (4), 237 (2004)
6. K. Kataoka, Frontier nanomedicine developed by nanotechnology (in Japanese), Proceedings of Nanomedicine, (2004)
7. K. Takagi, Nanotechnology and manufacturing equipment (in Japanese), *Fragrance Journal*, **31** (8), 595 (2004)
8. Y. Bando, Research of nanomaterials using state-of-the-art electron microscopes (in Japanese), *Kagaku to Kogyou*, **57** (6), 595 (2004)
9. International Symposium on Fusion of NANO and BIO technologies (2003)

Chapter 2
Latest Mixing Technologies for Cosmetics Production

Akihito Shundo

2. 1 Introduction

Mixing is indispensable in the production of cosmetics. For example, skin-care milky lotion and cream is produced by heating, dissolving and mixing oil phase ingredients, in which uniform mixing is crucial. Mixing is also necessary to dissolve thickeners and surfactants, to emulsify oil and water phases, and to uniformly cool the entire system after emulsification. Also in the production of lipsticks and foundations, it is important to disperse pigments homogeneously in oil and wax to develop color better, in which mixing is important for the dispersion. This chapter describes prospective of new mixing technologies in cosmetics production.

2. 2 Conventional mixing technologies

Generally, mixing used in the production of cosmetics can be broadly classified by purpose into macromixing, which mainly involves moving liquids for mixing, dissolving, and exchanging heat, and micromixing, which requires strong shearing force to produce emulsions, dispersions and fine particles. Low speed stirring machines are mainly used in macromixing. There are small-blade stirrers for stirring relatively low viscous liquids by rotating turbine blades and propeller blades at a low rotational speed, and large-blade stirrers for mixing viscous liquids by rotating large wings, such as anchor wings and ribbon wings, along the inner wall of a mixing tank.

In the latter micromixing, it is necessary to apply a strong shearing force on the liquid to process, and high-speed mixers are needed. There are high-speed mixers of the turbine stator type, such as T. K. HOMO MIXER (figure 1 (a)), and the dissolver type, such as T. K. HOMO DISPER (figure 1 (b)). When the liquid is low viscous, micromixing causes the entire liquid to flow, thus the process also involves macromixing.

The mixers should be chosen by considering the viscosity and flow characteristics of the liquid and previous and subsequent processes. It is also necessary to set mixing conditions, including the speed of rotation, duration of process, heating, cooling, and the order of adding ingredients. In actual processes, the viscosity of the products such as creams is low at the time of emulsification (at the high temperatures) becomes very high exceeding 10 000mPa·S as the temperature drops when emulsification finishes. Therefore composite mixers called vacuum emulsifiers such as T.K.COMBI MIX (figure 2(a)) and T.K.AGI HOMO MIXER (figure 2(b)), in which low-speed large-wing stirrer and a high-speed mixer are combined are used in production process.

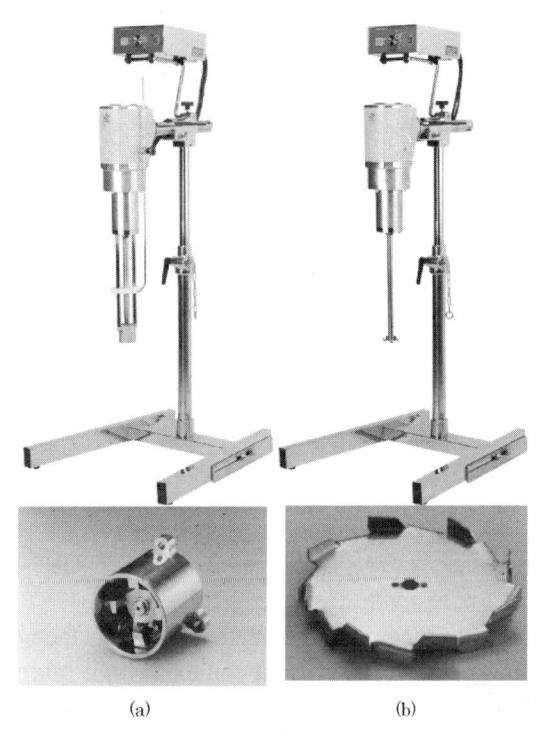

Figure 1 (a) T.K. HOMO MIXER MARK II and (b) T.K. HOMO DISPER

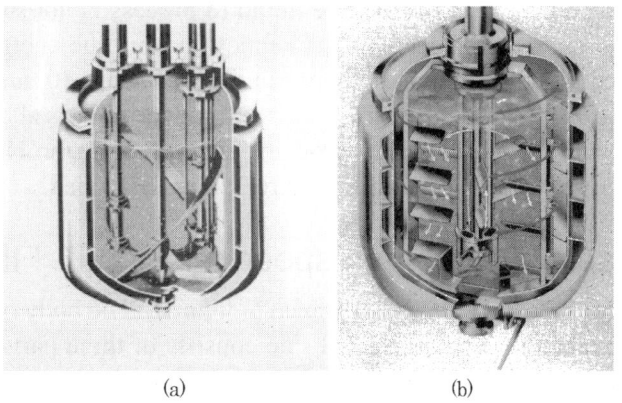

Figure 2 (a) T.K. COMBI MIX and (b) T.K. AGI HOMO MIXER

2. 3 Recent trends of cosmetics and demands for mixers

Cosmetics are required to be safe, stable, comfortable to use (touch and fragrance), and effective. These are essential. Besides these, cosmetics are increasingly demanded to have physical, physiological and psychological efficacies[1]. Cosmetics with such characteristics are produced by adding nanoparticles of titanium oxide and the zinc oxide to scatter ultraviolet rays and preparing them into emulsions smaller than 100 to 200 nm to improve penetration into the skin. These require high mixing technologies for emulsifying and dispersing microparticles. Mixing in such a nanodomain needs large mechanical force to be applied and emulsifiers and dispersants must be carefully selected because the particles are very small, have large solid

Figure 3 T.K. FILMICS® FM-80-50 type

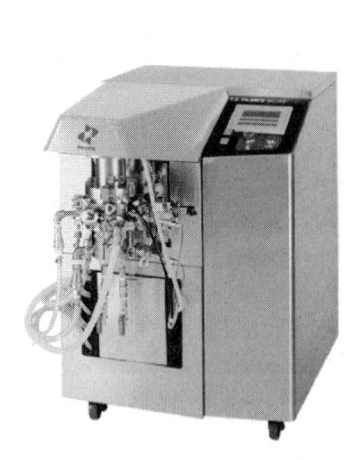

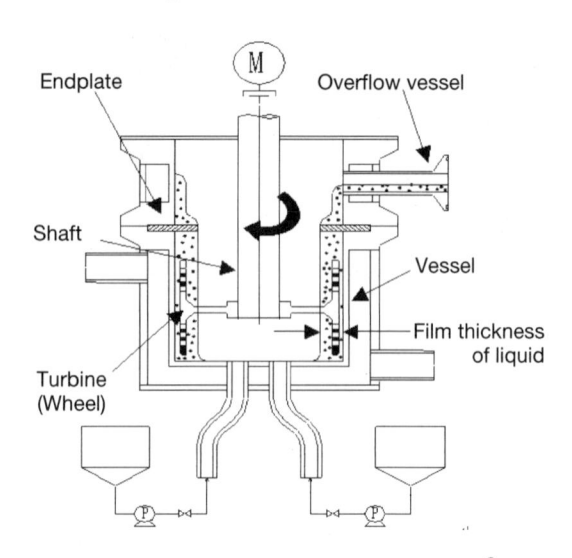

Figure 4 Basic structure of T.K. FILMICS®

to liquid or liquid to liquid interfaces, are in a state of high surface energy, and are thus very unstable. In other words, high-speed mixers that can exert high mixing energy are demanded. However, there is a limit in the mixing energy that can be exerted by conventional mixers.

Energy is limited not only in low-speed stirrers but also in high-speed mixers. The faster the blades are rotated to increase energy, the faster the blades expel the liquid, increasing cavitation. The impact force produced when cavitations are crashed also contributes to emulsification and dispersion, but when sufficient amount of the liquid to process is not supplied to the blades at which cavitation occurs, the blades idle and cannot transmit the energy to the liquid. The limit peripheral speed of conventional high-speed mixers is about 30 m/sec (the value varies depending on the viscosity of the liquid to process). Mixing at speeds above this value cannot transmit sufficient energy to the liquid to process. To resolve this problem, a thin-film spin system high-speed mixer "T.K. FILMICS®" (figure 3) was developed.

2. 4 Thin-film spin system high-speed mixer "T.K. FILMICS®"

The basic structure of "T.K. FILMICS®" is shown in figure 4. The body of revolution consists of two parts: a shaft and a turbine. The fixed side consists of three parts: a vessel, endplate, and overflow vessel. The structure is simple and consists of few parts. The process involves the turbine rotating the liquid to process, which is either contained in the vessel or supplied from the bottom, and the resultant centrifugal force pressing the liquid to the inner wall of the vessel and the end plate. When the velocity reaches a certain value, the liquid forms a cylindrical thin film that rotates at a high speed along the entire inner surface of the container and rotates receiving the centrifugal force determined by the velocity^2. The liquid on the wall is subjected to a centrifugal force of 6700G at a peripheral speed of 50 m/sec. This prevents idling caused by cavitation, which occurs under the gravitation of the earth, and enable large energy to be applied on liquids to process.

2. 5 Principles and effects of T.K. FILMICS®

Let us see how the liquid in the vessel flows in such a mixer. A schematic view of flows within the rotating thin film is shown in figure 5. For example, in FM-80-50 type, a turbine that has an outer diameter slightly smaller than the inner diameter of a vessel of 80 mm rotates at 12,600 min^{-1} the maximum inside the vessel. The turbine accelerates the liquid and form a rotating thin film. The peripheral speed (figure 5 (1)) of the turbine reaches 50 m/sec when the film is formed. The resultant centrifugal force forms a high-dimensional compressed energy field, which is free of cavitation and resultant idling. The top of figure 6 is a horizontal cross section when the liquid forms a rotating film on the top, and the bottom shows the speeds of the turbine and liquid (the distance from the center is plotted on the X axis, and the speed is plotted on the Y axis). The speed of the liquid is almost equal to the speed of the turbine almost to the wall. Within a very small clearance, the speed changes by 50 m/sec. This sudden change in speed (shear speed) causes strong shear stress to develop throughout the wall surface of the vessel[3]. This expands droplets into thin films, tearing them apart in an instant; and the centrifugal force presses aggregates onto the wall, which rolls and crushed by the rotational flow. This is the mechanism of emulsification and dispersion by the thin-film spin system high-speed mixer. It can produce emulsions and dispersions of the submicron to nanometer order, which cannot be made by conventional high-speed mixers due to insufficiency in shear speed. In addition, the liquid inside the thin film passes through the holes of the turbine by the rotation of the turbine, forming jet streams toward the walll of the vessel (figure 5 (2)), separates into upward and downward streams flowing on the wall surface (figure 5 (3)), and returns to the inside of the film (figure 5 (4)). The circulating flow and rotational flows that flow over 200 times minute (figure 5 (1)) apply uniform energy on the entire liquid within the film.

When particles receive strong shear force, the particles emulsify and disperse. At the same time, they become unstable as the interfacial energy increases, and they become prone to uniting and re-aggregating. To prevent this, the particles must be stabilized by covering with chemical agents (surfactant *etc.*) while they are at a high energy state. In conventional high-speed mixers, liquids in process are subjected to high shear force only when the liquids pass the mixing blades during emulsification and dispersion, and they return to an ordinary energy level soon. However, for emulsification and dispersion of submicron to nanometer order,

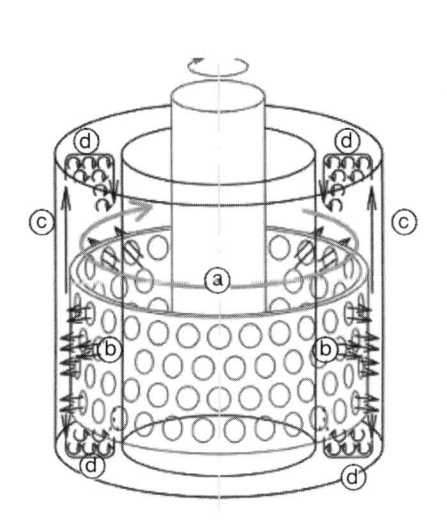

Figure 5 Flows in the rotating film

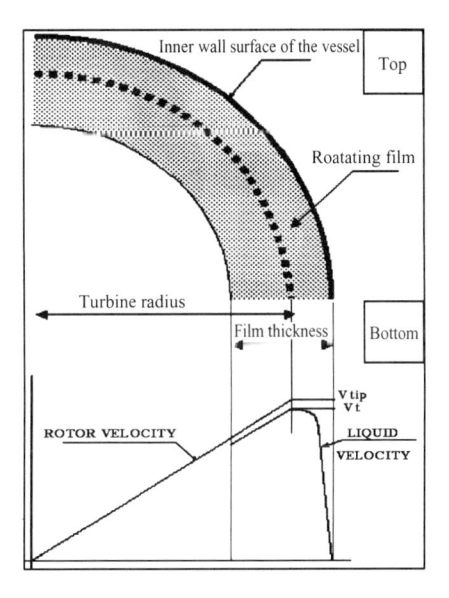

Figure 6 Horizontal cross section and speed distribution of the rotating film

instant application of energy is insufficient for stabilizing the particles, which coalesce and aggregate again. The rotating film produced in T.K. FILMICS is always at high density and uniform energy condition, and the high energy level is kept much longer than in conventional mixers. This enables chemical agents to cover the emulsified and dispersed particles while they still have high energy levels.

2.6 Examples of applications

T.K. FILMICS® can exert high-density and uniform mixing energy as described above. The machine has been used for various products that could not be produced otherwise. Some of interesting applications are described below.

2.6.1 Emulsification of liquid paraffin: sharp particle size distribution and particle size control

Particle size distribution of an emulsion of liquid paraffin prepared using T.K. FILMICS® is shown in figure 7. In this experiment, 250 mL of liquid paraffin was mixed for 1 minute at three different peripheral speeds, and the resultant particle size was measured. The particle sizes differed by speed still showing sharp distribution with M.D.=3.316 μm at 30 m/sec, M.D.=1.086 μm at 50 m/sec and M.D.=0.605 μm at 70 m/sec. The particle size distribution was kept uniform because energy was uniformly applied, and the particle size was controlled by changing the energy level.

2.6.2 Solubilization of lotion

Solubilization of lotion is shown in figure 8. The pre-mixture before processing had a particle size distribution with three peaks and M.D. = 426 nm. The distribution was improved to M.D. = 33.5 nm when processed using a high-pressure homogenizer, but still large particles of 0.3 to 2.0 μm remained. When processed using FILMICS® at 50 m/sec for 1 minute, the distribution was improved to M.D. = 37.0 nm, which was slightly larger than that when the high-pressure homogenizer was used, and no large particles remained. The transparency of the samples is compared in figure 9 The sample that was processed using the high-pressure homogenizer, which contained large particles, was turbid; while that processed using FILMICS® was transparent. There is a possibility that the amount of surfactants can be reduced from that in conventional mixers at least in some kind of surfactant at certain mixing ratios.

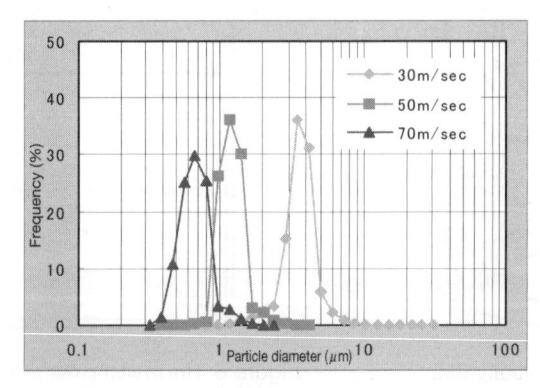

Figure 7 Emulsification of liquid paraffin

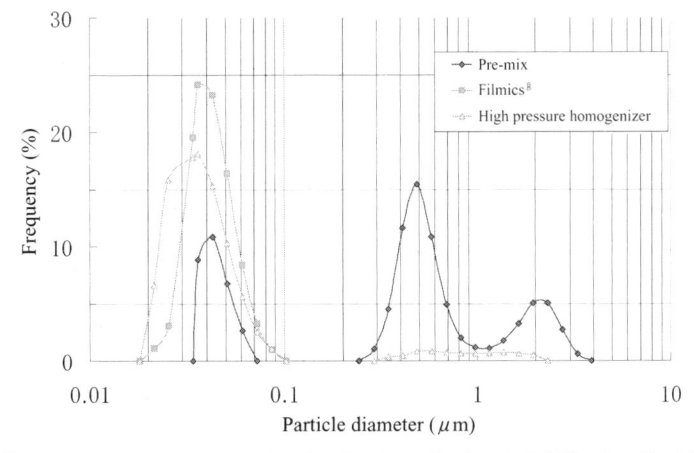

Figure 8 Changes particle size distribution of lotion (solubilization liquid)

Figure 9 Lotion samples (solubilization liquid)
(left) High pressure homogenizer processing,
(right) T.K. FILMICS® processing

2.6.3 Liposome

Liposomes were prepared using egg yolk lecithin so as to contain calcein as a model drug, and a comparative examination was performed using a high pressure homogenizer and T.K. FILMICS® to examine what changes occur by application of different mechanical energy levels. The results are shown in figure 10. Almost no change occurred in enclosure ratio by prolonging the process time in T.K. FILMICS®, but in the high pressure homogenizer, the enclosure ratio dropped sharply as the number of passes increased. The high-pressure homogenizer seemed at a glance to have produced particles of small mean diameter but it was because many liposomes were destroyed and mere pieces of lipids were measured. This agreed with the enclosure ratio of calcein. The high-pressure homogenizer crashed particles by the impact force produced by collision and caused them to rearrange. On the other hand, T.K. FILMICS® transformed liposomes flat by applying shear stress, causing them to separate into small liposomes, and thus the release of the drug was small. The machine is thus likely to be applicable to producing lotions that contain liposomes enclosing water-soluble active ingredients.

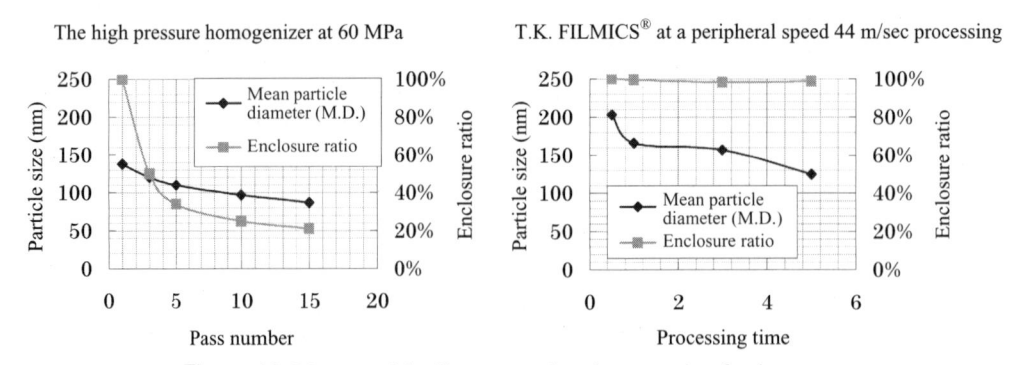

Figure 10 Mean particle diameter and enclosure ratio of a drug for a high pressure homogenizer and T.K. FILMICS®

2. 6. 4 Fine particles in cream containing polymer thickener

The results of an emulsification experiment of a cream that contained polymer thickener are shown in figure 11. When processed using a high-pressure homogenizer at 70 MPa for 1 pass, the M.D. was reduced to 150 nm. However, the polymer was cut into pieces, resulting in sharp drops in viscosity, and the cream separated into layers on the next day. On the other hand, when processed in T.K. FILMICS® at 56 m/sec for 1 min, the resultant particles were as large as M.D. = 400 nm, but the resultant cream was uniform, stable, and highly viscous because the polymer thickener remained intact. Even 20 months after processing, the particle diameter and sensory evaluation remained the same showing good long-term stability. T.K. FILMICS® applies high energy but do not cut polymers. Of course, particle sizes of emulsions containing polymer thickeners can be controlled by changing the peripheral speed, and thus can control the characteristics of resultant cream, such as the touch, spread and luster.

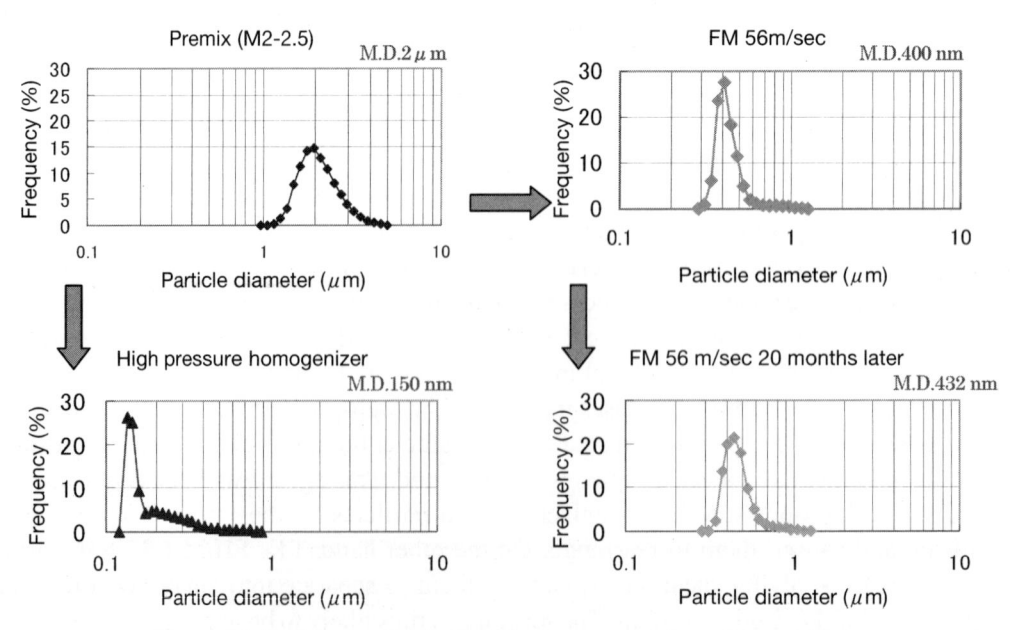

Figure 11 Particle size distribution of a cream containing polymer thickener

2.7 Conclusions

The characteristics of T.K. FILMICS® are summarized below:
- It can design emulsion and dispersion particles of nanometer sizes, which was difficult with conventional high speed mixers.
- It keeps the particle size distribution very sharp.
- It can be used for mass processing because it is a consecutive processing machine that can be controlled in units of second.
- It can be easily switched to batch operation.
- It is simple, compact, and easy to wash because it consists of few parts.
- It is easy to scale up.
- It cuts only a small proportion of polymers.
- It is medialess dispersion machine (but it should be limited to crashing of aggregates and should not be used for crushing primary particles and crystals).
- It is effective for stabilizing interfaces, and thus reduces the amount of surfactants and prevents re-aggregation.

The applications described here are mere examples, and may not apply to all liquids. There are liquids for which T.K. FILMICS® is not appropriate due to their physical properties. T.K. FILMICS® has produced results that have been difficult using conventional high speed mixers or other machines, including unexpected results. Further studies are needed to collect data of each material and integrate knowledge. Active participation of those who are involved in designing and producing micro particles is essential for the development of novel technical fields.

References

1. M. Masuda, E. Takasu, Encyclopedia of Cosmetics, p.122, MARUZEN Co., Ltd. (2005)
2. H. Shibuya, Perfection of Cosmetics, p.251-219, JOHOKIKO Co., Ltd. (2006)
3. H. Asa, Mixing technology of the 21st century, p.56-57, Kogyo Chosakai Publishing, Inc. (2000)

Chapter 3
Development of Microchannel Emulsification Technology

Mitsutoshi Nakajima and Isao Kobayashi

3. 1 Introduction

Emulsions are dispersions of immiscible twoliquids with one of them as droplets stabilized by the action of emulsifier and are used in various products, such as cosmetics, foods and pharmaceuticals. Colloid mills and high pressure homogenizers are commonly used for the production of emulsions. The machines break up the disperse phase into fine particles driven by mechanical shear stress and are highly productive, but cannot precisely control the particle size and are prone to producing polydispersity. The physical properties of an emulsion depend on droplet size and its distribution (coefficient of variation). In polydisperse emulsions consisting of non-uniform droplets a phenomenon called Ostwald ripening occurs, which involves big droplets becoming larger while small droplets becoming smaller. Monodisperse emulsions consisting of uniform droplets are thus advantageous. The membrane emulsification method for producing monodisperse emulsions was developed by Nakajima *et al.* (Miyazaki Prefecture Industrial Technology Center)[1]. It involves sending a liquid that serves as the disperse phase into the continuous phase through a porous membrane of a uniform pore size, and can prepare monodisperse emulsions relatively easily. Micro processing techniques have made remarkable progress and contributed to rapid development of electronics. Micro processing methods that utilize photofabrication has enabled production of micro production systems called MEMS (micro electro mechanical systems) with integrated microstructures[2]. MEMS has been applied in chemistry to develop μTAS (micro chemical and biochemical analysis system)[3] and microreactor[4]. A material commonly used in the micro processing technique is single-crystal silicon substrate, which is prepared by polishing the surface clean and flat by mechanical and chemical polishing, forming a metal membrane on the surface by surface oxidation and deposition, printing the mask pattern onto the substrate, developing photosensitive material (resist), and performing etching (wet etching, high density plasma etching)[2].

Microchannel (MC) is a very small channel of a micrometer scale prepared using a micro

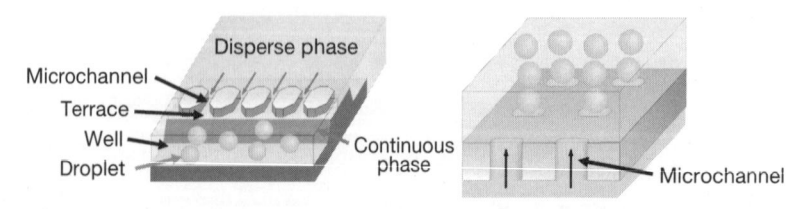

Figure 1 Grooved (left) and straight-through (right) microchannel arrays

Chapter 3 Development of Microchannel Emulsification Technology

processing technique. Typical structures include T- and Y-shape microchannels and microchannel arrays, which consist of a number of microchannels. The authors have microfabricated unique microstructures on silicon and other substrates and applied to produce emulsions. This chapter describes microchannel emulsification technology (figure 1) that uses microchannel arrays and is being developed by the authors.

3.2 Manufacture of monodisperse emulsion using microchannel arrays

Kikuchi *et al.* developed a microchannel-array blood rheometer[5]. Several thousand microchannels of a diameter size of 5 μm are arranged as a capillary model on a substrate. The authors applied the grooved microchannel (MC) array plate for the production of emulsion droplets[6-8]. We set the microchannel substrate in the module and filled the entire space inside the module with an aqueous phase containing surfactant. When the pressurized to be dispersed phase (e.g. vegetable oil) passed through the microchannels spread in the shape of a disk on the terrace, flowed from the tip of the terrace into the well containing the continuous phase, and transformed into droplets of a uniform size. Oil-in-water droplets (O/W) of a uniform size of about 20 μm were produced when microchannels of 5 μm were used. Investigation using grooved microchannels of different structures showed that the terrace at the channel exit played the crucial role in the preceding droplet formation. When there was no terrace, droplets were formed, but the timing of droplet separation was not constant, resulting in droplets of various sizes. With the terrace, the liquid (oil), which spread in the shape of a disk by the terrace structure, flowed smoothly from the terrace tip into the well phase by the Laplace pressure based on interfacial tension, enabling stable production of droplets of a uniform size (figure 2)[9-12]. We found that the size of the droplets can be controlled by the microchannel structure, such as the channel size and the length of the terrace, and clarified the effects of viscosity[9,13-15] and interfacial tension[16] on the size of droplets. In other words, microchannels of different sizes are likely needed to produce various sizes of controlled-size droplets. The sizes of droplets can also be adjusted for a narrow range by controlling the viscosity ratio between the disperse phase and the continuous phase. We also found that there is a supply speed of oil at which droplets of a uniform size are produced and supply speeds at which the oil outflows continuously and that the speeds are determined by the capillary number (Ca) calculated from the ratio of the interfacial tension to the viscosity of the disperse phase. The limit speed at which the disperse phase starts to outflow is higher at larger interfacial tension and lower viscosity of the disperse phase, and thus the productivity of emulsion is higher.

Surface oxidation of the substrate was found effective for stable production of oil-in-water (O/W) emulsions. Monodisperse water-in-oil (W/O) emulsions, in which vegetable oil or hexane

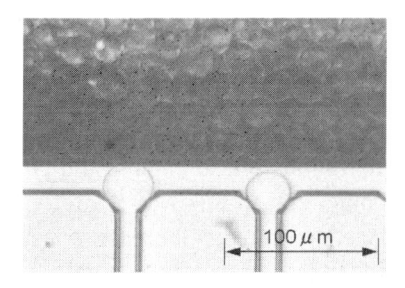

Figure 2 Monodisperse droplets produced from a grooved microchannel array

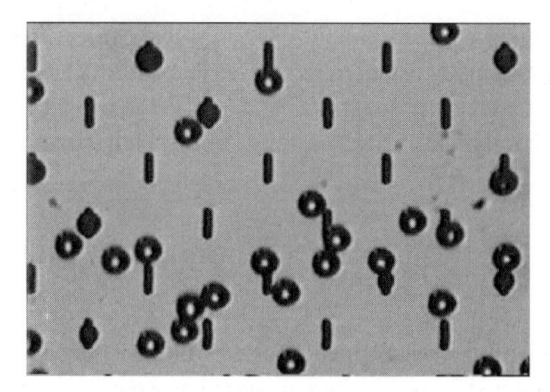

Figure 3 Monodisperse droplets formed from rectangular microchannels (longer side: 35μm, shorter side: 10μm) microfabricated on a silicon plate

was used as the oil phase, could be produced by making the substrate hydrophobic using the silane coupling method[17,18]. In production of emulsions, selecting an appropriate surfactant is important[19]. When an anionic surfactant or a nonionic surfactant of high emulsification ability is used, stable production of O/W emulsion droplets is possible. When cationic surfactant is used, electrostatic gravitation acts between the surfactant and a surface-oxidized silicon substrate, which is negatively charged, and the oil phase spreads as if wetting the substrate, resulting in difficult droplet formation. Anionic surfactant shows a strong electrostatic repulsion against the substrate and can prevent the wetting. The substrate surface is also likely to maintain hydrophilicity when nonionic surfactants are used. O/W emulsions can also be stably produced using proteins that have emulsification properties[20,21].

Because the productivity of the grooved microchannel array was low, we developed straight-through microchannels, which have through holes of rectangular and circular cross sections. The circular straight-through microchannels could not produce monodisperse droplets because the droplets continuously expanded from the exits. On the other hand, the microchannels with rectangular through holes resulted in smooth formation of droplets and was found effective for producing monodispersed emulsions (10 mL/h the largest using a substrate of 1 cm^2) (figure 3)[22]. The cross-section of the holes was important, and the longer side was found to be at least 3 times longer than the shorter side[23]. We have prepared straight-through microchannels of different sectional sizes, with which monodisperse droplets of 5–50 μm can be produced[24–26].

3. 3 CFD simulation and analysis of droplet production

The microchannel emulsification was modeled and analyzed using computational fluid dynamics (CFD). The CFD simulation results showed that the separation of droplets from a channel depended on the ratio between the longer and shorter sides of the channel. In other words, a droplet was detached from the channel when the longer side was sufficiently long, but the droplet expanded when the longer side was short because the disperse phase flowed out continuously[27–29]. The simulation of droplet formation from a rectangular microchannel is shown in figure 4. The figure shows that a neck, at which a droplet was to separate from the oil phase, was formed as the continuous phase entered the channel, and produced a droplet. Controlling the shape of the channel exit in this way is prerequisite for the continuous phase to enter the channel, the Laplace pressure to increase conspicuously, and a droplet to be formed spontaneously. This is a technology that was actualized by controlling the microspace structure.

(a) 15.0 ms (b) 36.7 ms (c) 38.7 ms

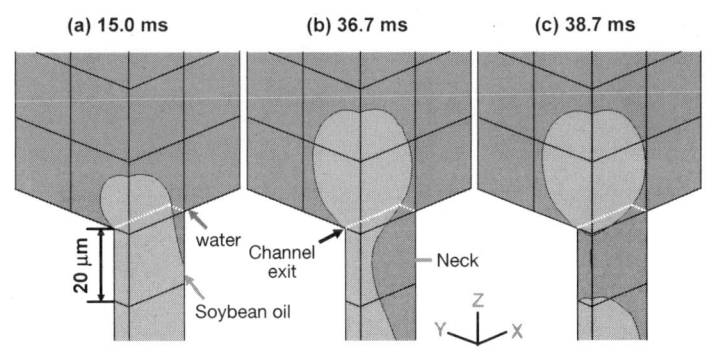

Figure 4 Simulation of droplet formation from rectangular straight-through microchannels (longer side: 40μm, shorter side: 10μm) (Droplets of 35μm were formed.)[28]

3.4 Development of the silicon asymmetric straight-through microchannels and non-silicon straight-through microchannels

The microchannels with a rectangular cross section were shown to be effective in efficiently producing monodisperse emulsions when a highly viscous oil phase, such as vegetable oil, was used. However, it was experimentally shown to be less feasible for stable production using low viscous oil phases, such as hexane, than the grooved microchannel array. To overcome the drawback, we developed a new asymmetric straight-through microchannel. As shown in figure 5, the new silicon microchannel array has an asymmetric structure in which thin channels (holes) are connected to slits. With this asymmetric straight-through microchannels, emulsions were stably produced from low viscous oil phases, such as hexane and decane[30].

Besides these silicon microchannel arrays, other substrate materials have been developed to reduce cost and improve durability and operability. A stainless steel grooved microchannel array was tested and found effective in preparing a monodisperse O/W emulsion[31].

The structure of the grooved microchannel array was reproduced on an acryl plate. A monodisperse W/O emulsion could be produced using the acryl plate[32,33]. This was because the acrylic plate is hydrophobic unlike a silicon substrate. We are also conducting a collaboration to develop through-hole array on an acrylic plate, and have produced an array of straight-through microchannels that has a precision as high as that of silicon substrates. As we have

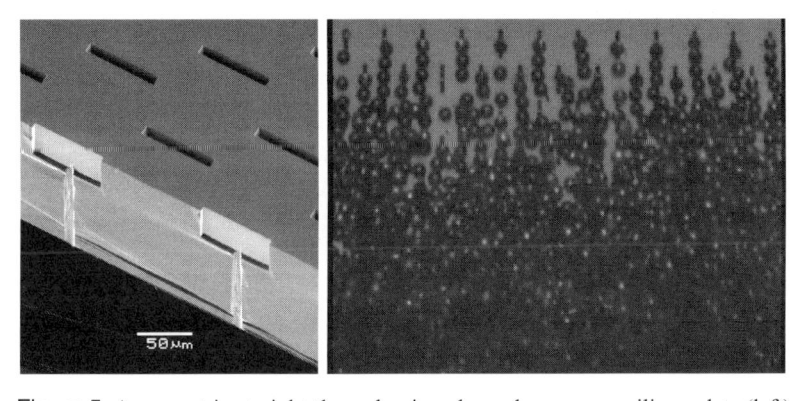

Figure 5 Asymmetric straight-through microchannel array on a silicon plate (left) and successful droplet formation (right)

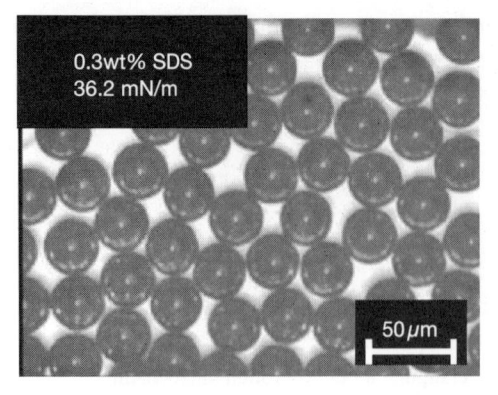

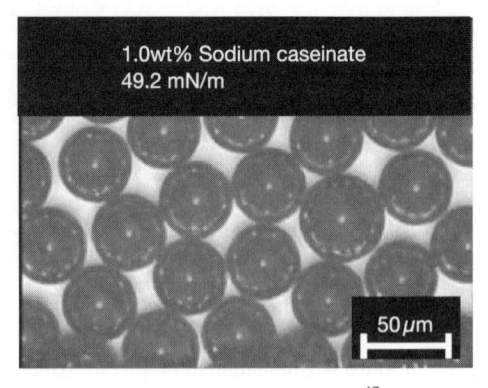

Figure 6 Monodisperse microbubbles prepared using microchannel emulsification[47] (Surfactants: sodium dodecyl sulfate (left), sodium caseinate (right))

reported, it can produce monodisperse W/O emulsions[34].

3. 5 Application of microchannel emulsification

The microchannel emulsification technologies for producing monodisperse emulsions can be applied for preparing microparticles of polymers and microcapsules. Applications that have been made are listed below. Please see the references for their details.

Monodisperse fine particles can be produced by preparing monodisperse droplets using the microchannel emulsification and solidifying the resultant droplets. We have prepared solid lipid microspheres mainly consisting of high melting point natural oils and fats[35], polymeric microspheres by polymerizing divinyl benzene[36,37], albumin microbeads[38], and gelatin microbeads[39,40], showing that monodisperse fine particles can be prepared. We have investigated preparation of gelatin and acacia coacervate microencapsules[41], monodisperse oil droplets using lysolecithin[42], and multiple emulsions[43,44] and downsizing of particles using microchannel emulsification and solvent evaporation methods[45]. Emulsion droplets are inherently spherical, but we have shown that they can be prepared in disks by precisely controlling the structure of microchannels[46].

Microchannel emulsification has been mainly used for liquid-liquid disperse systems. We have investigated the use for preparing microbubbles, and the results are shown in figure 6. Although the stability has not been thoroughly investigated, microbubbles of 30–50 μm and a coefficient of variation of less than 10 % could be produced by dissolving an appropriate surfactant in the water phase[47].

3. 6 Conclusions

This chapter outlined the microchannel emulsification technology, which is based on micro processing techniques. The characteristics of the technology are: (i) unlike shearing by mechanical mixing, the disperse phase is spontaneously transformed into droplets in microchannel emulsification by the interfacial tension dominant on a micrometer scale, (ii) monodispersed emulsions consisting of uniform droplets of 3–100 μm can be prepared using a grooved microchannel array that has a terrace structure, (iii) spontaneous-transformation-based droplet formation occurs with rectangular through microchannels and asymmetrical through microchannels, enabling monodispersed emulsion of 3–50 μm in particle size to be effectively produced, which is the smallest size available today using micro processing techniques, and (iv) microparticles of a uniform size and microcapsules can be produced using monodisperse droplets prepared by microchannel emulsification.

Approximately 50 prototype microchannel emulsification equipments have been utilized in chemical manufacturers, *etc.*, and investigations are being made toward commercial production. The studies on microchannel emulsification were partly funded by the Promotion of Basic Research Division of BRAIN and the Nanotechnology Project of MAFF. The authors thank Dr. Yuji Kikuchi, Dr. Takahiro Kawakatsu and Dr. Shinji Sugiura and many others for their cooperation.

References

1. T. Nakashima, M. Shimizu, M. Kukizaki, *Adv. Drug Delivery Reviews*, **45**, 47-56 (2000)
2. H. Fujita, "Introduction to micro and nanomachine technology", pp.9-224 (2003)
3. T.Kitamori, S. Shoji, Y. Baba, H. Fujita *ed.*, "Technology and application of micro-chemical chip", MARUZEN Co., Ltd., pp. 1-352 (2004)
4. W. Ehrfeld, V. Hessel, H. Lowe, Microreactors, WILEY-VCH (2000)
5. Y. Kikuchi, H.E. Kikuchi, *SPIE*, **4265**, 14 (2001)
6. T. Kawakatsu, Y. Kikuchi, M. Nakajima, *JAOCS*, **74**, 317-321(1997)
7. T. Kobayashi, M. Nakajima, J. Tong, T. Kawakatsu, H. Nabetani, Y. Kikuchi, A. Shono, K. Satoh, *Food Sci. Technol. Res.*, **5**, 350-355 (1999)
8. M. Nakajima, *Japan Journal of Food Engineering*, **5** (2), 71-81 (2004)
9. S. Sugiura, M. Nakajima, S. Iwamoto, M. Seki, *Langmuir*, **17**, 5562-5566 (2001)
10. S. Sugiura, M. Nakajima, M. Seki, *Langmuir*, **18**, 5708-5712 (2002)
11. S. Sugiura, M. Nakajima, N. Kumazawa, S. Iwamoto, M. Seki, *J. Phys. Chem. B.* **106**, 9405-9409 (2002)
12. S. Sugiura, M. Nakajima, T. Oda, M. Satake, M. Seki, *J. Colloid Interface Sci.*, **269** 178-185 (2004)
13. T. Kawakatsu, G. Tragardh, Y. Kikuchi, M. Nakajima, H. Nabetani, T. Yonemoto, *J. Surfactants Detergents*, **3**, 295-302 (2000)
14. T. Kawakatsu, G. Tragardh, Ch. Tragardh, M. Nakajima, N. Oda, T. Yonemoto, *Colloids and surfaces A: Physicohemical Eng. Aspects*, **179**, 29-37 (2001)
15. I. Kobayashi, K. Uemura, M. Nakajima, *Colloids Surf. A: Physicochem. Eng. Aspects,* **296** 285-289 (2006)
16. S. Sugiura, N. Kumazawa, S. Iwamoto, T. Oda, M. Satake, M. Nakajima, *Kagaku Kogaku Ronbunshu*, **30** (2), 129-134(2004)
17. M. Kawakatsu, H. Komori, M. Nakajima, Y. Kikuchi, T. Yonemoto, *J. Chem. Engineering Japan.* **32**. 241-244 (1999)
18. S. Sugiura, M. Nakajima, M. Ushijimax, K. Yamamoto, M. Seki, *J. Chem. Eng. Japan*, **34**, 757-765 (2001)
19. J. Tong, M. Nakajima, H. Nabetani, Y. Kikuchi, *J. Surfactants Detergents*, **3**, 285-293 (2000)
20. M. Saito, L. Yin, I. Kobayashi, M. Nakajima, *Food Hydrocolloids*, **19** 745-751 (2005)
21. M. Saito, L-Y. Yin, I. Kobayashi, M. Nakajima, *Food Hydrocolloids*, **20** (7), 1020-1028 (2006)
22. I. Kobayashi, M. Nakajima, K. Chun, Y. Kikuchi, Fujita H., *AIChE J*, **48**, 1639-1644 (2002)
23. I. Kobayashi, S. Mukataka, M. Nakajima, *J. Colloid Interface Sci.*, **279**, 277-280 (2004)
24. I. Kobayashi, M. Nakajima, S. Mukataka, *Colloids Surfaces A*, **229**, 33-41(2003)
25. I. Kobayashi, S. Mukataka, M. Nakajima, *Ind. Eng. Chem. Res.*, **44**, 5852-5856 (2005)
26. I. Kobayashi, T. Takano, R. Maeda, Y. Wada, K. Uemura, M. Nakajima, Microfluid Nanofluid, DOI 10.1007/s10404-007-0167-2, 2007
27. I. Kobayashi, S. Mukataka, M. Nakajima, *Langmuir*, **20**, 9868-9877 (2004)

28. I. Kobayashi, S. Mukataka, M. Nakajima, Experimental and CFD studies, *Langmuir*, **21**, 5722-5730 (2005)
29. I. Kobayashi, K. Uemura, M. Nakajima, *J. Chem. Eng. Japan*, **39** (8), 855-863 (2006)
30. I. Kobayashi, S. Mukataka, M. Nakajima, *Langmuir*, **21**, 7629-7632 (2005)
31. J. Tong, M. Nakajima, H. Nabetani, Y. Kikuchi, Y. Maruta, *J. Colloid Interface Science*, **237**, 239-248 (2001)
32. H. Liu, M. Nakajima, T. Nishi, T. Kimura, *Nippon Shokuhin Kogaku Kogaku Kaishi*, **5** (4), 259-266 (2004)
33. H. Liu, M. Nakajima, T. Kimura, *JAOCS*, **81**, 705-711(2004)
34. I. Kobayashi, S. Hirose, T. Katoh, Y. Zhang, K. Uemura, M. Nakajima, Proc. HARMST2007, pp.91-92 (2007)
35. S. Sugiura, M. Nakajima, J. Tong, H. Nabetani, M. Seki, *J. Colloid Interface Sci.*, **227**, 95-103 (2000)
36. S. Sugiura, M. Nakajima, H. Ito, M. Seki, *Macromolecular Rapid Communications*, **22**,773-778 (2001)
37. S. Sugiura, M. Nakajima, M. Seki, *Ind. Chem. Eng. Res.*, **41**, 4043-4047 (2002)
38. T. Kawakatsu, N. Oda, T. Yonemoto, T. Nakajima, *Kagaku Kogaku Ronbunshu*, **26**, 122-125 (2000)
39. S. Iwamoto, K. Nakagawa, S. Sugiura, M. Nakajima, *APPS Pharm Scitech*, **3** (3), article 25 (2002)
40. S. Iwamoto, K. Nakagawa, M. Nakajima, H. Nabetani, *Kagaku to Seibutsu*, **43** (6), 410-415 (2005)
41. K. Nakagawa, S. Iwamoto, M. Nakajima, A. Shono, K. Satoh, *J. Colloid Interface Sci.*, **278**, 198-205 (2004)
42. J. Tong, M. Nakajima, H. Nabetani, *Euro. j. Lipid. Sci. Technol.*, **104**, 216-221(2002)
43. S. Sugiura, M. Nakajima, K. Yamamoto, S. Iwamoto, T. Oda, M. Satake, M. Seki, *J. Colloid Interface Sci.*, **270**, 221-228 (2004)
44. I. Kobayashi, X. Lou, S. Mukataka, M. Nakajima, *JAOCS*, **82**, 65-71 (2005)
45. I. Kobayashi, Y. Iitaka, S. Iwamoto, S. Kimura, M. Nakajima, *J. Chem. Eng. Japan*, **36**, 996-1000 (2003)
46. I. Kobayashi, K. Uemura, M. Nakajima, *Langmuir*, **22**, 10893-10897 (2006)
47. M. Yasuno, S. Sugiura, S. Iwamoto, M. Nakajima, *AIChEJ*, **50**, 3237-3233 (2004)

[Emulsification]

Chapter 4
Fast, Simple and Easy Production of O/W Fine Emulsion

Tohru Okamoto

4.1 Introduction

Like ordinary colloids, there are two methods for preparing emulsions: the condensation and dispersion methods. The condensation method involves forming and growing deposits in a uniform solution or oil at a solubilizable state. The dispersion method separates big dispersoids into fine particles by applying shear force. The dispersion method is more widely used than the other.

In the dispersion method, emulsion is prepared by forming water-oil interfaces. Because water-oil interfaces have interfacial tension, work is needed to form fine dispersion particles each surrounded by water-oil interface. In nanoemulsion, which consists of nano-size emulsion particles, the total area of the interface is much larger than in ordinary emulsions. Therefore, for efficient preparation of nanoemulsion, the work load for emulsification should be reduced by lowering the interfacial tension, and emulsification equipment of high shear power should be used to overcome the interfacial tension and form new interfaces. Well known methods for preparing micro emulsions, such as the inverse phase emulsification method, HLB temperature emulsification method and D phase emulsification method, involve emulsification at low interfacial tensions. Development of emulsification equipment that have strong shear power, such as high pressure homogenizers, has facilitated preparation of emulsions of sub-micrometer levels. Even with these methods and equipment, formulations and production conditions still need to be carefully optimized to prepare emulsions of several tens of nm levels.

On the other hand, the condensation method is advantageous in terms that it does not need generation of nanoparticles and mechanical shear power. However, it has several disadvantages: the formulation is limited and emulsification process is difficult to control.

This article describes methods and concepts for solving these issues and efficiently producing nanoemulsion.

4.2 Preparation of fine emulsion using techniques of surface chemistry

Emulsion is formed when a three-ingredient system consisting of surfactant, oil and water is stirred. The conditions of the resultant emulsion duffer greatly depending on preparation process even when the same formulation is used. Figure 1 is a typical phase diagram of a surfactant-oil-water system. Let us investigate the effects of preparation process for producing an emulsion of the final composition ●.

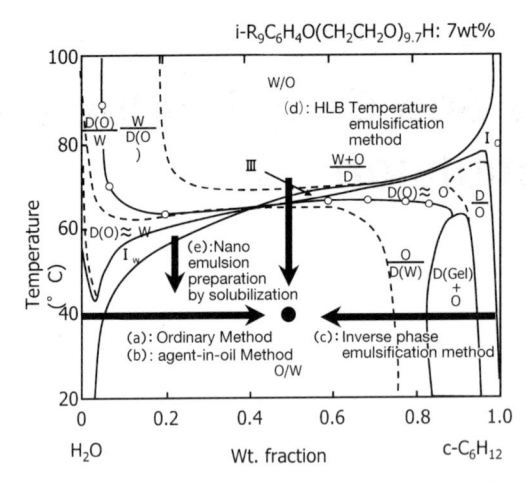

Figure 1 Comparison of emulsion processes and phase diagram of surfactants-oil-water systems

Hydrophilic surfactant is used for producing O/W emulsion. Preparation process (a), which involves dissolving hydrophilic surfactant in the water phase and adding the oil phase into the water phase by stirring, seems natural, but finer emulsion particles can be obtained by dissolving the surfactant in the oil phase and adding the oil phase to the water phase (the agent-in-oil method, (b)). This is because in the unbalanced state before the oil and water phases reach the distributive balance, the interfacial tension is lower when the oil phase is distributed and diffused into the water phase than when a surfactant adsorbs from the water phase, to which the surfactant is more affinitive than the oil phase, to the interface[1]. From this, it can be understood that reduction of the interfacial tension of the oil-water interface is important in the dispersion method for obtain the fine emulsion.

4.2.1 Inverse phase emulsification method[2]

The inverse phase emulsification method prepares O/W emulsion of fine particles by dissolving surfactant in oil and adding the water phase while stirring (figure 1 (c)). When water is gradually added to oil containing surfactant, reverse micelles are formed first, followed by formation of the lamella liquid crystal phase and O/D emulsion, which change into the region of O/W emulsion. The series of process occurs continuously. At the liquid crystals phase, interfacial tension become small, and fine particles are formed. The process resembles the agent-in-oil method at the point of dissolving a surfactant to oil, but inverse phase emulsification uses the phase change of the surfactant when water is slowly added.

4.2.2 HLB temperature emulsification method[3-5]

The HLB temperature emulsification method for preparing fine emulsion is based on the phenomenon that the hydrophile lyophile balance (HLB) of nonionic surfactants varies by temperature (figure 1 (d)). Let us investigate the generation process by following the arrow. At high temperatures, nonionic surfactant is hydrophobic and produces W/O emulsion. At lower temperatures, the surfactant is more hydrophilic. At temperatures below the phase inversion temperature (PIT, HLB temperature), emulsions change to O/W emulsions. Around the phase inversion temperature, which is the intermediate temperature range between the temperatures of W/O and O/W emulsions, three phase regions of the surfactant, water and oil phases appear.

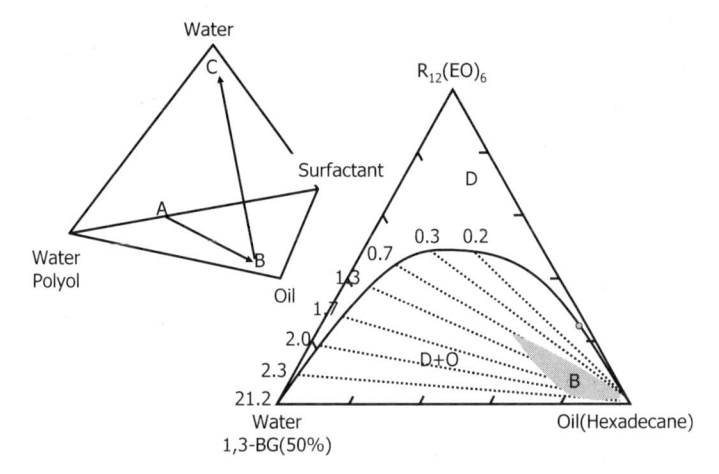

Figure 2 Emulsification process in the D phase emulsification method
and the oil phase and interfacial tension of D phase

The oil-water interfacial tension is remarkably low in this temperature range, and even weak stirring power can result in fine emulsions of the submicrometer level. In this temperature range, fine emulsions are easily formed, but on the other hand, emulsion particles coalesce fast and are unstable. Therefore once emulsion is formed, it is preferable to cool the emulsion to shorten the stay time in this temperature range, and a surfactant that has the guaranteed temperature far from the HLB temperature should be used.

4.2.3 D phase emulsification method [6–8]

The D phase emulsification method uses polyhydric alcohol as the fourth component together with oil, water and surfactant to prepare fine O/W emulsions. The preparation process is shown in figure 2. The method involves dispersing oil in a solution containing surfactant, water and polyol, which results in O/D gel emulsion (figure 2(A→B)), and adding water to the gel emulsion to form fine O/W emulsion particles (figure 2(B→C)). When oil is added to the water phase of the composition shown as A in the figure, the oil is dissolved until the solubilization limit and separates above the limit and forms cloud. When oil is added, the system becomes transparent gel (region B in figure 2). This is because the refractive index between dispersoids and the disperse phase, which is a thick water solution of polyol and surfactant, is reduced. The numerical values in figure 2 denote the interfacial tension between the D phase and oil. The change of the continuous phase from water to D phase lowers the interfacial tension of oil, resulting in fine emulsion particles.

4.3 Preparation of nanoemulsion using the condensation method

4.3.1 Preparation of nanoemulsion using the solubilization region

Nanoemulsion particles of diameters not exceeding 50 nm can be produced by solubilizing a surfactant-oil-water system and quickly cooling the resultant microemulsion of one-solution phase to the room temperature to shift the system into the two phase region on a phase diagram. Figure 3 shows a solubilization phase diagram of a polyoxyethylene (8) hexadecyl ether (C16E8)-oil-water system. In this method, nanoemulsion is prepared along the arrow in the figure. The region between a solubilization limit curve and cloud point curve is where oil is solubilized in large quantities and is microemulsion. Nakajima *et al.* showed that nanoemulsions of particle sizes of droplets could be easily prepared by quickly cooling microemulsions to the room temperature to make the systems shift to the conceptual two phase region[9-11]. At around the solubilization limit temperature, two-phase emulsions are unstable and are difficult to maintain the size of microemulsion. Therefore it is necessary to reduce the retention time at this temperature by quickly cooling and using oil and surfactant that are advantageous in keeping stability. Oil that is little soluble to water and surfactant that has a large molecular size are preferable if the HLB is the same.

A nanoemulsion that has a particle diameter of micelle is obtained when the turbidity of a one-phase O/W microemulsion around the solubilization limit temperature is kept in the two phase region. Because the size of solubilization micelles is determined by the weight ratio between surfactant and oil, the size of nanoemulsion particles is controlled by controlling the weight ratio (figure 4). In the figure, which shows a liquid paraffin system, the particle size

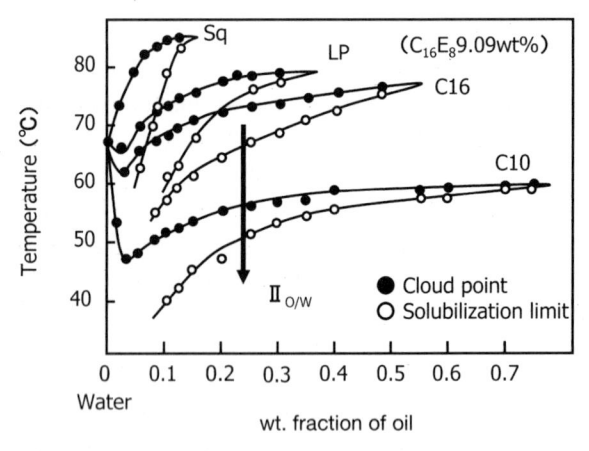

Figure 3 Solubilization phase diagram of a system of Polyoxyethylene (8) Hexadecyl Ether ($C_{16}E_8$), oil and water[9]
(SQ: Squalane, LP: Liquid paraffin, C16: Hexadecane, C10: Decane)

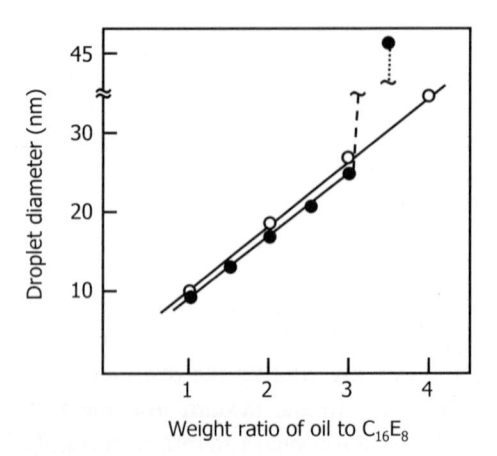

Figure 4 Relationship between the particle size and the weight ratio of oil and Polyoxyethylene (8) Hexadecyl Ether ($C_{16}E_8$)[9]
○: Hexadecane (C16), ●: Liquid paraffin

deviated from the line at oil / surfactants = 3.5 and over because not all oil can be dissolved at this ratio. Let us investigate the differences between this method and the other aforementioned emulsification methods shown in figure 1 (figure 1 (e)). Unlike the other methods, which involve dispersion, this method is not affected by mixing conditions because the procedure takes place in the solubilization region in which oil droplets are spontaneously formed although there is a limit in the quantity of oil that can be used in the formulation.

4. 3. 2 Wax nano dispersion

As a concrete example of nanoemulsion preparation using the condensation method, wax nano dispersion is described. Carnauba wax and Candelilla wax are widely used in haircare products because they make the hair shiny and strongly keep the hair in shape. However, the waxes have high melting points are easy to separate out, and the use has been limited to sticks and creamy formulations.

Matsuura et al. solubilized wax in a mixture system consisting of a nonionic surfactant of HLB of 9 to 12 and an amphoteric surfactant of the betaine type and found that ultra fine wax dispersion of around 50 nm could be prepared by rapid cooling[12, 13]. Formulas of nonionic surfactants used for solubilizing 10 % Carnauba wax in 10 % nonionic surfactant and 5 % amphoteric surfactant (cocoyl amide propylmethyl glycine) are shown in table 1 together with the results of solubilization. The properties after cooling are shown in table 2. Carnauba wax was solubilized in all nonionic surfactants of HLB grade 9 regardless of straight or branched chain. The particle size of the solubilization state was maintained during the cooling process for shifting the system into the two-phase region when nonionic surfactants that had long alkyl chains of 22 or 18 carbons were used. On the other hand, when surfactants that had short alkyl chains were used, the system became cloudy during the cooling process. This was because

Table 1 Structure of nonionic surfactants and the solubility of wax at a temperature above the melting point[12]

Structure of hydrocarbon chain of nonionic surfactants		HLB of surfactants								
		5	6	7	8	9	10	11	12	13
Straight-chain type	C12					○	○			
	C16				×	○				
	C18				×	○				
	C18 (oleyl)					×	○			
	C22	×				○				×
Branched-chain type	Isostearyl				×		○	×		
	Octyldodecyl						○	×		
	decyltetradecyl					×	○			
Bulky-chain type	Hydrogenated Castor Oil	×	×	×	×	×	×	×	×	

Table 2 Structure of nonionic surfactants and their appearances and properties at the room temperature[12]

Structure of nonionic surfactants	HLB	Appearance	Melting Point (°C)	Freezing Point (°C)
C22 (EO) 10	9	Transparent	61.5	45.3
C18 (EO) 8	9	Transparent	60.7	42.1
C16 (EO) 7	9	Turbid	63.3	61.5
C12 (EO) 5	9	Turbid	64.2	62.8

particles were unstable against coalescence in the two-phase region around the solubilization limit temperature and surfactants of large molecular sizes (big hydrophilic group) were advantageous in keeping stability.

4.3.3 Nanoemulsion preparation using polyoxyethylene · polyoxypropylene random copolymer dimethyl ether

The aforementioned method involved preparing an one-liquid phase emulsion at a high temperature in which HLB is balanced and quickly cooling the emulsion to shift it into the two-phase region. Miyahara *et al.* reported a method of preparing nanoemulsion of around 50nm in size by producing microemulsion using polyoxyethylene · polyoxypropylene random copolymer dimethyl ether[14], which was a new solution solvent, at room temperature and diluting with water[15-17].

Figure 5 shows the phase equilibrium at 25°C of a system consisting of 70 % POE (7) oleylether and Polyoxyethylene (9mol) · polyoxypropylene (2mol) random copolymer dimethyl ether (EPDME(9/2)) mixture, liquid paraffin and water. When EPDME was used, microemulsion was produced in both the microemulsion region Wm and the region where the oil content was higher than that in the lamellar liquid crystal phase (Lα). When the microemulsion of the constitution shown with "a star" in the figure was mixed and dispersed in water, nanoemulsion of 54 nm size was produced at the room temperature by stirring slowly at about 400 rpm. Figure 6 shows the structure of microemulsion of high oil inside phase. EPDME (9/2) attaches to the hydrophilic groups of microemulsion, extending the interface. This decreases the association number of polyoxyethylene surfactant and enlarges the occupation area per molecule. The spaces among surfactant molecules are filled with EPDME. An objective of preparing emulsion is to mix water and oil at an arbitrary ratio, and it is desirable to emulsify a large amount of oil using a little amount of surfactant to extend the freedom of formulation. The special microemulsion phase with an expanded curvature in figure 3 was formed instead of taking a lamella structure in which HLB was balanced likely because EPDME (9/2) occupied the hydrophilic groups of the surfactant, resulting in an expanded oil-water interface area. This enabled a larger quantity of oil to be included than in an ordinary microemulsion phase.

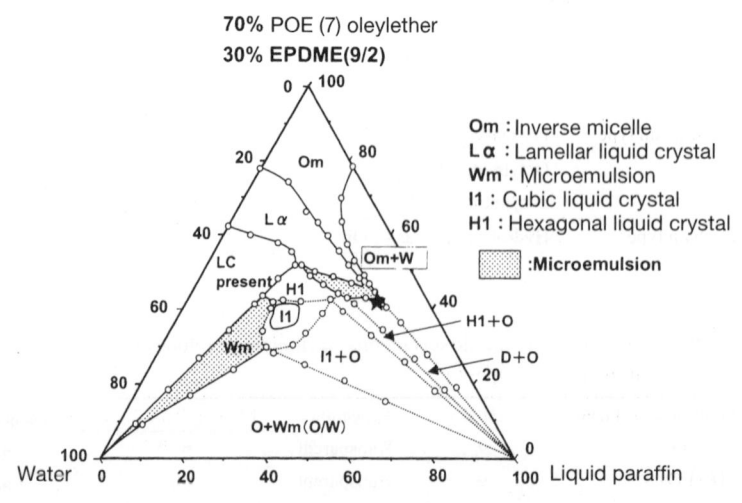

Figure 5 Phase equilibrium diagram (25°C) of a POE (7) oleylether (70%) +EPDME (9/2) (30%)-liquid paraffin-water system[17]

⌇ : Surfactant ● : **EPDME(9/2)** molecule

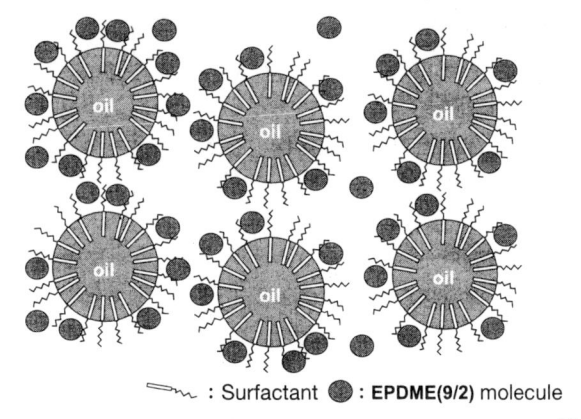

Figure 6 Structure of microemulsion of high inside phase[17]

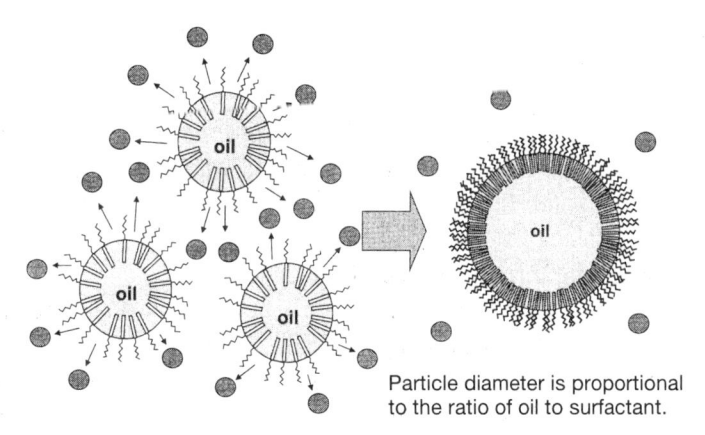

Particle diameter is proportional to the ratio of oil to surfactant.

Figure 7 Condition changes by dilution of microemulsion of high inside phase into water [17]

Microemulsion of high inside phase is densely filled with oil, which forms cubic liquid crystals, and is thus highly viscous and difficult to handle. At a composition in which the surface tension is almost 0, the association number of surfactant rises, sometimes resulting in highly viscous phase, such as lamella liquid crystals, particularly at the room temperature. During a dilution process, oil particles unite when the phase shifts from one liquid phase to two liquid phase, and coarse particles can be generated if the dispersibility is bad. The microemulsion prepared using EPDME is low in adsorption of surfactant and has a unique structure in which EPDME moves freely (figure 6). Thus, the microemulsion can keep fluidity and low viscosity despite of its high inside phase ratio and can be dispersed in water smoothly. When the microemulsion is dispersed in water, EPDME diffuses in the water phase, and the emulsion becomes particles of the size determined by the ratio of surfactant to oil (figure 7).

4. 3. 4 Preparation of nanoemulsion from uniform solutions

A colloid dispersion system can be prepared by decreasing the solubility of molecularly dissolved dispersoids by some means and condensing and depositing. The nanoemulsion preparation method is described below.

Silicone oils of low molecular weights dissolve in ethanol uniformly. Dissolving surfactant

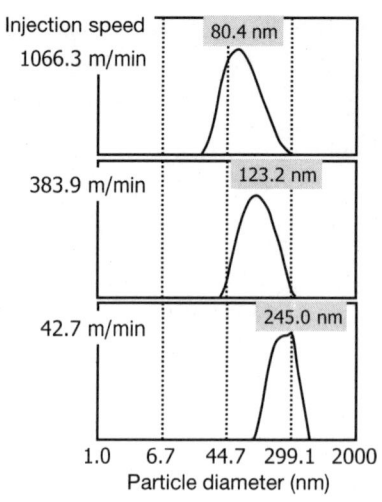

Figure 8 Injection speed of ethanol phase and emulsion particle diameter

into the ethanol solution and injecting the solution into water at a high speed result in a nanoemulsion of silicone oil[18]. When the alcohol phase in which silicone oil is dissolved mixes with water uniformly, the solubility of silicone oil falls, and silicone is deposited, separates, and forms large-size particles. Choosing appropriate surfactants at this time and preventing aggregation by increasing the injection speed of the ethanol phase to water result in successful nanoemulsion preparation. Figure 8 shows injection speeds and particle size distribution when an ethanol solution containing 0.2 % dimethyl polysiloxane and 0.2 % POE (20) hydrogenated caster oil in 10 % ethanol was injected into the water phase. Particle sizes were smaller at higher injection speeds, and the particle size was 80.4 nm at an injection speed of 1066.3 m/min. The manufacturing method that involves dissolving surfactant and oil in alcohol before adding to water is widely used to produce solubilization system lotions. However, manufacturing conditions are difficult to control because overdosing of oil to exceeding the solubilization limit causes oil particles to unite during the oil depositing process and oil slicks and coarse oil drops to form. The coalescence of oil particles can be prevented by increasing the speed of injecting the alcohol phase into the water phase and preventing the collision of particles immediately after their generation. Since the speed of injection can be controlled even with an ordinary solvent delivery pump, this preparation method of fine emulsion is simple, easy and energy-saving and needs no special device. It should be noted, however, that unlike the solubilized region method, which produces particles of a uniform size before shifting to a two-phase system and thus the particle size of the resultant nanoemulsion is quite uniform, nanoemulsions produced using this method show a wide particle size distribution because oil condenses and oil particles grow during the process of particle formation.

4. 4 Nanoemulsion preparation using high pressure homogenizer

Because preparation of emulsions by the dispersion method needs work to form a new interface, it is effective to use an emulsifier of high shear power. Figure 9 shows the relationship between the particle size of emulsions and the weight ratio of surfactants and oil in a system that contained liquid paraffin and POE (30) oleylether (C18E30) at a total concentration of 30 %, when the weight ratio of surfactants and oil and the emulsification pressure of high pressure homogenizer were changed. Small emulsion particles were produced when the emulsification pressure of the high pressure homogenizer was increased. New interfaces were generated when the sizes of oil drops were reduced, and a surfactant in the water phase adsorbed to the new interface and stabilized the emulsion. That is to say, the condition in which all of the surfactant molecules in the formulation were adsorbed is regarded as a smallest particle diameter of the emulsion. It is difficult to enlarge the interface area (reduce the particle size) further without affecting the stability. Thus, the particle size at this condition is the smallest particle diameter.

The dashed line in figure 9 shows the minimum particle diameter at each formulation calculated from the area occupied by a single molecule of surfactant. The line shows the ideal emulsification state in which surfactant acts most efficiently. The figure shows that the minimum particle diameter can be easily achieved (because the target particle diameters are big) in systems of high oil contents in relation to that of surfactants, but increasing surfactant contents to reduce particle size results in actual particle sizes detaching from the dashed line. Increasing the shear force reduces the detachment, but emulsification particles smaller than 100 nm are difficult to prepare without changing the formulation because there is a limits in increasing mechanical power.

4. 4. 1 Preparation of nanoemulsion using water soluble solvents

The use of water soluble solvents has been reported as an efficient method for preparing nanoemulsion using a high pressure homogenizer[11]. Figure 10 shows the relationship between the concentration of water soluble solvent and emulsification particle diameter in an emulsified O/W emulsion consisting of liquid paraffin (20 %), $C_{18}E_{30}$ (10 %) and water that contained water soluble solvents (70 %), when the emulsions were prepared using a high pressure homogenizer. The particle sizes of emulsion were smaller at higher concentration of water soluble solvents in both systems that used glycerol and 1,3-butanediol as water soluble solvents (figure 10), of which the trend was more notable in 1,3-butanediol.

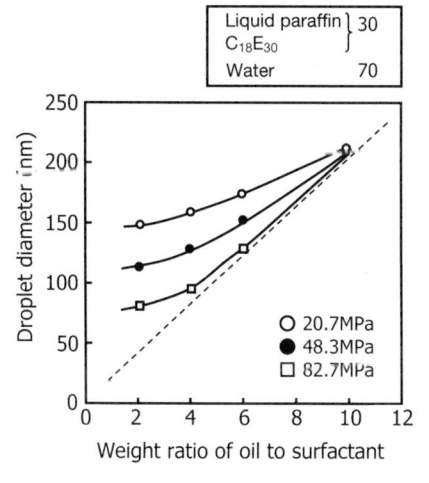

Figure 9 Relationship between emulsion particle size and the surfactant to oil weight ratio at various emulsification pressures. (The dashed line shows the smallest particle diameter calculated from the adsorption onto the cross section area of the oil-water interface of surfactants.)

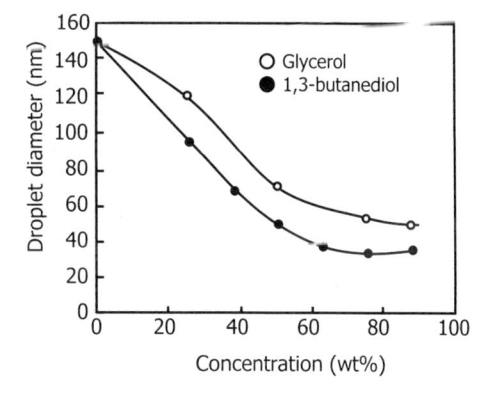

Figure 10 Relationship between the emulsion particle diameter and the concentration of water soluble solvent in the water phase (20.7 MPa) [11]

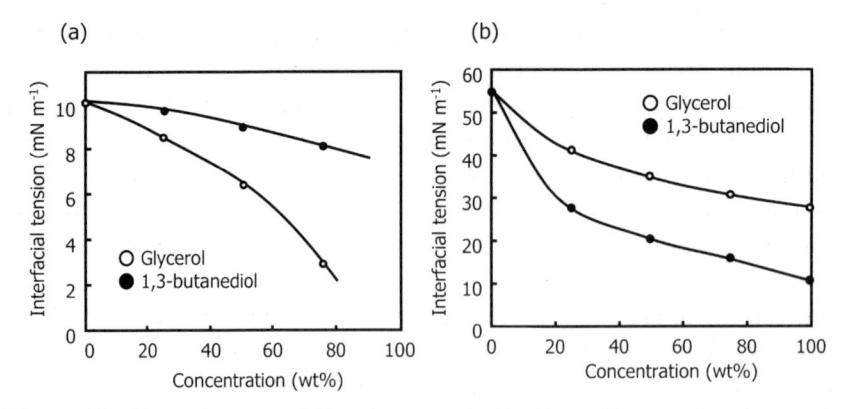

Figure 11 Effects of waster soluble solvents on the liquid paraffin-water interface tension
(a) Interfacial tension of liquid paraffin-1 wt% $C_{18}E_{30}$ water solution,
(b) Interfacial tension of liquid paraffin-water

The mechanisms of particle size reduction by water soluble solvents have been investigated in terms of their effects on interfacial tension. In ordinary emulsification, the action of an aqueous surfactant solution and the oil phase in reduction of the oil-water interfacial tension is considered as an index for expressing the ease of emulsification. However, the decrease in interfacial tension caused by glycerol was larger than that by 1,3-butanediol in a system prepared by adding water soluble solvents into an aqueous solution of a high surfactant content, and correlation could not be determined between the decrease and size reducing effects (figure 11(a)). On the other hand, 1,3-butanediol was effective for reduction of the interfacial tension when surfactants were not adsorbed, supporting the result of fine particle preparation (figure 11(b)). The actions of water soluble solvents while preparing nanoemulsion using a high pressure homogenizer depended on their properties to decrease the interfacial tension at a condition in which only a small amount of surfactant is adsorbed on the oil-water interface. When emulsion particles are divided into smaller particles in a high pressure homogenizer, only small amounts of surfactant molecules are adsorbed on the newly formed interface because the speed of the division is very fast. Therefore, to lower the energy for dividing emulsion particles and facilitate preparation of fine particles, it is effective to reduce the interfacial tension (dynamic interfacial tension) at low adsorption quantities. Interfacial tension plays an important role in preparing fine emulsion particles, but it should be noted that its action differs between preparation using an ordinary emulsifier and a high pressure emulsifier.

The use of water soluble solvents is effective for preparing nanoemulsion particles, but too high concentrations of water soluble solvents adversely affect the process and cause increases in particle diameters. This is because water soluble solvents such as 1,3-butanediol destabilizes emulsion at high temperatures. High pressure emulsifiers generate heat, rising and destabilizing emulsions, which become prone to causing associations of particles. This suggests that both stabilization and preparation of fine particles should be considered when preparing nanoemulsions.

4. 4. 2 Nanoemulsion of surfactant-amphiphile-oil-water system (milky lotion and cream)

Nanoemulsion of particle size of 20-30nm can be prepared using an optimum formulation and a high pressure homogenizer at the ideal emulsification condition (the condition in which most surfactant molecules are adsorbed on the interface (dashed line in figure 9)). It is not necessary to consider the influence of surfactant molecules not used for emulsification and remained in water needs when the surfactant dissolves in water in a form of micelles, but for surfactants that form gel and/or interact with other ingredients, non-used surfactant molecules may affect the appearance and/or properties of resultant products. Effects of an ideal emulsification condition on the stability of the products are exemplified below.

Milky lotion and lotion for cosmetics consist of amphiphiles, surfactant, oils, and water systems. Amphiphiles and surfactants form α gel in the water phase. α gel has a structure similar to lamellar liquid crystal, in which surfactants and amphiphiles are arranged with their hydrophobic groups forming a hexagonal structure. The structure forms a network, increases the viscosity, solidifies the entire system, and keeps the emulsions stable. α gel is also a key ingredient in giving emolliating performances and unique moisturizing feeling to the cosmetics. Because giving viscosity and efficacies as cosmetics are linked, producing non-viscous products of the same efficacies were deemed difficult, but the difficulty was resolved by preparing nanoemulsion using ideal emulsification processes[19, 20].

The properties and appearances of emulsions prepared using stearyl trimethyl ammonium chloride (STAC) as a cationic surfactant, stearyl alcohol ($C_{18}OH$) as fatty alcohol and dimethyl polysiloxane at various preparation methods are shown in figure 12. The emulsion having STAC and fatty alcohol at a mole ratio of 1:3 and an ordinary process (figure 12(a)) formed α gel, which made the product viscous. Processing this product using a high pressure homogenizer at temperatures above the melting point of the gel produced less viscous product (figure 12(b)), which returned to cream when kept at high temperature. On the other hand, the low viscosity could be maintained when prepared into emulsion particles smaller than about 30 nm as shown in figure 12(c).

Figure 13 shows the particle sizes of emulsions prepared using dimethyl polysiloxane and their DSC charts. When the particle size was big, transition peak was assigned to the α-gel peak of a cationic surfactants-fatty alcohol-water system. However, α gel peak was smaller in emulsions of smaller particle sizes and appeared on the low temperature side. From the relationship between the phase transition enthalpy of α gel and particle size, the ingredients of α gel were likely to have adhered to new interfaces produced by the reduction of particle sizes, decreasing the phase transition enthalpy of α gel. This caused gradual decreases of dispersed α gel particles in the water phase, preventing reconstruction of the network and increasing

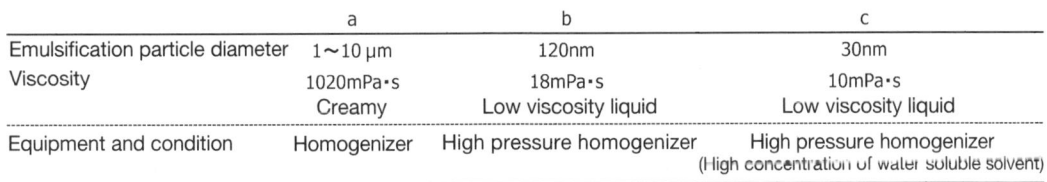

	a	b	c
Emulsification particle diameter	1~10 μm	120nm	30nm
Viscosity	1020mPa·s Creamy	18mPa·s Low viscosity liquid	10mPa·s Low viscosity liquid
Equipment and condition	Homogenizer	High pressure homogenizer	High pressure homogenizer (High concentration of water soluble solvent)

Figure 12 Changes in property and appearance of milky lotion accompanying size reductions of emulsion particles

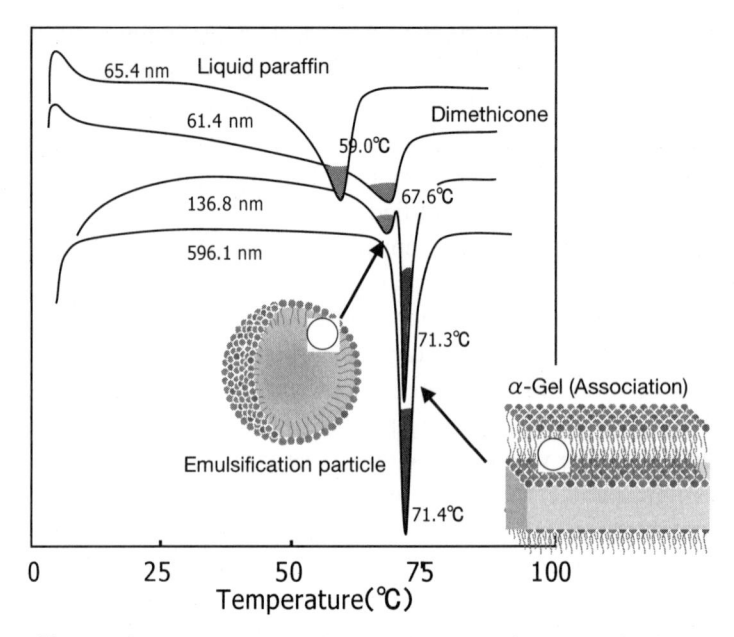

Figure 13 Phase transition behavior and particle diameters of cationic surfactants-fatty alcohol-water system emulsion

in viscosity. This phenomenon is likely to involve reducing the size of emulsion particles, making almost all surfactant and amphiphiles in the water phase to adsorb to the interface, and thus stabilizing the system. Thus, reducing the size of emulsion particles up to an ideal emulsification state is indispensable for keeping the stability of a system.

4.5 Conclusion

Nanoemulsion was once believed to be difficult to prepare, but development of high-performance emulsifiers and progress in interface chemistry have facilitated the preparation. There are still rooms for improvement, such as developing simple methods and enabling use of free formulations. Today, Green Sustainable Chemistry is attracting attention. Technologies for preparing emulsions of fine particles using less energy and less surfactant will be increasingly noticed.

References

1. T. Mitsui, S. Nakamura, F. Harusawa, Y. Machida, *Kolloid Z.Z.Polym.*, **259**, 227 (1972)
2. H. Sagitani, *J. Am. Oil Chem. Soc.*, **58**, 738 (1968)
3. K. Shinoda, H. Saito, *J. Colloid Interface Sci.*, **26**, 70 (1968)
4. K. Shinoda, H. Saito, *J. Colloid Interface Sci.*, **30**, 258 (1969)
5. K. Shinoda, H. Saito, *J. Colloid Interface Sci.*, **32**, 647 (1970)
6. H. Sagitani, T. Hattori, K. Nabeta, M. Nagai, *Chem. Soc. Jpn.*, **1983**, 1399 (1983)
7. H. Sagitani, Y. Hirai, K. Nabeta, M. Nagai, *J. Jpn. Oil Chem. Soc.*, **35**, 102 (1986)
8. H. Sagitani, *J. Dispersion Sci. Technol.*, **9**, 115 (1988)
9. T. Tomomasa, M. Kawauchi, H. Nakajima, *J. Jpn. Oil Chem. Soc.*, **37**, 1012 (1988)
10. T. Tomomasa, M. Kawauchi, H. Nakajima, *J. Soc. Cosmet. Chem. Jpn.*, **23**, 288 (1990)

11. H. Nakajima, *Surface*, **36**, 39 (1998)
12. E. Matsuura, A. Noda, Y. Shinojima, T. Ohmura, Y. Nakama, M. Yamaguchi, Y. Kumano, The 21th IFSCC international congress proceedings, Berlin (2000)
13. E. Matsuura, A. Noda, Y. Shinojima, T. Ohmura, Y. Nakama, M. Yamaguchi, Y. Kumano, The collection of points of the 21th IFSCC international congress proceedings, Berlin, 34 (2000)
14. T. Ohmori, Y. Yamamura, K. Nakahara, R. Miyahara, K. Hosokawa, K.Maruyama, T. Okamoto, H Kakoki, *Journal of Oleo Science*, **55** (7), 365-375 (2006)
15. R. Miyahara, K. Watanabe, T. Ohmori, Y. Nakama, *Journal of Oleo Science*, **55** (8), 403-411 (2006)
16. R. Miyahara, K. Watanabe, T. Ohmori, Y. Nakama *Journal of Oleo Science*, **55** (9), 473-482 (2006)
17. R. Miyahara, *Fragrance Journal*, **34** (10), 13 (2006)
18. T. Okamoto, T. Teshigawara, M. Matsumura, The collection of lectures of the 84th Chem. Soc. Jpn. spring annual convention, 288 (2004)
19. T. Okamoto, S. Anzai, H. Nakajima, The substance of the 48th DMCIC, 112 (1995)
20. T. Okamoto, S. Anzai, H. Nakajima, The substance of the 48th DMCIC, 654 (1995)

Chapter 5
Technologies for Stabilizing Emulsions

Sadanori Ban

5. 1 Introduction

Emulsification is a technology for mixing oil and water to produce materials of high functions and added values, and is used in various fields such as for cosmetics, pesticides, medical supplies and foods. There are two methods for preparing emulsions: the condensation method, which uses the characteristics of surfactants (emulsification using the surface chemistry), and the dispersion method, which involves the use of machines such as mixers and mills (mechanical method), which are commonly used combined. Emulsification process is thermodynamically unstable, and emulsions are prone to separation. In other words, almost all emulsions separate into two phases someday unless in very special cases. There is a misunderstanding that "there are methods for producing ever stable emulsions" because there are emulsion products that do not separate into two phases in several years. However, stable emulsification denotes producing emulsions that have a relatively long life (are slow to separate) by use of a special device and other measures, and just means "producing relatively stable emulsions".

This chapter describes the basics of emulsification, including the types of emulsions, preparation methods, stabilization, and thermodynamic properties of emulsification.

5. 2 Types of emulsions

Emulsion is a system consisting of one liquid (dispersed phase) being dispersed in another liquid (disperse medium), and is classified by the combination of liquids. Emulsions can be broadly classified into oil-in-water (oil in water) and water-in-oil (water in oil) type. There are also oil-in-water-in-oil and water-in-oil-in-water types, which are called multiple emulsions[1]. A three phase emulsion[2] was recently developed, which has an independent phase around droplets. A schematic diagram of the emulsification types is shown in figure 1. The type is determined by the kind and density of surfactant, the volumetric ratio between oil and water, temperature, and other factors. Bancroft[3] discovered oil-soluble emulsifying agents tend to form W/O emulsions

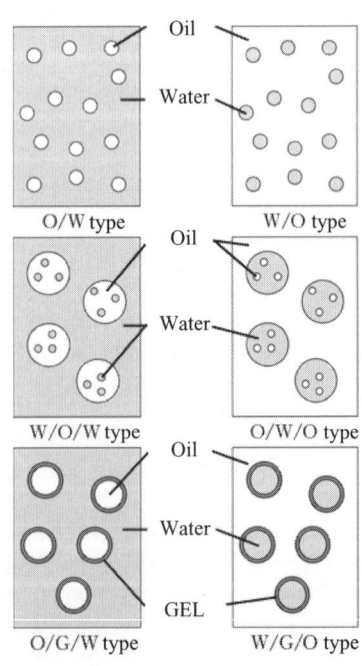

Figure 1 Schematic diagram of emulsion types

and water-soluble emulsifying agents tend to form O/W emulsion (Bancroft Rule). This rule applies in general.

There are some methods for determining whether an emulsion is O/W emulsion or W/O emulsion. A simple method involves dripping the emulsion of question in a beaker containing water. If the emulsion diffuses in water, the emulsion is O/W emulsion. If it floats on water, it is W/O emulsion. A viscous emulsion can be tested by adding water. If it is diluted by water, it is O/W emulsion; and it is W/O emulsion if it cannot be diluted. There is another method that involves measuring electrical conductivity. O/W emulsion show high conductivity while W/O emulsion show low electrical conductivity. The method requires addition of a small amount of salt if the surfactant used is nonionic (because electricity is not transmitted through a system that does not contain ions).

5.3 Preparation of emulsions

As described in the introduction, a third material is necessary to stably disperse water in oil or disperse oil in water, which are mutually immiscible. The most popular material is amphipathic materials, such as surfactants. There are also polymers (proteins, gelatin, *etc.*), fine powder of insoluble solids (clay, silica, *etc.*), and self-organized association (gel phase, *etc.*). Pickering emulsion is formed by selective adhesion of insoluble particles to oil and water surfaces (figure 2). Droplets can also be stabilized, even temporarily, by adding simple inorganic electrolytes such as potassium thiocyanate[4]. The difficulty of emulsification exists in what third materials (emulsifying agents) and emulsification method to be used. The two major emulsification methods are described below in detail.

5.3.1 Emulsification using surface chemistry (Condensation method)

In emulsification that uses surface chemistry, fine-particle emulsions are produced by using the least amount of mechanical energy, devised mixing methods and additives other than emulsifying agents as well as emulsifying agents. Kitahara[5] pointed out that the key is to produce a uniform state (transparent, gel, *etc.*) during the emulsification procedure, from which droplets are separated by changing temperature or adding a poor solvent, such as water and oil. The process is schematically shown in figure 3.

The phase inversion method (figure 3(1)) involves adding a small quantity of water to oil to make W/O emulsion, into which water is further added to change the phase (phase inversion) from lamellar liquid crystal to oil-in-D (detergent) phase emulsion and to O/W emulsion in

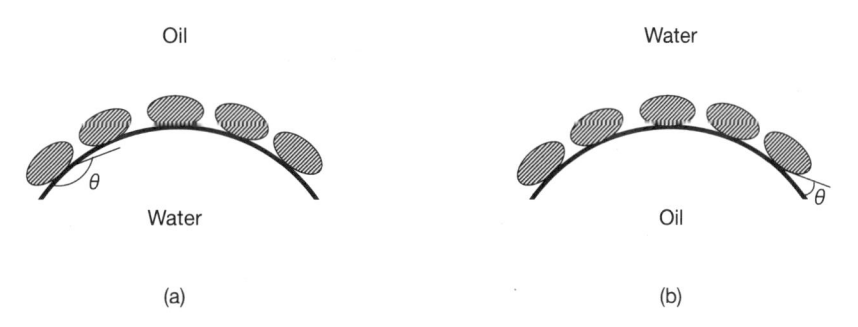

Figure 2 Example of the Pickering emulsion
Emulsion is stabilized by selective wetting of particulates by the solvent. (a) W / O emulsion formed by selective wetting of particulates by oil. (b) O / W emulsion formed by selective wetting of particulates by water.

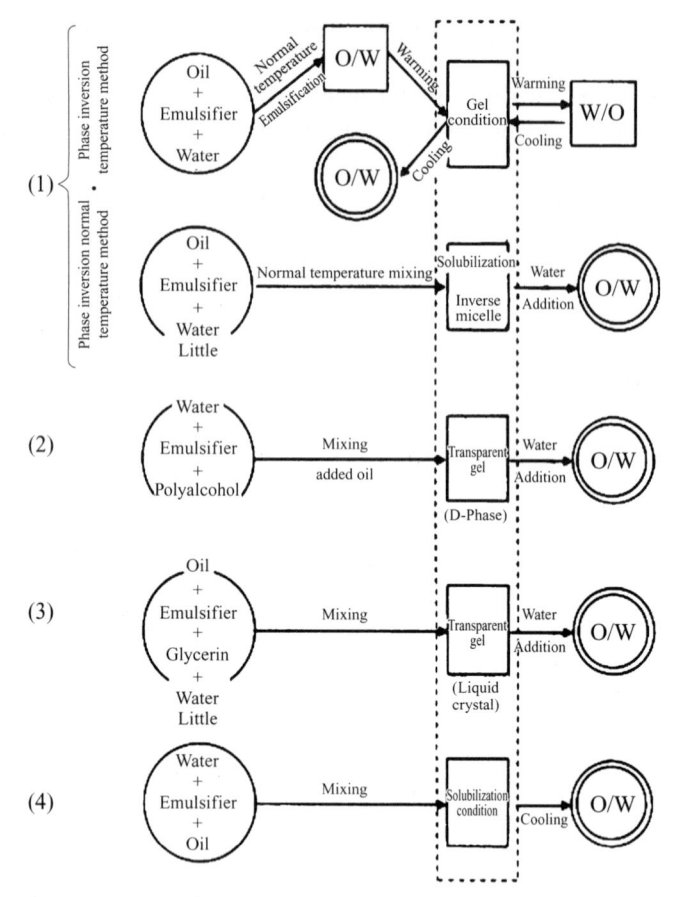

(1) Phase inversion method, (2) D-phase emulsification method, (3) Liquid crystal emulsification method, (4) Solubilization conversion emulsification method
◎ denotes a product. Those within the dashed-line rectangle are in uniform conditions.

Figure 3 Schematic illustration of physicochemical emulsification methods[5]

order to produce fine emulsion droplets[6]. Simultaneously, the surface tension between the surfactant and oil in the oil-in-D phase emulsion domain is reduced. There are also the D phase emulsification method[7], which uses the high-order structure of surfactants and polyalcohol or water, the liquid crystal emulsification method[8] and the gel emulsification method[9,10].

5.3.2 Mechanical emulsification (Dispersion method)

Mechanical energy can be applied either by swing mixing (stirring), grinding mixing (milling) or the two combined using appropriate emulsifiers. Emulsifying agents assist the machines in producing fine droplets and stabilize the resultant emulsion.

Recently, highly powerful emulsifiers have been developed and put on market, which can emulsify materials that have been deemed difficult to emulsify. The machines can produce emulsification particles smaller than 100 nm. It is also easy to produce liposomes in mass. The machines include high pressure homogenizers, high speed high shear type mixers and ultrasonic emulsification devices.

5. 4 Destruction and stability of emulsions

Stability is the most important topic of emulsion products. Creaming, flocculation, coalescence and Ostwald ripening are destabilization (destruction) processes of emulsions. The schemes of the four processes and their mutual relationships are shown in figure 4 [11].

5. 4. 1 Destruction processes of emulsions
5. 4. 1. 1 Creaming

Creaming is the migration of oil droplets in O/W emulsions, under the influence of density difference, to the top of sample while the oil droplets remain separated. Equation (1) of Stokes can be used to express the surfacing of oil droplets, but this expression is just an approximation because oil droplets are not rigid bodies and their relative gravity is usually large in emulsions.

$$u = \frac{2a^2(\rho-\rho_0)g}{9\eta} \qquad (1)$$

where, u is sedimentation speed, a is the radius of the particle, ρ and ρ_0 are the density of the particle and the medium, g is acceleration of the gravity, and η is the viscosity.

In creaming, droplets just assemble and are separated from each other by films of water. Thus, the droplets can be dispersed again by stirring lightly. However, flocculation is easy to occur because the oil droplets are close to each other.

The best method to prevent creaming is to reduce the particle size (a). This is one of the reasons for producing emulsions of a small particle size.

5. 4. 1. 2 Flocculation

Universal gravitation acts among colloidal (emulsion) particles, and the phenomenon that particles come in contact with each other is called flocculation. To prevent flocculation, it

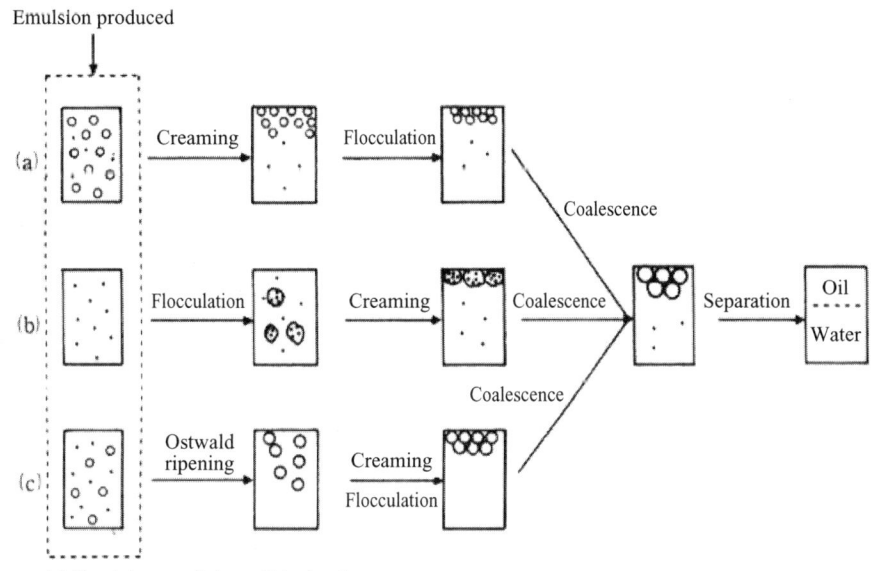

(a) Emulsion consisting of big droplets
(b) Emulsion consisting of minute droplets
(c) Emulsion consisting of minute droplets of various sizes and soluble liquid droplets.

Figure 4 Destruction process of emulsion[11]

is necessary to increase the repulsive force between particles using electric charge (electric double layer interval) or use the repulsive force of the layer (thick adsorption layer interval) of polymers adsorbed around particles. Emulsions of large zeta potential and lower salt concentration in the continuous water solution phase are strong against flocculation. Because destabilization of emulsions progresses from flocculation to coalescence, the two have been investigated together rather than flocculation alone.

5.4.1.3 Coalescence

The process in which droplets stick to and unite with each other, losing the adsorption films separating them, is called coalescence. Coalescence is the most critical event in the emulsion destabilization process. Coalescence among droplets advances, resulting in the oil phase separating from the water phase. To prevent coalescence, flocculation, which occurs prior to coalescence, should be prevented, and the adsorption film should be made strong enough to not allow adsorbed molecules dissociate or move within the film.

5.4.1.4 Ostwald ripening

When the size of fine particles separated from a supersaturated solution is not uniform, small particles gradually disappear, and large particles increase in size. This phenomenon is called Ostwald ripening.

In an O/W emulsion, even a small solubility to water of the oil phase results in a successive reaction of dissolution from oil droplet → diffusion in the water → separation to the big droplets. Thus, Ostwald ripening is hard to occur when the oil phase is little soluble to water and the water phase is viscous[11].

5.4.2 Stabilization of emulsions

Stable emulsions can be prepared by referring the methods described above. The methods differ slightly between O/W and W/O emulsions. The differences in stability between O/W and W/O emulsions are shown in table 1. As shown in the table, W/O emulsions do not have electric repulsion around the particles and thus do not solvate (O/W emulsions hydrate) and are more unstable than O/W emulsions. Of possible methods for stabilizing W/O emulsions, three are shown below:

(1) Reduce the size of particles and their distribution width,
(2) Protect emulsion particles by
   • forming an adsorption layer consisting of polymers and/or
   • protecting adsorption using fine particles, and
(3) Make the medium to take a structure of
   • gel or liquid crystal formation, or
   • high internal ratio emulsion (the ratio of the water phase is emulsion more than 80 %).

The measures (1)–(3) cannot solve all destabilization problems, but improve the stability of emulsions. These can also be used for O/W emulsions.

Table 1 Stability comparison between O/W emulsion and W/O emulsion

	O/W Emulsion	W/O Emulsion
Adsorption of emulsifier	Yes	Yes
Electric double layer	Yes	Very little
Solvation	Yes	Very little

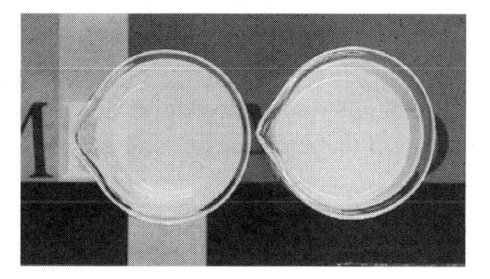

Before dilution   After dilution (×100)

Figure 5 Example of discriminating emulsion

5. 4. 3 Evaluation methods of emulsions

As described in 5. 4. 2, the particle size and its distribution are the keys for making stable emulsions. When emulsion particles can be observed under a microscope, microscopic observation should be the first act to be taken for evaluating the emulsion. When droplets are difficult to see due to polymers and other additives, it is usually effective to dilute the emulsion before observing under a microscope. Particularly, it is easy to detect unevenness of particle size by observing under a microscope while diluting the emulsion to 1/10 and then to 1/100 and so on. An example is shown in figure 5. When two kinds of emulsions of similar average particle diameters were diluted to 1/100, they appeared different. The beaker on the left was more transparent than the other. Because emulsions that contain large particles look turbid, the left beaker should be more uniform in particle size than the other. Particle size distribution measuring instruments and transmissivity meters can also be used.

Coalescence can be prevented by preventing flocculation of emulsion particles (by protecting the particles). The risk of flocculation can be assessed by measuring zeta potential. The larger the zeta potential, the more unlikely that the emulsion undergoes flocculation. Particle size distribution, transmissivity and zeta potential are sometimes measured by diluting samples. To prevent the dilution process from breaking emulsion particles, the same solvent used for the dispersing medium (such as emulsifying agent and buffer) should be used.

Instability of emulsions can also be estimated by measuring the viscosity and viscoelasticity.

Besides these measurements, a station test must be conducted. The test involves stationing an emulsion sample and observing its states (whether it separates into layers, *etc.*). It is also necessary to check that the sample wets the wall of the container uniformly. There are also centrifugation tests and accelerated station tests, which involve keeping samples in cycles of high and low temperatures (repetitive temperature change test). However, these methods do not shorten a station test.

5. 5 Thermodynamics of emulsions

The thermodynamics of emulsions are investigated by comparing a system in which two immiscible liquids are mixed and one is dispersed in the other and a system in which the two liquids are mixed and separated in two phases. figure 6 shows a system in which oil (O) is dispersed in water.

In the non-dispersed system (System I), the free energy of the entire system is:

$$G^{I} = G^{I}{}_{O} + G^{I}{}_{W} + G^{I}{}_{OW} = H^{I}{}_{O} - TS^{I}{}_{O} + G^{I}{}_{W} + G^{I}{}_{OW}$$

$G^{I}{}_{O}$ and $G^{I}{}_{W}$ are the free energy of the oil and water bulks, and $G^{I}{}_{OW}$ is the excess free energy related to the interface between oil and water. Because interfacial tension (γ_{OW}) is the excess free energy of the interface per unit area, interfacial tension can be expressed as:

$$G^{I}{}_{OW} = \gamma_{OW} \times A^{I}$$

In a system in which oil is dispersed in water (System II, an O/W emulsion), the free energy of the system is:

$$G^{II} = G^{II}{}_{O} + G^{II}{}_{W} + G^{II}{}_{OW} = H^{II}{}_{O} - TS^{II}{}_{O} + G^{II}{}_{W} + G^{II}{}_{OW}$$

Because the free energy of the bulk does not change by dispersion process, the term that reflects the increase in order of the system caused by the formation of the oil droplets (entropy) cannot be ignored[12]. Thus, changes (ΔG) in free energy caused by emulsion formation are as shown below:

$$\Delta G = G^{II} - G^{I} = G^{II}{}_{OW} - G^{I}{}_{OW} - T(S^{II}{}_{O} - S^{I}{}_{O})$$
$$S^{I}{}_{O} \doteqdot 0$$

Therefore,

$$\Delta G = \gamma_{OW} \Delta A - TS^{II}{}_{O}$$

ΔA is the increase in surface area from the non-dispersed system. $S^{II}{}_{O}$ is the term showing the disorder that cannot be converted into heat (temperature). When an ordinary expression of Gibbs' free energy ($\Delta G = \Delta H - T\Delta S$) is used to express emulsification, the enthalpy term (ΔH) shows the force (heat) that acts between particles making them to assemble, and the entropy term (ΔS) shows the degree of discrepancies among particles. $TS^{II}{}_{O}$ is a positive value but can be ignored in ordinary emulsions because it is much smaller than $\gamma_{OW}\Delta A$. Thus, the change (ΔG) in free energy by emulsification is always a positive value, showing that emulsions are thermodynamically unstable. Therefore, O/W emulsions separate into two phases of oil and water. However, there are dynamically stable cases, known as spontaneous emulsification[13]. This long-known phenomenon involves spontaneous emulsion (dispersion)

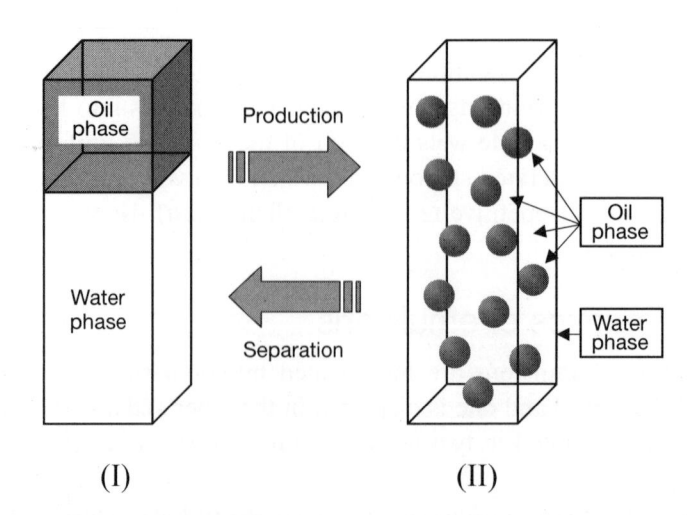

Figure 6 Model of O/W emulsion production and separation

formation between oil and water just by coming in contact with each other, and is believed to occur because the interfacial force becomes partly negative[12]. Recently, Nishimi[14] and Miller found that minute emulsification droplets were spontaneously formed at the room temperature when the HLB of an oil-water interface changed spontaneously while forming intermediate-phase microemulsion. This is an ultimate emulsification method in surface chemistry.

5. 6 Conclusions

Everett wrote a book[15] on the stability of gold sol (colloid). He mentioned that Faraday made gold colloidal particles in 1856 and 1857, and contributed to studies on colloid. The "gold colloid" of Faraday exists in the Royal Institution of Great Britain and is a dispersion liquid still purple even after the period of 150 years (as of 1998, when the author checked the color, figure 7). Faraday showed a method for protecting gold colloid using gelatine, which is a natural polymer. The same applies to emulsions, and protecting droplets (preventing coalescence) is essential for stabilizing emulsions. The size of droplets and the structure formation of the medium should also be controlled.

Future emulsification technologies should be environment friendly and reduce CO_2 emissions. For example, methods that involve operation of high pressure homogenizers at "low emulsification pressures" and "reduced pass number" are demanded. They will not only reduce the production costs but also prolong the life of emulsification devices. Improved emulsification technologies (for producing high functional emulsions with high added values) should be developed by utilizing the knowledge of surface chemistry and high-performance emulsification devices.

Figure 7 Gold colloid of Faraday (Photograph taken by the author in 1998)

References

1. S. Matsumoto, Y. Kita, D. Yonezawa, *J. Colloid Interface Sci.*, **57**, 353(1976)
2. K. Tajima, *Hyomen*, **37**, 611 (1999)
3. W. D. Bancroft, *J. Phys. Chem.*, **17**, 514 (1913)
4. T. Sasaki, T. Hanai, T. Mitsui, "Science of Emulsion", p.79, Asakura Publishing Co., Ltd. (1968)
5. H. Kitahara, *Pharm Tech Japan*, **12**, 1127 (1996)
6. H. Sagitani, M. Takenouchi, *J. Oleo Sci.*, **30**, 38 (1918)
7. H. Sagitani, T. Hattori, K. Nabeta, M. Nagai, *Nippon Kagaku Kaishi*, 1983, 1399
8. T. Suzuki, H. Takei, S. Tamazaki, *J. Colloid Interface Sci.*, **129**, 491 (1989)
9. Y. Suzuki, Patent Gazette, Syowa 57-29213 (1982)
10. Y. Tabata, Patent Gazette, Syowa 60-9853(1985)
11. H. Kitahara, "The basics of interface and colloid science", p.143, Kodansha Co., Ltd. (1994)
12. S. Komura, T. Araki, H. Kodama, R. Morikawa, J. Fukuda "An introduction to Soft matter", p.138, Springer Japan KK. (2002)
13. J.T. Davies, E. K. Rideal, "Interfacial phenomena", p. 364, Academic Press (1961)
14. T. Nishimi, C.A. Miller, *Langmuir*, **16**, 9233 (2000)
15. D.H. Everett, *KOROIDO KAGAKU NO KISO*, Kagaku-Dojin Publishing Company, INC (1992)

Chapter 6
Control Technologies for Emulsion Films

Yuji Sakai

6. 1 Introduction

Skincare cosmetics are used to keep the skin healthy and beautiful. Most skincare cosmetics form films when applied on the skin. The thin films, or so called cosmetic films, manifest effects acting to the skin. However, the cosmetic films have little been discussed because their effects were difficult to assess and visualize. Recent progress of measurement equipment have enabled cosmetic films, particularly emulsion films or films of emulsion, to be assessed and visualized, and studies have been reported on the relationship between the states and functions of emulsion films[1-4]. This chapter gives a brief summary of the studies.

6. 2 Compatibility between moisturizing and occlusive functions[1-3]

6. 2. 1 Emulsion film structures to increase both moisturizing and occlusive functions

Dry and rough skins have little moisture, and a large amount of water loss from the skin. To prevent this, skincare cosmetics, particularly emulsion-type cosmetics, are widely used. Emulsion cosmetics are made of hydrophilic parts consisting of water and polyol and hydrophobic parts consisting of oil and wax *etc*. It can supply water to the skin continuously for hydrophilic parts contain much water. This phenomenon is generally called "moisturizing function"[5]. Lipophilic parts prevent water from transpiring from the skin. This phenomenon is generally called the "occlusive function"[5,6]. Therefore, emulsion lotions have both moisturizing and occlusive functions. For example, a rise in the amount of glycerin in an emulsion causes an increase of the moisturizing function because the hydrophilic parts expand. On one hand, when oil is increased, the occlusive function is increased because the lipophilic parts expand.

Obviously, the lipophilic parts are narrowed when the hydrophilic parts of an emulsion expand, and the hydrophilic parts are narrowed when the lipophilic parts expand. Thus, the moisturizing and occlusive function compete with each other, and it is hard to increase both functions at the same time. It has been reported that the competitive relationship between the moisturizing and occlusive function cannot be broken just by simply adding oil that has polarity or holds water or other polyols[5]. This point is regarded as the function limit of emulsion lotions. Therefore, when an emulsion of enhanced occlusive function is used, a toner of an enhanced moisturizing function should be used at the same time.

To solve this problem, we investigated the conditions and structures of emulsion films. Because it is the emulsion films, and not emulsion, that show the aforementioned functions and are in contact with the skin, We thought it is important investigate this part. We hypothesized

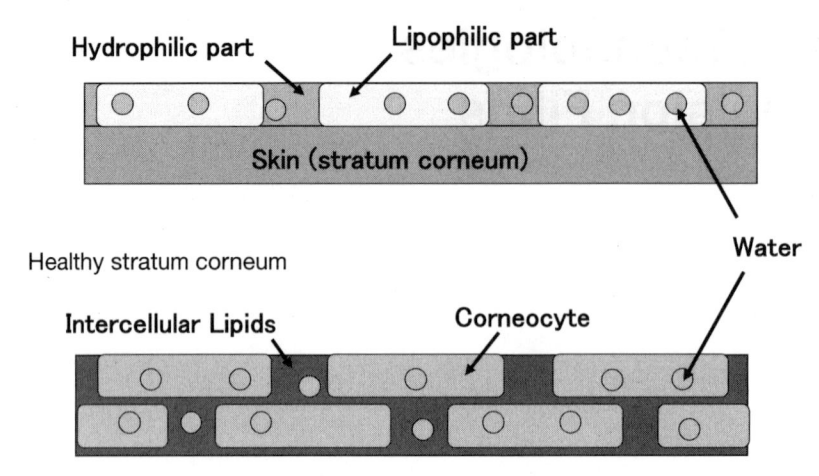

Figure 1 Healthy stratum corneum and emulsion films having enhanced moisturizing and occulsive function

that modeling an emulsion film that resembles healthy stratum corneum (figure 1) would enable me to increase both the moisturizing and occlusive function of the emulsion film at the same time. Healthy stratum corneum can maintain water well as it contain much moisturizing ingredients in the corneocytes, which means that the moisturizing function is high. Also water inside the skin is hard to evaporate because intercellular lipids form a strong structure, meaning that the occlusive function is enhanced. Corneocytes of the stratum corneum were regarded to be the lipophilic parts of emulsion films, and intercellular lipids were regarded to be the hydrophilic parts of emulsion films. Dispersing a large amount of water in the lipophilic parts so that the corneocytes contain much water and strengthening the hydrophilic parts so that intercellular lipids are strong were hypothesized to be effective for the coexistence of moisturizing and occlusive function. Therefore, in this study, the conditions and structures of emulsion films were investigated to develop widely used O/W emulsions in which both water holding capacities and occlusive function are increased, which were difficulty to coexist at the same time.

6. 2. 2 Dispersions of water into the lipophilic parts

First, the time historical changes in the quantity of water contained in emulsion films were investigated using two different surfactants (figure 2). In both surfactants, the quantity of water in emulsion films decreased with time. The moisture content was always higher in the emulsion that contained polyglycerin surfactants than that with polyoxyethylene surfactants. This was likely attributable to the difference in hydration power between polyglycerin chain (hydroxyl) and polyoxyethylene chain[7]. A method was developed for dispersing water in the lipophilic parts of emulsion films by adding polyglycerin surfactants. A method that involved dispersing water into oil was used because it was difficult to measure emulsion films directly. A test of polyoxyethylene surfactants resulted in easy dispersion of water into oil (squalane) even when lipophilic surfactants and polyoxyethylene-25-stearate (25ES, HLB15) of high hydrophilicity were used (figure 3). On the other hand, the use of highly hydrophilic decaglyceryl monostearate (10GMS, HLB14.5) alone resulted in formation of hydrate crystals in oil. This was likely attributable to the aforementioned difference in hydration power between the polyglycerin

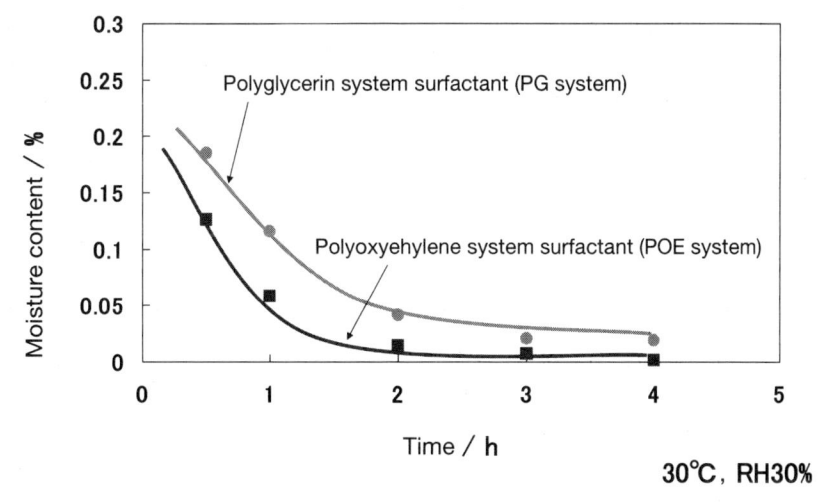

Figure 2 Changes of moisture contents in different emulsion films

Figure 3 Dispersion of water into squalane by dispersants (water 1.5 wt%) : (a) 1 wt% 25ES, (b) 1 wt% 10GMS, (c) 1 wt% 10GMS, 2 wt% Cetanol, (d) 2 wt% Cetanol

chain (hydroxyl) and polyoxyethylene chain. Addition of other additives was investigated, in which cetanol was found to be effective in dispersing water.

6. 2. 3 Structure reinforcement of hydrophilic parts

Next, a method for strengthening the structure of the hydrophilic parts of emulsion films was investigated in order to enhance the occlusive function of emulsions. However, because it is difficult to measure the properties at the hydrophilic parts of emulsion films directly, investigations were made on the hydrophilic part of emulsion particularly the structure of the continuous phase. Two emulsions of the same formulation were prepared using two different emulsification methods. One of the emulsions was prepared using shearing stress, and the structure of the hydrophilic parts was weak, resulting in fluiding emulsion. The other emulsion was prepared using the D phase emulsification method[8]. The structure of the hydrophilic parts was strong, and the emulsion did not fluid.

The occlusive function of both emulsions were measured using the cup method (figure 4). The cup method measures the performance in blocking water transpiration by sealing the top of an acrylic cup that contained a certain amount of water with filter paper, applying a sample to the filter paper, incubating at 30°C and R.H.30%, and measuring the amount of water loss during the period of incubation. Because the amount of water loss is in a linear relationship with time, the inclination of the line was defined as Water Loss Index (WLI) (figure 5). The emulsion with strong structure was found to be always low in transpiration and showed high blocking effects (figure 6). Systems consisting of only surfactants have been reported to show high water transpiration at very low viscosity (high fluidity)[9], supporting this experiment.

Formation of the continuous phase of emulsions was tested using polyglycerin system surfactants, which were chosen based on the previous examination. It was found difficult to form structures such as liquid crystals in the continuous phase using polyglycerin system surfactants. In the prescription of cosmetics, there is a method for forming structures in the continuous phase of O/W type emulsions, which involves increasing the amounts of surfactants, oil, solid oil, higher alcohol and thickening agents. However, the increases may adversely affect the quality, safety and the comfort of use of the resultant emulsion as cosmetics. Appropriate hydrophilic surfactants to be added to a base consisting of decaglyceryl monostearate (10GMS) and glycerol monostearate (GMS) were investigated based on a hypothesis that highly hydrophilic

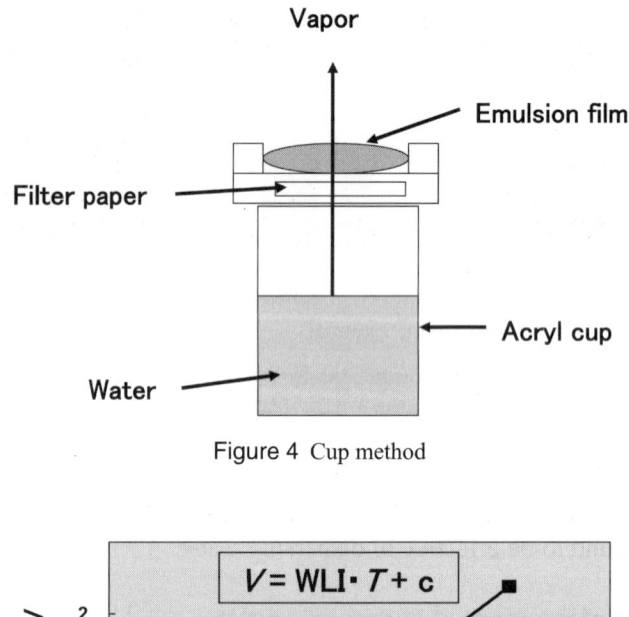

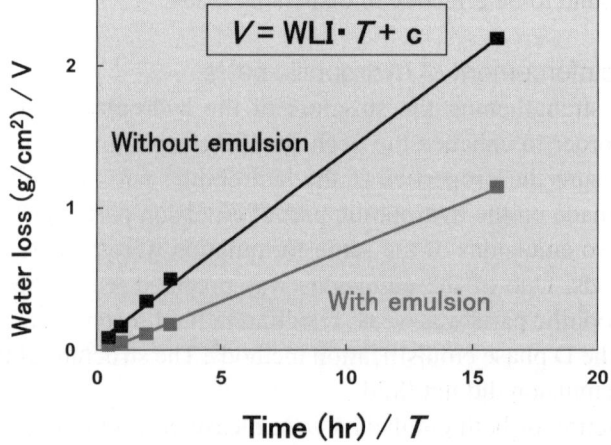

Figure 4 Cup method

Figure 5 Water loss from the acrylic cup method and difinition of the water loss index (WLI)

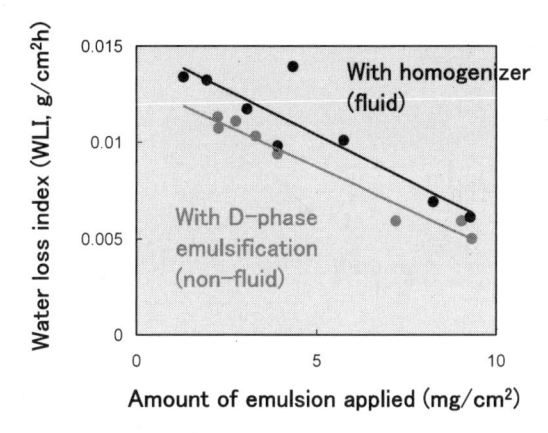

Figure 6 Water loss from emulsions prepared using different emulsification methods (preparation in 25ES)

Table 1 Effects of hydrophilic surfactants on the continuous phase of emulsions (a) Non fluid, (b) Fluid

Hydrophilic surfactant	10GMS	(25ES)
Without hydrophilic surfactant	2	1
Sucrose stearate	2	1
PEG-45 stearate	2	1
PEG-100 hydrogeneted castol oil	2	1
PEG-150 distearate	2	2
PEG-150 stearate	2	1
PPS	1	1

0.5wt% Hydrophilic surfactant, 1.6wt% 10GMS, 2wt% GMS, 3wt% butylene glycol, 3wt% cetyl alcohol, 14wt% squalane, 8wt% trioctanoin, water(the residue).

surfactants of a high HLB value facilitate formation of structures in the continuous phase. The use of polyglycerin-13-polyoxybutylene-14-stearylether (PPS) was found to reduce the fluidity of emulsions (table 1) at parts shown in figure 7.

To investigate the causes, the structure of the emulsion was subjected to Small Angle X-ray Scattering. A scattering peak at 62 Å was found only in the PPS-combined system (figure 8). Because the molecule chains of PPS were 130 to 180 Å, this peak should be of GMS origin and not of PPS origin. It seemed that PPS protected the liquid crystal structure of GMS.

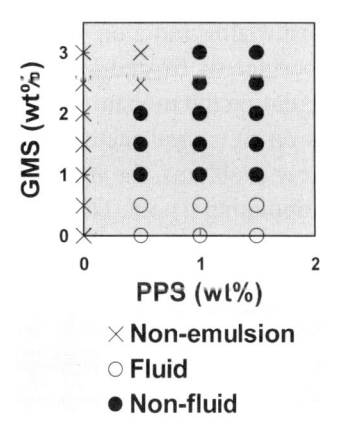

Figure 7 Non-fluid emulsion area (rigid hydrophilic part) (1.5 wt% 10GMS)

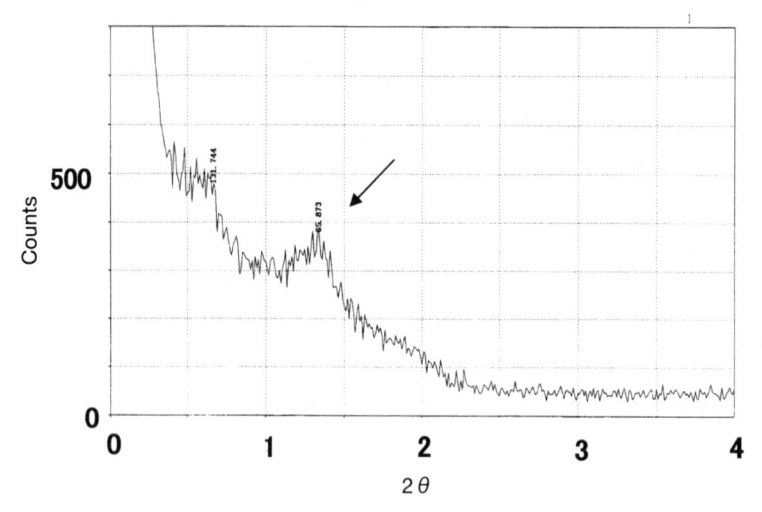

Figure 8 Small angle X-ray scattering spectrum of an emulsion
(1.6 wt% 10GMS, 2.5 wt% GMS, 1.5 wt% PPS)

6. 2. 4 Evaluation of moisturizing and occlusive functions

Emulsion films of two kinds of emulsions shown in table 2, which were Emulsion A developed in this study and a similar but old-type Emulsion B in which the results of this study were not reflected, were prepared and tested for moisturizing and occlusive function. As described above WLI, which is used as an index to show occlusive function, depends on the quantity of emulsion applied on the filter paper. I newly defined Intrinsic Films-Occlusive index (IFO) as follows:

$$IFO=(x/y)^{1/z} \qquad (1)$$

  x: the WLI without emulsion (g/cm²·h)
  y: the WLI with emulsion (g/cm²·h)
  z: Amount of emulsion on filter paper per surface area (g/cm²)

The moisturizing function of emulsion films were evaluated by applying emulsion samples on the surface of frosted glass, weighing after leaving at 30°C and relative humidity 30%, and calculating the moisture content. This method had several problems: the moisture content depends on time, and factors other than the properties of emulsion films were reflected. To solve these problems, the moisture contents of emulsion films were monitored over long time and extrapolating 0 hour. This was defined as Intrinsic Film-Moisturing index (IFM) as follows (figure 9):

$$IFM = W \cdot \exp(aT) \qquad (2)$$

  W: Water content in emulsion film
  T: Time after application of emulsion
  a: Constant

The resultant IFM of the conventional (B) and new (A) emulsion films were 14.4 and 17.5, respectively (figure 10), showing an improvement of 22%. The IFOs of the conventional (B) and new (A) emulsion films were 1.11 and 1.46, respectively, showing an improvement of about 32%. Both the moisturizing and occlusive function were successfully improved at the same

Table 2 Composition of two different emulsion types
(A) Experimental emulsion A, D phase emulsification
(B) Conventional emulsion B, homogenizer

Ingredient	a	b (wt%)
GMS	1.5	2.0
25ES	0.0	2.0
10GMS	2.0	0.0
PPS	0.5	0.0
butylene glycol	3.0	3.0
cetyl alcohol	3.0	0.0
squalane	14.0	17.0
trioctanoin	8.0	8.0
glycerin	5.0	5.0
water	63.0	63.0

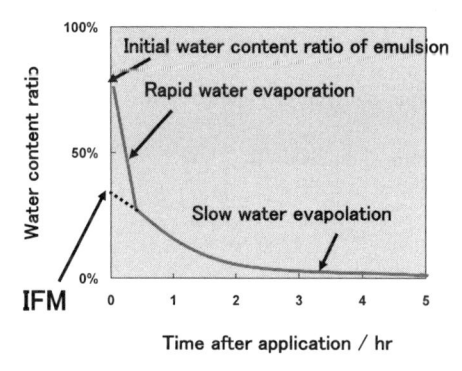

Figure 9 Water content ratio of an emulsion film

Figure 10 IFO (at various quantities of squalane and glycerin) and IFM of a new emulsion and a conventional emulsion

time. Like in conventional emulsions, increasing the proportion of oil resulted in increases in occlusive function but drops in moisturizing function; and increasing glycerin resulted in rise in moisturizing function and drops in occlusive function.

6.2.5 Observation of the conditions of the emulsion films

The film conditions of the new (A) and conventional (B) emulsions were observed under a laser microscope to investigate the results from the view point of makeup film conditions. The new emulsion film showed droplets of emulsion finely spreading over the film (figure 11). However, no such a structure was observed on the film of the conventional emulsion B. A three-dimensional analysis of the conditions (figure 12) revealed steeper and more densely formed ascents and descents on the surface of the new emulsion film than in the traditional emulsion B film (table 3). Thus, the surface area of the former was larger than the latter. The investigation showed that the emulsion films differed greatly in both structure and condition.

6.2.6 Continuous using evaluation

Finally, the effects to the skin (stratum corneum) of the new emulsion were confirmed. The new emulsion was subjected to continuous use evaluation, in which 44 female volunteers

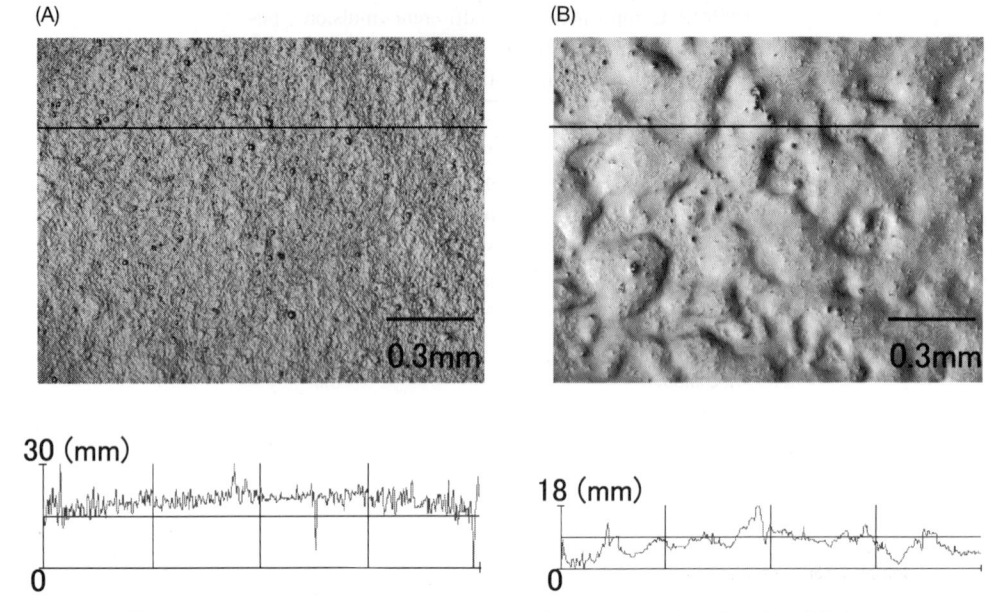

Figure 11 Laser microscope images of emulsion films and sectional undulations
(30 minutes after application)
(a) Experimental emulsion A, (b) Conventional emulsion B

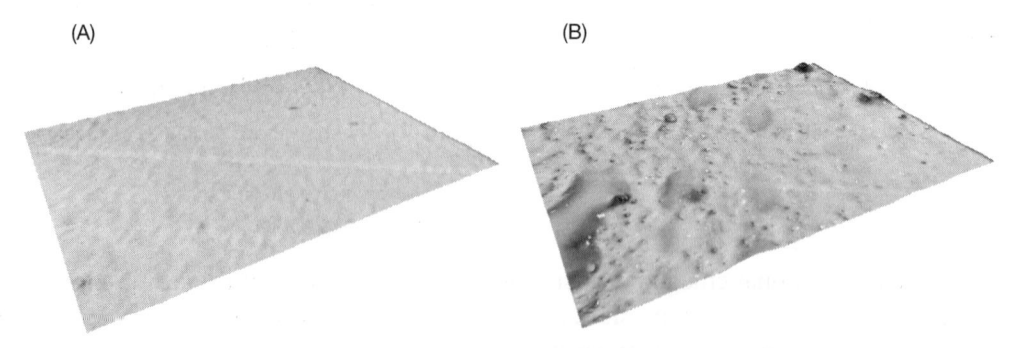

Figure 12 Three-dimensional image of emulsion films (30 minutes after application)
(a) Experimental emulsion A, (b) Conventional emulsion B

Table 3 Three-dimensional analysis values of emulsion films
(A) Experimental emulsion A, (B) Conventional emulsion B

	A	B
Area (mm^2)	1.51	1.51
Surface area (mm^2)	7.85	3.03
Surface area / area	5.21	2.01
Average height of undulation (mm)	4.08	2.83
Average space of undulation (mm)	14.2	19.5

tested the trial product for about one month. Corneocyte specimens were sampled using a tape stripping before and after the use of the emulsion. The corneocyte were observed under a microscope, and their conditions were evaluated based on stratified abrasion, which is an index of moisturizing function, and arrangement of the cells, which is an index of occlusive function[10]. Stratified abrasion was graded in five, ranging from very good (score: 1) to very bad (score: 5). The arrangement of the cells was also evaluated by giving scores ranging from very good (score: 1) to very bad (score: 4).

The resultant mean score in stratified abrasion significantly decreased from 2.75–2.32 (improvement) (figures 13 and 14). The mean scores in cell arrangement also decreased from

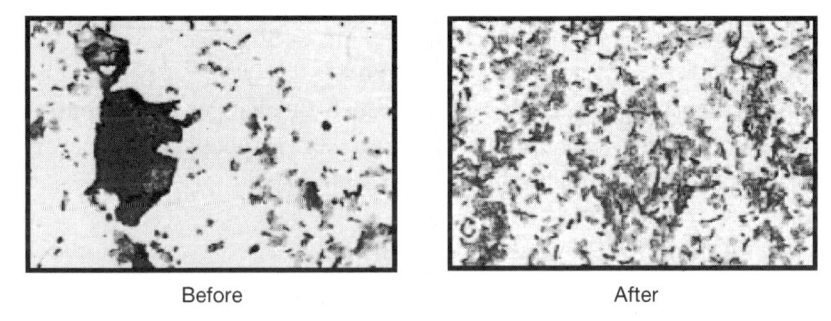

Figure 13 Changes in stratified removal of the stratum corneum by continuous use of the emulsion (18 times)

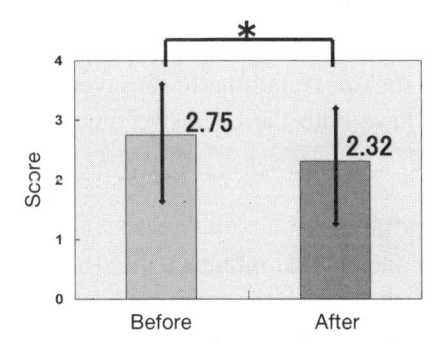

Figure 14 Average changes in stratified removal of the stratum corneum by continuous use of the emulsion

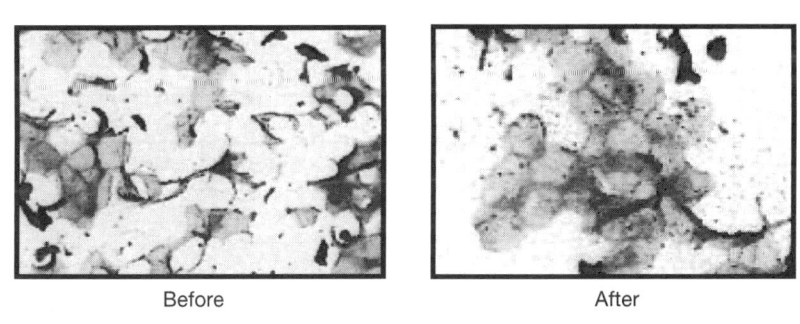

Figure 15 Changes in the arrangement of corneocytes by continuous use of the emulsion (90 times)

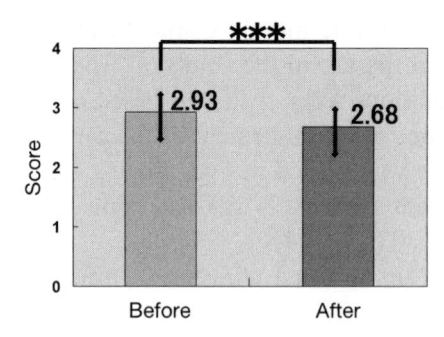

Figure 16 Average changes of the corneocyte arrangements by continuous use of the emulsion

2.93–2.68 (improvement) (figures 15 and 16). Because the changes in the control group, who did not use the new emulsion, were drops of 0.09 and 0.02, respectively, the new emulsion was shown to be effective in improving the stratum corneum.

6. 3 Improvement of sensory evaluation [4]

6. 3. 1 Emulsion film structures of enhanced occlusive functions and improved sensory evaluation

In general emulsions, the occlusive function can be enhanced by increasing the oil content. However, the increase makes the emulsions "sticky" and "rough", simply reflecting the sensory evaluation of oil. To resolve the problem, investigations were conducted focusing on emulsion films once again. For example, a two-layer film, in which the upper layer consists of hydrates gel to control the stickiness and the lower layer is a three-dimensional film with barrier performances, should have both the occlusive function and high sensory evaluation (figure 17).

6. 3. 2 Investigation of surfactants

As described above, one of the factors that influence the structure of emulsion films greatly is surfactants. Thus, screening of surfactants was conducted. Polyoxyethylene system surfactants, polyglycerol fatty acid esters and acrylates/c10-30 alkyl acrylate crosspolymer produced uniformly gelatinous emulsion films and did not form a double film structure consisting of a hydrate gel film and an oil film. This is likely because the lipophilic groups of the surfactants are big resulting in no separation of water and oil (The oil-water interface is strong, too). Also,

Figure 17 Design of an emulsion film (two-layer structural films)

Figure 18 Chemical structure of propylene glycol alginate

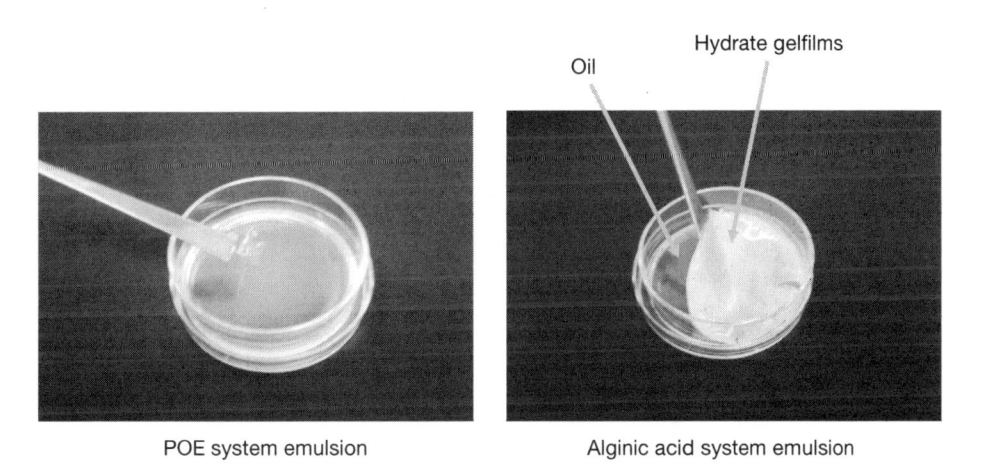

Figure 19 Film conditions of alginic acid system emulsions

hydrate gel film was not formed because the surfactants have low molecular weights and few hydroxyl groups. Surfactants that have small lipophilic groups and many hydrophobic groups, form hydrate gel films, and can emulsify were searched for. Propylene glycol ester alginate was found (propylene glycol alginate, figure 18 and 19) to satisfy all the criteria. Propylene glycol ester alginate could also emulsify polar oil, silicone oil and fluorine system oils, which are difficult to emulsify.

6. 3. 3 Stability of emulsion prepared using propylene glycol alginate

Since propylene glycol alginate has small lipophilic groups, it was difficult to prepare strong films on the surface of emulsified oil drops (interface films), and the coagulation and union of oil drops, which occurred along the passage of time and at high temperatures, could not be prevented. Coagulation and union of emulsion particles are usually controlled by increasing viscosity by adding carbomers and polysaccharides. However, this widely method was not effective. Thus, a new stabilization method was tested, which involved bridging emulsion particles instead of increasing the viscosity, to physically prevent union of droplets. Bridging agents are needed to form bridges between polymers and emulsion particles and between the polymers. As a result of the screening, sodium alginate and calcium chloride were acceptable in bridge structure formation and was effective for emulsion stabilization. The concrete process involves adsorbing alginic acid on the surface of oil drops after emulsification with propylene

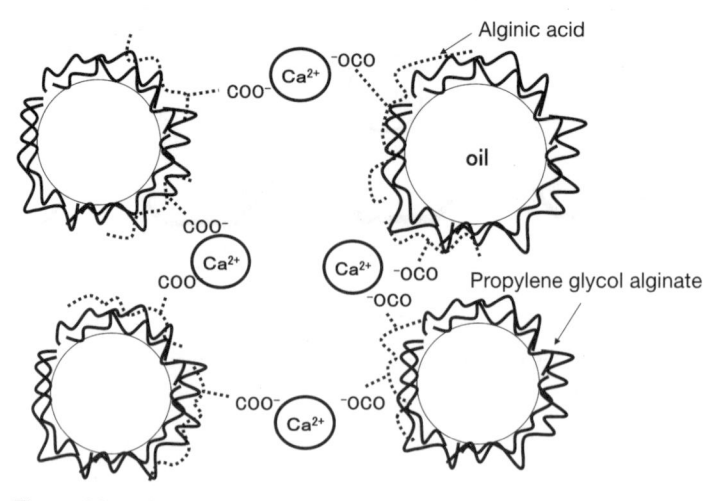

Figure 20 Stabilization mechanism of alginic acid system emulsions

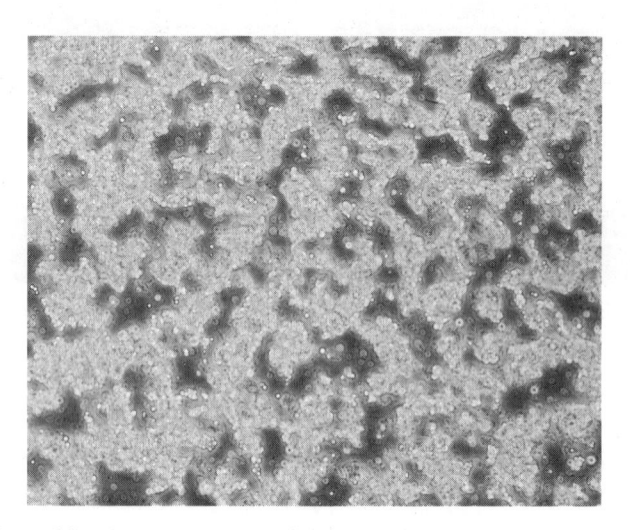

Figure 21 Microscopic image of alginic acid system emulsions (×200)

glycol alginate and connecting the alginic acid with alginic acid on the surface of other oil drops with calcium chloride. The process controlled the movement of oil drops and prevented cohesion and union (figure 20). As a result of bridge formation, the emulsion oil drops were non-uniformly dispersed to a degree that is not seen in conventional stable emulsions (figure 21).

6. 3. 4 Conditions of emulsion films and sensory evaluation

A microscopic study of the alginic acid system emulsion applied on a slide glass showed formation of no emulsion particles (figure 22). This showed that the bridge structures were broken when the thin film was formed and the emulsion particles were totally broken. In emulsions prepared using conventional POE system surfactants, emulsion particles were observed even after emulsion films were formed. This was likely because the orientation of the surfactant at the interface of the emulsion particles was high, resulting in a strong interfacial film, and some emulsion particles remained unbroken even after the formation of the thin film. Laser microscopic observation showed that the new emulsion formed a surface much smoother than conventional emulsion films (figure 23). This was because a hydrate gel film was formed above an emulsion film. In a sensory evaluation, the new emulsion obtained high scores in "non stickiness" and "feeling smoothness" (figures 24 and 25). This was because the hydrate gel film, which formed as an upper layer of the emulsion film, covered the bad feeling attributable to oil. Particularly, even emulsions of high oil contents obtained high scores in these items, which is a notable point.

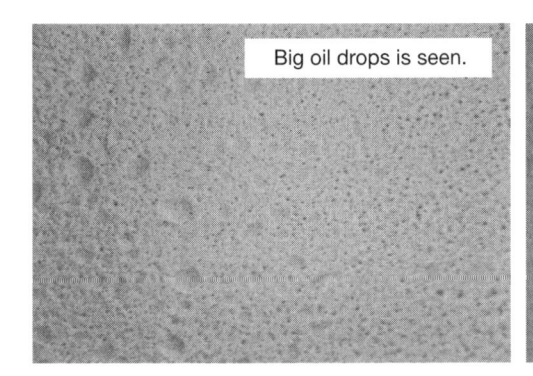

Figure 22 Transmission microscopic images of emulsions (×175)

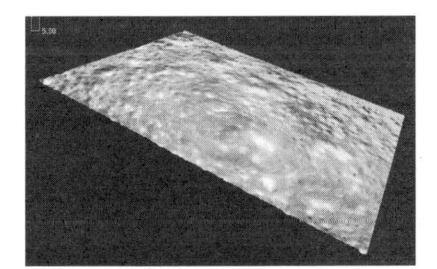

Figure 23 Laser microscope images of the emulsion films

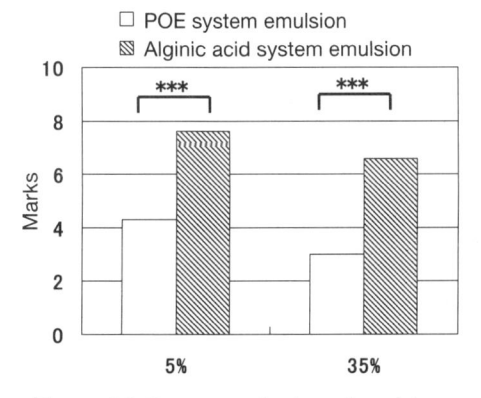

Figure 24 Sensory evaluation of emulsion (non stickiness)

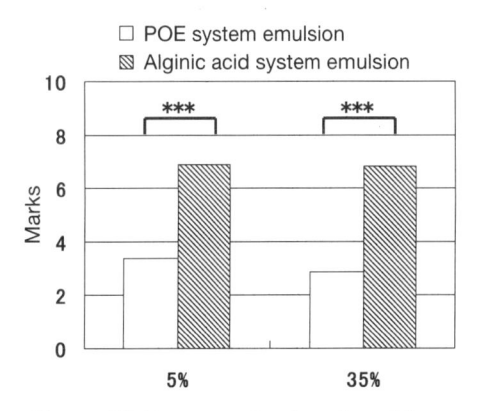

Figure 25 Sensory evaluation of emulsion (smoothness)

References

1. Y. Sakai, M. Suzuki, *IFSCC Magazine*, **9**, 23-28 (2006)
2. Y. Sakai, M. Suzuki, *J. Soc. Cosmet. Chem. Japan*, **40**, 95-104 (2006)
3. Y. Sakai, M. Suzuki, *Fragrance Journal*, **10**, 25-31 (2005)
4. M. Suzuki, Y. Sakai, 24th IFSCC Congress Osaka Japan, Poster 202 (2006)
5. S. Nishiyama, H. Komatsu, M. Tanaka, *J. Soc. Cosmet. Chem. Japan*, **16**, 136-143 (1983)
6. A. M. Kligman, *Cosmetics and Toiletries*, **93**, 27-35 (1979)
7. M. Fukuda, K. Shinoda, *J. Jpn. Oil Chem. Soc.*, **48**, 587-594 (1999)
8. H. Sagitani, Y. Ikeda, Y. Ogo, *J. Jpn. Oil Chem. Soc.*, **33**, 156-161 (1984)
9. Y. Aoki, Y. Sumida, 22nd IFSCC Congress Edinburgh, Podium 38 (2002)
10. N. Kashibuchi, Y. Muramatsu, *J. Soc. Cosmet. Chem. Japan*, **23**, 55-57 (1989)

[Powders and Fine Particles]

Chapter 7
Application of Composite Nanoparticles to Cosmetics

Toyokazu Yokoyama and Hiroyuki Tsujimoto

7.1 Introduction

Active attempts are being made to use nanoparticles in cosmetics. Nanoparticles are generally defined as very small particles of diameters smaller than 100 nm, although the definition varies by field. In this chapter, the term "nanoparticle" is used to also include particles smaller than 400 nm, which are smaller than conventional submicron particles, different from the submicron particles in properties and handling methods, and smaller than the wavelength of visible light.

The ultra-small particles passes light when they are properly dispersed in a solution, and thus can serve as filler that maintains the transparency of the dispersion, and can also express various functions even only with a small quantity. It is known that nanoparticles are highly absorbable and modification of their surfaces is highly effective because nanoparticles are very small and have a large specific surface area[1].

Major build-up methods for producing nanoparticles are broadly classified into vapor-phase methods and fluid-phase methods as shown in table 1, each of which can be subclassified

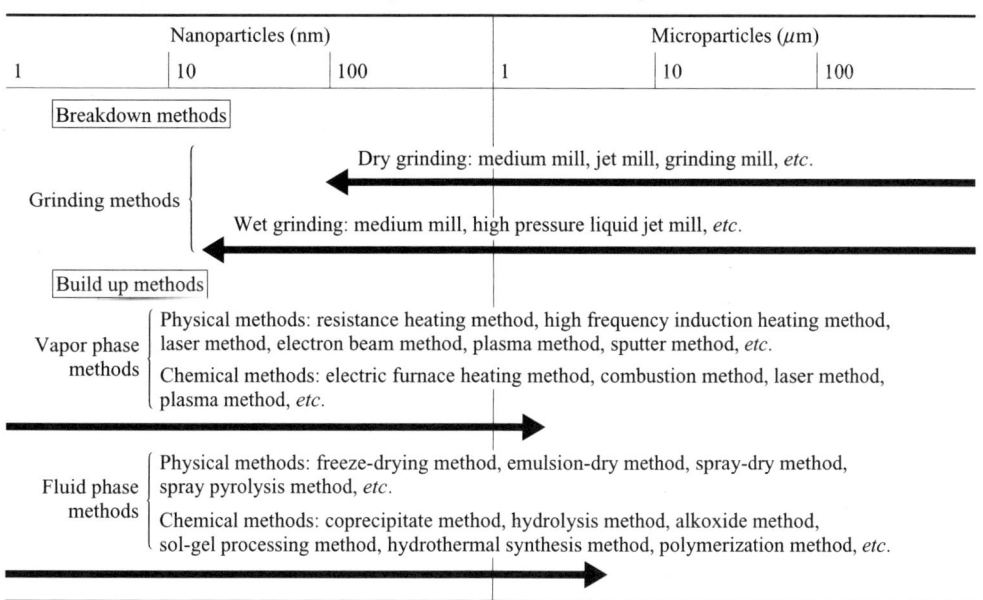

Table 1 Preparation methods of microparticles

	Nanoparticles (nm)			Microparticles (μm)		
1	10		100	1	10	100

Breakdown methods

Grinding methods
- Dry grinding: medium mill, jet mill, grinding mill, *etc.*
- Wet grinding: medium mill, high pressure liquid jet mill, *etc.*

Build up methods

Vapor phase methods
- Physical methods: resistance heating method, high frequency induction heating method, laser method, electron beam method, plasma method, sputter method, *etc.*
- Chemical methods: electric furnace heating method, combustion method, laser method, plasma method, *etc.*

Fluid phase methods
- Physical methods: freeze-drying method, emulsion-dry method, spray-dry method, spray pyrolysis method, *etc.*
- Chemical methods: coprecipitate method, hydrolysis method, alkoxide method, sol-gel processing method, hydrothermal synthesis method, polymerization method, *etc.*

into those that involve chemical reactions and those that involve only physical processes, such as evaporation, condensation, and crystallization. Composite nanoparticles, in which nanoparticles are composed in a nanometer level, can also be produced using these methods and devising operation methods and conditions. Nanoparticles can be classified into inorganic and organic substances, which differ in key points when applied to cosmetics. This chapter describes preparation of inorganic and organic nanoparticles, methods of composition, and some possible and actual applications to cosmetics.

7.2 Preparation of inorganic oxide nanoparticles and applications

7.2.1 Preparation of inorganic oxide nanoparticles using the vapor-phase method

Many vapor-phase methods have been proposed, such as those that involve chemical reactions using laser, plasma, high-frequency waves, electric furnace heating or another energy source and those that involve only physical changes, such as dissolution and evaporation. The vapor-phase methods generally produce nanoparticles of higher purities than liquid-phase methods and are superior in processibility. However, the resultant concentration of nanoparticles is low, leading to low productivity and high product prices.

The Flash Creation Method (FCM) was developed as a technology to mass-produce nanoparticles of a high purity using the vapor phase methods[2]. FCM involves vaporizing and reacting a solution of a metal compound at a high temperature, which is in a plasma state, to produce particles that serve as nuclui and rapidly cooling and collecting the nanoparticles before they grow by grain growth and coagulation. The method can produce several kilograms of nanoparticles of a size of about 100 nm/h even with a pilot system, although the amount may vary depending on the materials and specifications used. In principle, FCM should be feasible for producing compounds other than oxides but has been exclusively used to produce nanoparticles of oxides.

7.2.2 Characteristics of inorganic composite nanoparticles

FCM is characterized for its ability to directly produce composite nanoparticles in the same process for producing nanoparticles by using two or more materials together as well as the ability to process diverse elements. The patterns of composite nanoparticles produced are broadly classified into: (1) solid solutions, in which oxides of two or more different elements are mutually dissolved, (2) surface coating type in which a certain oxide (nucleus) is covered by another oxide, and (3) fine particle dispersions, in which nanoparticles are dispersed in a matrix of another kind of oxide (figure 1).

The structure of resultant composite nanoparticles is determined basically by the combination of the materials used, but the addition rate and conditions of processing also alter the characteristics of the composite nanoparticles.

An electron micrograph of TiO_2-ZnO-CeO_2-SiO_2 composite nanoparticles, which were prepared to investigate the applicability of composite nanoparticles to cosmetics, is shown in figure 2 (a). The particles were prepared by micro-dispersing TiO_2, ZnO and CeO_2, which screen ultraviolet rays (UV) of different frequencies, in a matrix of SiO_2, which has good dispersibility. An element analysis, which was conducted at several micro domains of one nanoparticle, showed that the percentages of the elements were almost the same at the three points investigated (figure 2 (b)), suggesting uniform dispersion of the elements within the nanoparticle. Solar rays are classified by wavelength into infrared, which has the longest wavelength, visible light, ultraviolet rays, X rays, and gamma ray. Ultraviolet rays are further

Chapter 7 Application of Composite Nanoparticles to Cosmetics

(a) Compound-solid solution type
(b) Surface coating type
(c) Fine particles dispersion type

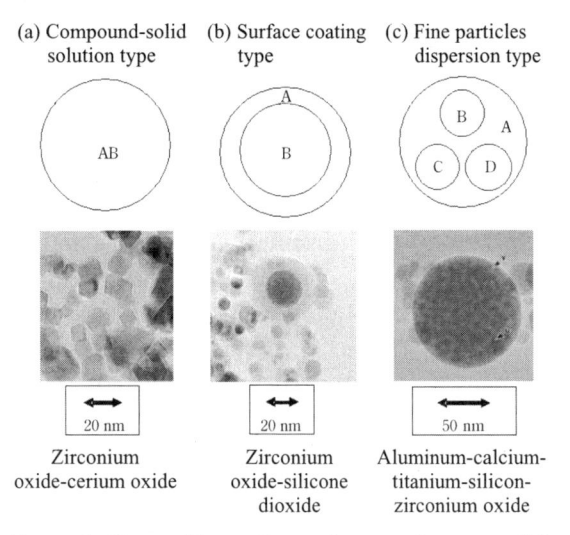

Zirconium oxide-cerium oxide

Zirconium oxide-silicone dioxide

Aluminum-calcium-titanium-silicon-zirconium oxide

Figure 1 Composition patterns of composite nanoparticles

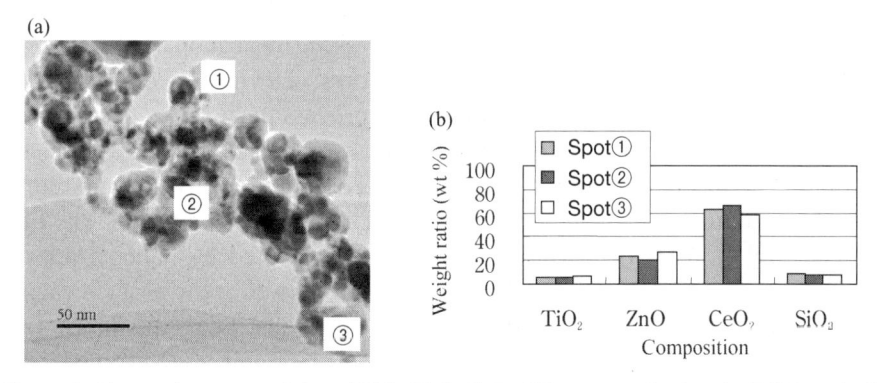

Figure 2 Composite nanoparticles of TiO_2-ZnO-CeO_2-SiO_2 prepared using the FCM method
(a) Electron micrograph
(b) Comparison of local element analysis at several different spots

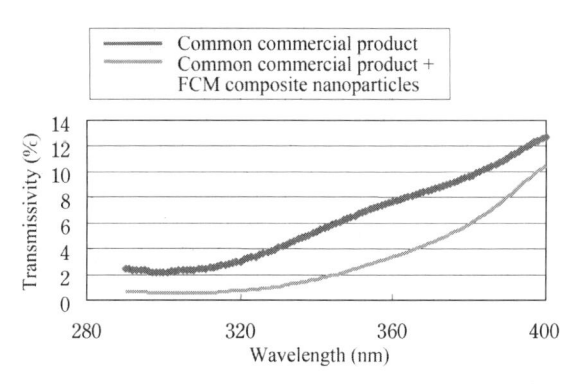

Figure 3 Effect of adding composite nanoparticles TiO_2-ZnO-CeO_2-SiO_2 on the wavelength dependence of ultraviolet ray transmissivity

divided at 320 and 280 nm into UV-A, -B and -C in a decreasing wavelength order.

UV-C is blocked by the ozone layer and hardly reaches the ground surface. Thus, the influences of UV-A and UV-B to the human body should be dealt. UV-B has shorter wavelengths than UV-A and thus a higher energy level and causes stronger adverse effects on the epidermis of the human skin. On the other hand, UV-A reaches the dermis, a deeper layer of the skin than the epidermis, and accounts for at least 90 % of the energy reaching the skin. Use of composite nanoparticles was tested to control the transmission of UV-A and UV-B. When composite nanoparticles were added to an ordinary commercial sunblock, the nanoparticles dispersed well and reduced the transmission of not only UV-B, which has short wavelengths, but also of UV-A, which has a large energy (figure 3). Composing oxides of different properties will enable articles of comprehensively new properties to be created.

7. 2. 3 Dry composition of inorganic oxide nanoparticles by mechanical techniques

Most of nanoparticles produced using the vapor-phase method exist coagulated as shown in the electron micrograph in figure 2. The size of nanoparticles is commonly determined by estimating the specific surface area by measuring nitrogen gas adsorption. The particle size estimated from specific surface area has agreed relatively well with the size observed under an electron microscope. Therefore, the coagulation of nanoparticles should have relatively few neckings. To dissociate the nanoparticles to make use of their original characteristics and improve handling of the powder and feeling of the skin, the nanoparticles were combined with particles of a micrometer-size, which serve as cores[3]. Devices used for processing apply strong mechanical force to a mixture of powder and process particles as designed, such as by (1) composing particles, (2) precisely mixing particles, and (3) making particles spherical (figure 4). Figure 5 is an example of mechanochemical bonding (MCB) of a mixture of PMMA particles of about 10 μm in diameter and fluoric resin of tens of nanometers. The figure shows that the surface of the PMMA particles was uniformly coated by nanoparticles of fluoric resin. In such a composite particle, the electrification properties of the particle change greatly. Furthermore, when nanoparticles of titanium oxide is used instead of the fluoric resin to coat the surface of the PMMA particles, the penetration speed of water into the powder filling layer increased notably from when the powders were simply mixed, showing a great improvement of hydrophilicity (figure 6)[4]. This was likely attributable to the improvement in the dispersibility of the nanoparticles. Such composite particles have an ultraviolet ray blocking effect while

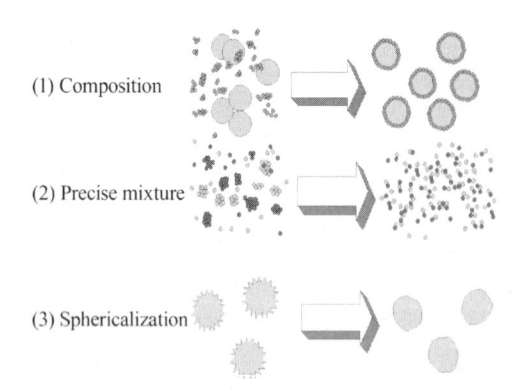

Figure 4 Particle processing functions of a mechanical particle composing device

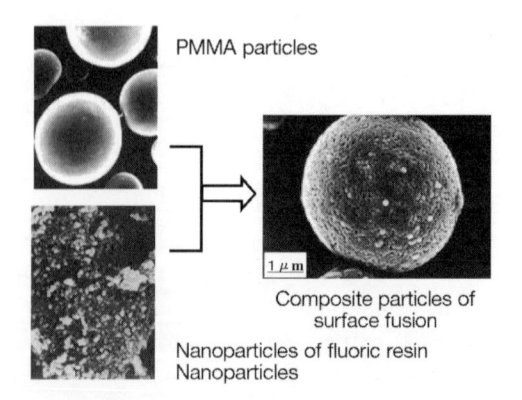

Figure 5 Example of composing nanoparticles of two different resins using the MCB technology

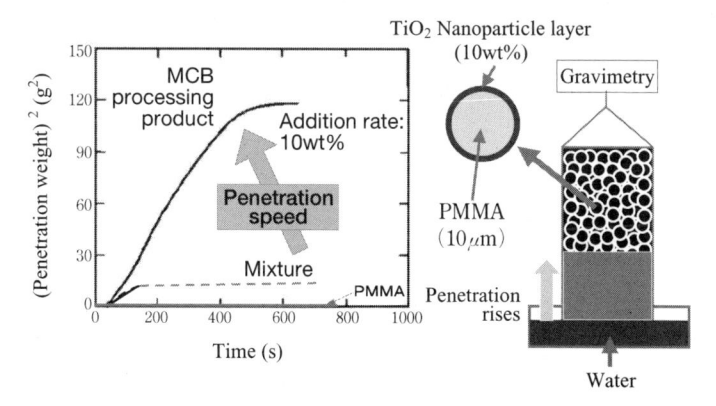

Figure 6 Changes in wetting characteristics by dry composition of nanoparticles

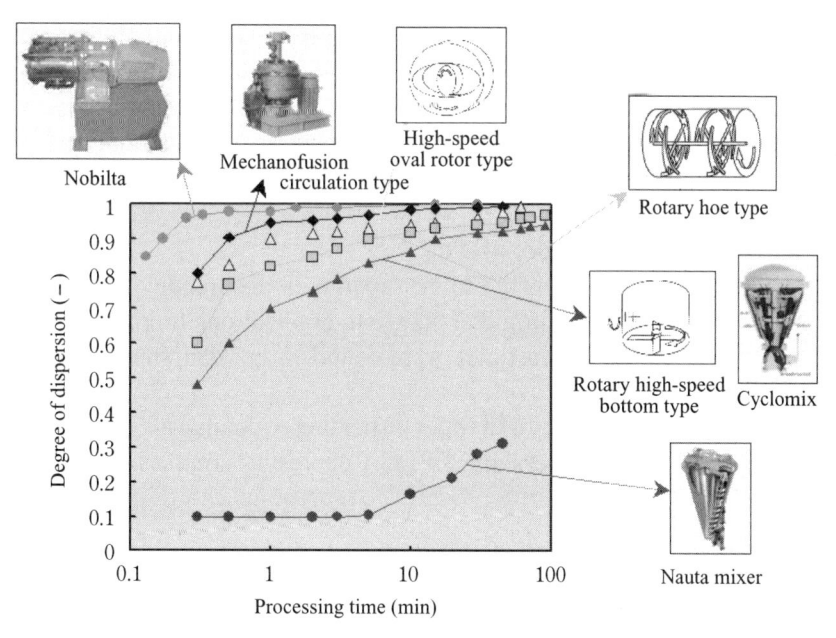

Figure 7 Comparison of mixing speed of various powder processors

maintaining the feeling of the micrometer-size PMMA, and are used in foundations in which titanium oxide is finely dispersed to block ultraviolet rays.

Various types of devices, which differ in the intensity and method of force application, are used for dry powder mixing and particles composing. There are some methods for evaluating the degree of nanoparticle composition, of which a widely used method involves mixing nanoparticles of different colors and measuring the optical reflection strength to evaluate the dispersion of the particles. The results are used to plan and design the necessary processes of powder processing. Mixing speeds of some powder processors evaluated by a method that is proposed by the Mixing Kneading and Forming Technical Group of APPIE (The Association of Powder Process Industry and Engineering, Japan) and involves use of white powder of calcium carbonate and red iron oxide powder are shown in figure 7[5]. Soft mixers such as the Nauta Mixers required low power and were superior for the overall mixing. However, for nanoparticle dispersion, MCB processors, which use high energy density, such as a high-speed particle composer Nobilta and Mechanofusion System, were effective. The devices are different not in mixing speed but also in the resultant degree of dispersion within a realistic time, which greatly affect the properties of the products.

7.3 Preparation and application of composite nanoparticles of biocompatible polymers

7.3.1 Preparation of PLGA nanoparticles by spherical crystallization

As an example of organic nanoparticles, this section describes PLGA (poly(lactic-co-glycolic acid)), which is a biocompatible copolymer recently developed in a DDS development project funded by NEDO (New Energy and Industrial Technology Development Organization). PLGA is absorbed into the bodies, degrades by hydrolysis, and is harmless to people. A technology has been developed to prepare PLGA into nanospheres that enclose effective ingredients. It is called spherical crystallization and involves dissolving PLGA in a mixed solution of acetone and alcohol, which is a good solvent, dripping it into a water solution of PVA, which is a poor solvent, causing the solvents to mutually diffuse, and crystallizing nanoparticles of PLGA by self-emulsification (figure 8). The nanoparticles are spherical and have a relatively uniform particle size of about 200 nm. After removal of acetone and alcohol and freeze-drying, nanoparticles enclosing drugs are obtained.

7.3.2 Application to whitening cosmetics

PLGA nanoparticles have been reported to be superior in absorbency through the intestinal wall compared to particles of a micrometer size, and applications to inhalants are also being investigated[7,8]. Recently, the nanoparticles were found to be also superior in percutaneous absorption[9].

Figure 9 compares the penetrability of a water dispersion of coumarin 6, which is a fluorescent substance, and that of a water dispersion of PLGA enclosing coumarin. The former showed

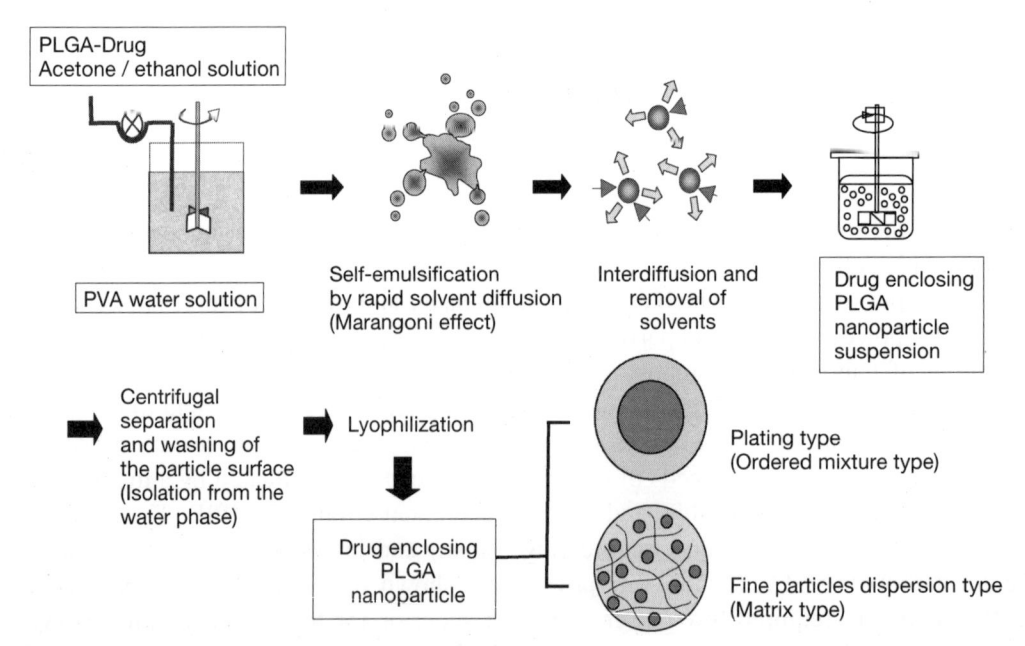

Figure 8 Preparation process of drug enclosing PLGA nanoparticle by the spherical crystallization technique

almost no penetration even at high concentrations, while the latter showed a large quantity of the substance penetrating into the skin and remaining within the skin even after 4 hours. Also in an experiment that enclosed fat-soluble provitamin C (VC-IP, ascorbyl tetraisopalmitate, a derivative of vitamin C) in PLGA nanoparticles, the quantity converted into ascorbic acid was much larger and the vitamin remained longer when enclosed in the nanoparticles than when the derivative was in a water dispersion.

The experiments suggested that the PLGA nanoparticles having penetrated the stratum corneum, were absorbed into the skin, and degraded by hydrolysis, slowly releasing VC-IP, which was deesterified into reduced-form vitamin (figure 10). Using these characteristics, high-performance cosmetics "NanoCrysphere" were developed in which PLGA nanoparticles enclosing derivatives of vitamin E (tocopheryl acetate), vitamin A (retinyl palmitate), and

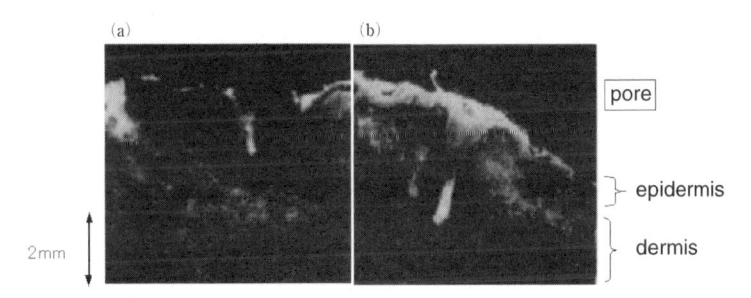

Figure 9 Comparison of penetrability of PLGA nanoparticles using excised human skin specimens (in 4 hours after application)
(a) Water dispersion of fluorescent substance coumarin 6
(b) Water dispersion of coumarin 6 enclosed in PLGA nanoparticles
Excised human skin specimen (35-years old woman, under an arm, divided into 6 small pieces)
Stratum corneum (from the skin surface-0.01 mm), epidermis (0.01-0.1 mm), dermis (0.1-1.7 mm)
(Photograph: Prefectural University of Hiroshima, Miwa laboratory)

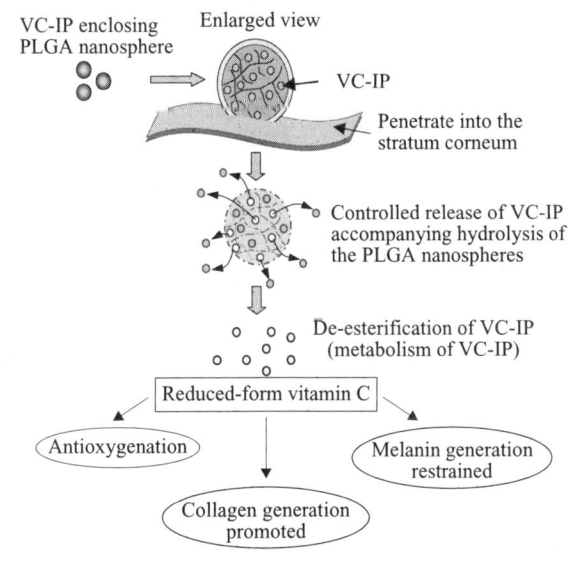

Figure 10 Vitamin C delivery system to deep parts of the skin by VC-IP enclosed PLGA nanoparticles

vitamin C are dispersed[10].

PLGA nanoparticles have an average particle diameter of 200 nm and are relatively large as a nanoparticle. Various possibilities are being investigated to explain the high penetrability of PLGA nanoparticles, such as presence of small particles, easy variability in the shape of the particles due to swelling, and stabilization due to drops in free energy.

The results of an experiment of the whitening effect of "NanoCrysphere" are shown in figure 11. Sample (a) is an electron micrograph of human skin melanocytes. In Sample (b), theophylline was applied to induce melanin production, and the cells were tanned. On the other hand, in the sample in which "NanoCrysphere" was applied before theophylline application, melanin production was suppressed by antioxygenation effect of vitamin C, showing the whitening action of the cosmetics.

VC-IP enclosed in PLGA nanoparticles has also been shown effective for controlling damages to DNA by UV rays. When skin cells receive UV irradiation, active oxygen is produced and damages DNAs in the cells. Damaged sites can be marked using fluorescent dye, with which the degree of damage can be compared. Damages to DNA by exposure to UV rays in excised human skin specimens treated and not treated by VC-IP enclosing PLGA nanoparticles are compared in figure 12. When UV rays were irradiated at an energy strength (UV-A: $200J/cm^2$, UV-B: $2J/cm^2$) equivalent to what a person receives a day (UV-A: $200J/cm^2$, UV-B: $2J/cm^2$), UV-A reached the epidermis and dermis and damaged DNAs over a wide area, and UV-B caused damages to DNAs in the epidermis. DNA damage was sharply mitigated in cells treated with PLGA nanoparticles containing VC-IP in advance. The effect was likely because the PLGA nanoparticles were highly absorbable and vitamin C that was slowly released when the particles were decomposed by hydration inactivated active oxygen.

7. 3. 3 Application to scalp care

There are many possible applications by using the superior percutaneous absorption

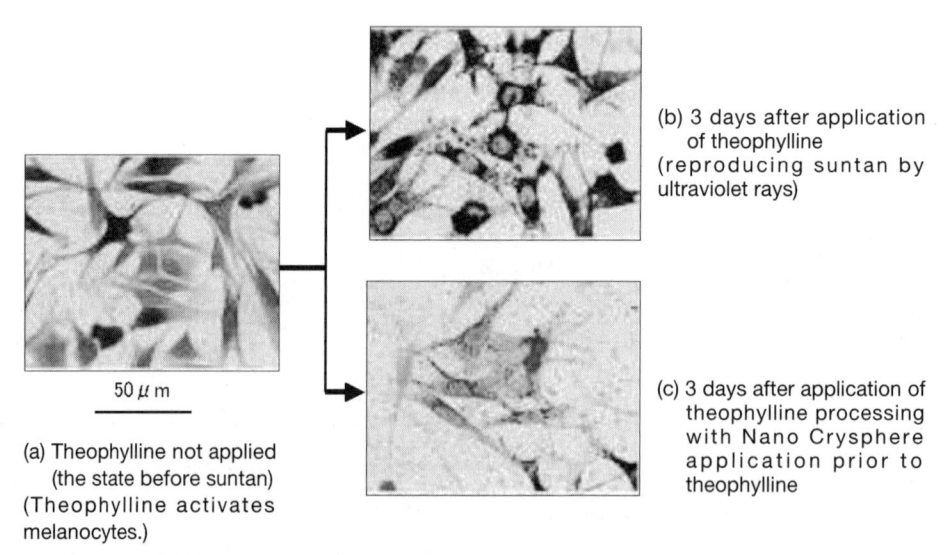

(a) Theophylline not applied (the state before suntan) (Theophylline activates melanocytes.)
50 μm

(b) 3 days after application of theophylline (reproducing suntan by ultraviolet rays)

(c) 3 days after application of theophylline processing with Nano Crysphere application prior to theophylline

(Concentration of the essence lotion in the cell culture medium: 0.01%)

Figure 11 Investigation of the whitening effect of NanoCrysphere (electron micrograph of the human skin melanocytes HMV-II)
(Photograph: Prefectural University of Hiroshima, Miwa laboratory)

characteristic of the biocompatible nanoparticles. Use for scalp care is one of promising applications. Human hair repeats a cycle of growing, degrading and resting, which is called the hair cycle. During the growth period, hair grows by hair matrix cells multiplying at the hair root. During the degradation period, the hair root gradually shrinks, and the activity stops during the resting period. Usually, the old hair falls off, and a new hair shaft is generated,

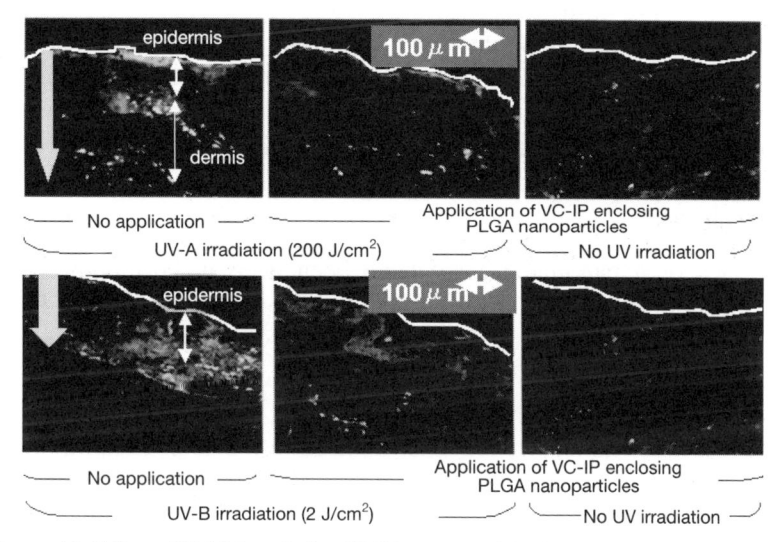

Figure 12 Effect of VC-IP enclosing PLGA nanoparticles in suppressing DNA cutting in excised human skin specimens
24 hours after ultraviolet irradiation, FITC/TUNEL method, dye (excised human skin specimens: 47-years old woman, auricular skin)
(Photograph & evaluation: Prefectural University of Hiroshima, Miwa laboratory)

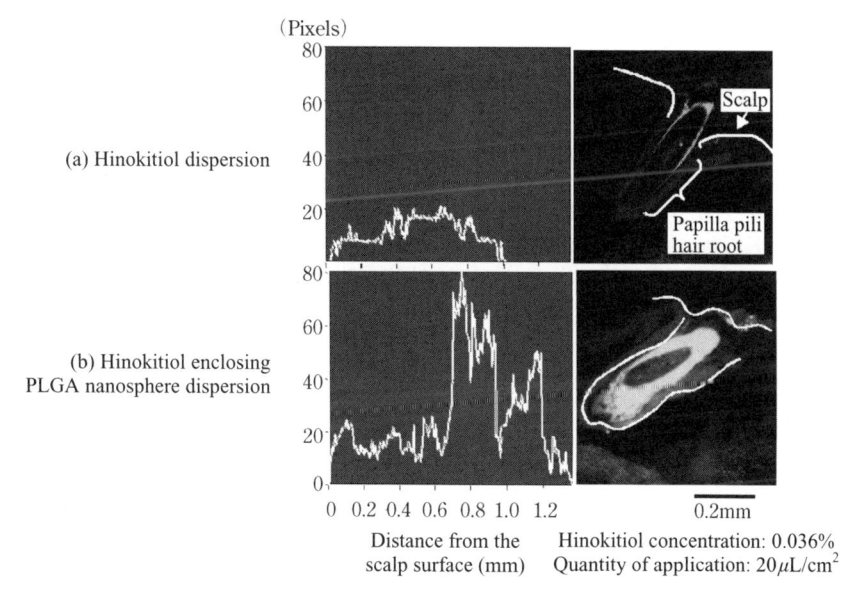

Figure 13 Penetrability of hinokitiol PLGA nanoparticles into the scalp
4 hours after application, 40-years old woman, the head
(Photograph: Prefectural University of Hiroshima, Miwa laboratory)

starting the hair cycle again. An important point is to effectively deliver an appropriate amount of ingredients needed for activating, promoting multiplication of and controlling hair matrix cells to the sites (hair papilla) that need the ingredients.

The effects of PLGA nanoparticles in increasing the penetrability of hinokitiol into the scalp are shown in figure 13. The experiment, which involved applying dispersion of either hinokitiol alone or hinokitiol enclosed in PLGA nanoparticles to excised human scalp specimens and quantifying the fluorescence of hinokitiol, showed that a far larger quantity of hinokitiol penetrated into the scalp when it was enclosed in the PLGA nanoparticles than it was simply dispersed, particularly near the hair follicles. This basic technology has been used to enclose and dissolve plant extracts and other ingredients in PLGA nanoparticles and the resultant products are now commercially available as new scalp care products called "NanoImpact".

7. 4 Conclusions

Some fine particles of a nanometer level have been used in cosmetics. Recently, nano-costmetics that actively use the unique characteristics of nanoparticles and nanostructures are attracting attention. As cosmetics are applied directly on the skin, their safety must be carefully and thoroughly examined. There are diverse nanoparticles that differ in material and property. Further application of nanoparticle technologies for preparing organic and inorganic nanoparticles, processing nanoparticles into composites, and evaluating the particles to cosmetics is expected.

References

1. Y. Kawashima, H. Yamamoto, H. Takeuchi, Y. Kuno, *Pharm. Devel. Tech.*, **5** (1), 77-85 (2000)
2. M. Kawahara, Text for the 37th Lecture Panel Discussion on Powder Engineering, p.11-18 (2003)
3. T. Yokoyama, *Convertech*, **374** (5), 42-45(2004)
4. T. Yokoyama, K.Urayama, M. Naito, T. Yokoyama, *KONA*, (5), 59-68 (1987)
5. M. Inoki, *KINOU ZAIRYO*, **24** (7), 77-86 (2004)
6. T. Yokoyama, H. Tsujimoto, Y. Kawashima, *Powder Science & Engineering*, **36** (10), 63-71 (2004)
7. H. Tsujimoto, K. Hara, C.C. Huang, T. Yokoyama, Y. Kawashima, *Foods & Food Ingredients Journal of Japan*, **210** (5), 437-448 (2005)
8. H. Tsujimoto, K. Hara, Y. Kawashima, *J. Soc. Powder Technol. Jpn*, **41** (11), 765-772 (2005)
9. H. Tsujimoto, K. Hara, C.C. Huang, T. Yokoyama, H. Yamamoto, H. Takeuchi, Y. Kawashima, N. Akagi, N. Miwa, *J. Soc. Powder Technol. Jpn*, **41**, 867-875 (2004)
10. H. Tsujimoto, *Drug Delivery System*, **7**, 405-415 (2006)
11. Hosokawa Micron Cosmetics, 「NanoCrysphere」, 「NanoImpact」, www.nanocrysphere.com

Chapter 8
Dispersion of Fine Particle Titanium Dioxide and Fine Particle Zinc Oxide

Takatsugu Yoshioka and Keiko Iwasaki

8. 1 Introduction

In Japan, the cosmetics market is generally said to have been matured, but sunscreen products are making a steady growth these years. This is because the influence to the skin by ultraviolet rays (UV) has been widely recognized. Functions that are demanded to sunscreen products include safety, enhanced block of UV, improved sensory evaluation (such as feeling pleasant and not remaining white when applied on the skin), and other additional functions, such as mosquito repelling and whitening of the skin. UV scattering agents and UV absorbents are added to sunscreen products either alone or together depending on purpose. To prevent UV ray absorbents from irritating the skin, there are (non-chemical) sunscreen products on the market that contain only UV scattering agents and still declare to block UV rays. However, UV scattering agents need improvements on dispersion technologies because the agents make the products apt to "remain whitish on the skin", "become hard to spread" etc. when added at high concentrations.

This chapter describes dispersion of fine particle TiO_2 and fine particle ZnO, which are typical UV scattering agents.

8. 2 The Present Condition of Sunscreen Products

Time historical changes in the amounts of shipment of cosmetic products and suncare products are shown in figure 1. Suncare products had made a gradual growth until 2004, when the sales jumped due to the hot summer. In 2006 when it was a cool summer, the growth leveled off, but the products accounted for about 1.6 % in the total sales of all cosmetics and are likely to grow more in the future.

To investigate the trends of recent sunscreen products, we investigated 130 sunscreen products on the market in Japan. The average SPF (Sun Protect Factor) value of the 130 items was about 40, and the average PA (Protection Grade of UV-A) was about 2.5. Products of SPF50 and 50+ or over exceeded the half of all products (figure 2), suggesting that consumers were highly conscious about blocking UV.

Products that contained only UV absorbents or UV scattering agents had relatively moderate ultraviolet ray defense effects. Products of high UV defense effects contained both UV absorbents and UV scattering agents (figure 3).

The SPF values of the products that contained only UV absorbents were only 30 the maximum (figure 4), but the products accounted for 10% of the sales in the market showing that the demands of consumers for products that do not remain white on the skin and feel pleasant to use were also high.

Of products that contained only ultraviolet scattering agents, more used TiO_2 and ZnO combined than TiO_2 alone. The average SPF value of the former was larger by about 10 than that of the latter (figures 3 and 4).

The demands for ZnO and TiO_2, which are used in sunscreen products, also showed the same trend. ZnO is needed not only to block UV-B, which causes erythema, but also to block UV-A, which is believed to cause photo-aging.

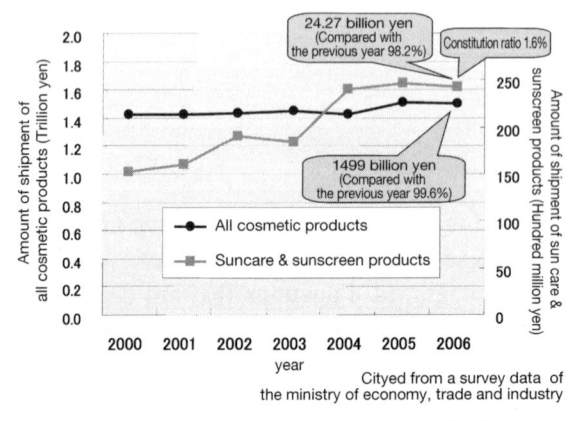

Figure 1 Changes in the amount of shipment of all cosmetic products and sun care products by year

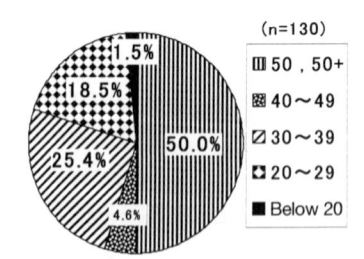

Figure 2 Percentages of SPF values of sunscreen products on the market

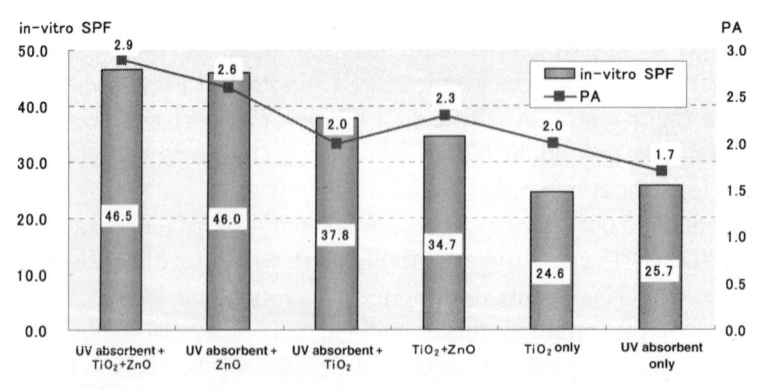

Figure 3 Mean SPF and PA values of sunscreen products on the market for each UV blocking agent

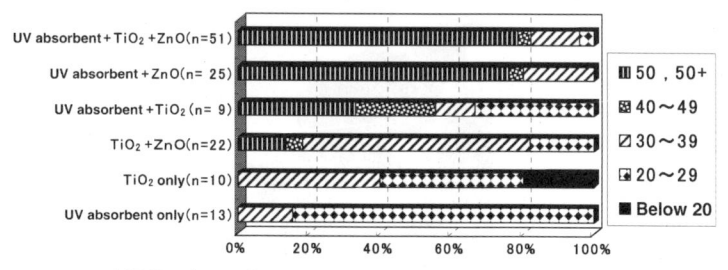

Figure 4 Percentages of SPF values of sunscreen products on the market for each UV blocking agent

8.3 Particle sizes and shapes of ultraviolet scattering agents

Major constituents of sunscreen products are ultraviolet scattering agents, which consist of TiO_2 for cutting UV-B and ZnO for cutting UV-A. To achieve high SPF values, it is necessary to mix these UV scattering agents in a good balance. However, including a lot of UV scattering agents makes the products to result in unnaturally pale makeup when applied on the skin. To produce sunscreen lotion that has a high SPF value and does not remain whitish but transparent on the skin, it is important to convert secondary aggregates into primary particles as much as possible.

Based on the relationship between particle diameter of fine particle TiO_2 and transmittance at each wavelengths, a particle diameter of around 20 nm was found to be suitable to achieve transparency and high blockage of UV-B; and a particle diameter of around 70–90 nm was suitable to block UV-A[1,2].

Crystals of TiO_2 have various shapes (figure 5). Typical forms are granules, spindles, needles, and sticks, the last two of which were developed to reduce the blue color appearance peculiar to TiO_2. For sunscreen products, the crystals are adjusted to sizes of 10–30 nm in order to cut UV rays best. The forms of bow ties and butterflies are effective for cutting off UV-A as their particle sizes are slightly big.

The refractive index of ZnO is 2.0 and is smaller than that of TiO_2. Thus, ZnO is easier to

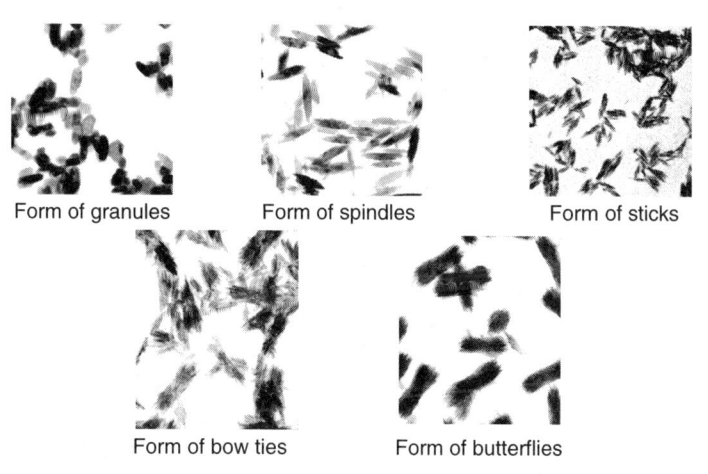

Figure 5 Forms of fine TiO_2 particles (power: ×100 000)
(Cited from the catalogue of Titankogyo Co., Ltd "fine particles TiO_2 ST-400 series for cosmetics")

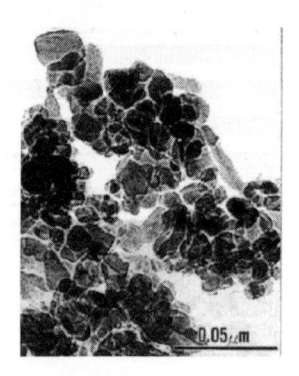

Figure 6 Forms of fine ZnO particles

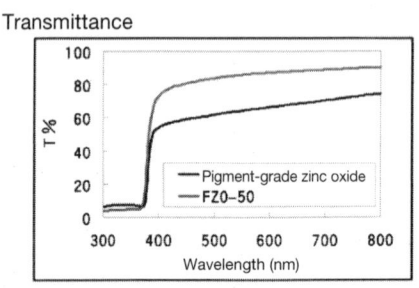

Liquid paraffin vaseline stearic acid system PWC3.3% Applicator 8mills

Figure 7 Transmittance curves of fine particle ZnO
(Cited from the catalogue of Ishihara Sangyo Kaisya Co.
"ultra fine particles ZnO FZO-50")

make transparent than TiO_2 of the same particle diameter[3]. Its UV screening ability is improved when combined with TiO_2 and/or UV absorbents. Crystals of ZnO come mainly in the form of granules (figure 6). Particles of 10–40 nm are widely used. The greatest absorption wavelength range is 370–380 nm (figure 7).

As described, the UV blocking performances of TiO_2 and ZnO are affected by the particle size and shape, and thus they should be carefully determined.

8. 4 Dispersion of Particle Titanium Dioxide and Particle Zinc Oxide

As described above, TiO_2 and ZnO can both block UV and produce transparent sunscreen lotions, but the effects depend on the dispersion conditions of the powders[4]. Particularly, TiO_2 tends to become aggregates when just included in products. When it becomes aggregates, the system becomes unstable and its UV defense effect makes a sudden fall.

As a technology for controlling dispersibility, we developed a high-performance functional UV scattering agent paste called "Cosmeserve® paste", which keeps fine particles stable in the base of high shear strength. The paste enables resultant products to not remain whitish on the skin.

We evaluated the performances of Cosmeserve® paste by preparing W/O sunscreen cream specimens (volumes of powder composition is 10.2%) (table 1) by (1) using Cosmeserve® paste, which is a dispersion of fine particles TiO_2 in the squalane base, and (2) adding each component separately, and comparing the performances of the resultant specimens. The UV

Chapter 8 Dispersion of Fine Particle Titanium Dioxide and Fine Particle Zinc Oxide

defense effect and light transmittance characteristics of the specimens are shown in table 2 and figure 8.

Table 2 and figure 8 showed the superior transparency of Cosmeserve® paste as well as its effective blockage of UV-A and UV-B and high transmittance in the visible range of over 400nm.

The results showed that the use of Cosmeserve® paste enabled fine particles of inorganic

Table 1 Evaluation of Cosmeserve® dispersion (W/O sunscreen cream formulation)

		Constituents	INCI	SC-WS-10R	SC-WS-10S
A	(1)	Abil EM-90	Cetyl dimethicone copolyol	2.30	2.30
	(2)	Cosmol 82	Sorbitan sesquioleate	0.70	0.70
	(3)	Salacos 913	Isotridecyl isononanoate	7.00	7.00
	(4)	Squalane	Squalane	8.00	8.00
	(5)	Silicone MP-1	Phenyl trimethicone	3.00	3.00
	(6)	Propylparaben	Propylparaben	0.05	0.05
	(7)	Silicone DN-196	(Dimethicone/Vinyl dimethicone) Crosspolymer/Dimethicone	2.00	2.00
B	(8)	Fine particles titanium dioxide	Titanium dioxide, Silica, Aluminum hydroxide, Stearic acid	10.20	—
	(9)	Squalane	Squalane	6.20	—
	(10)	Rheopearl	Dextrin palmitate	0.60	—
	(11)	Cosmeserve® paste	Titanium dioxide, Silica, Aluminum hydroxide, Stearic acid, Squalane, Dextrin palmitate	—	17.00
C	(12)	Purified water	Water	52.80	52.80
	(13)	1,3-butylene glycol	Butylene glycol	6.00	6.00
	(14)	Methylparaben	Methyl paraben	0.15	0.15
	(15)	Sodium chloride	Sodium chloride	1.00	1.00
			Total	100.00	100.00

Table 2 UV defense ability of Cosmeserve® paste

	In vitro SPF*
(1) Mixed as Cosmeserve® paste	17.7
(2) Mixed as powder	11.0

* Measured using SPF Analyzer SPF-290S plus

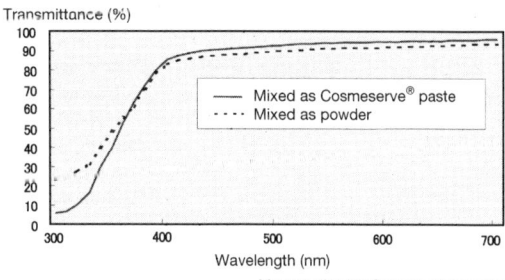

Figure 8 UV defense ability and transmittance curves of Cosmeserve® paste

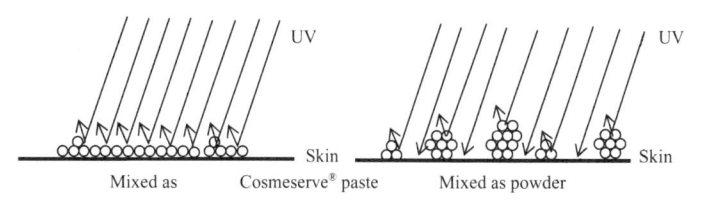

Measured using Spectrophotometer HITACHI U-3010 (Application thickness: 6 μm)

Figure 9 Schematic view of Cosmeserve® paste applied on the skin

powder to spread on the skin more uniformly and the specimen to be more transparent on the skin and have a higher SPF than adding each ingredient separately.

A schematic diagram of the specimen, which was prepared using Cosmeserve® paste, on the skin is shown in figure 9.

Cosmeserve® paste is available in several grades, which differ in disperse medium and the kind of fine particle powder. Cosmeserve® WP-UF (V) is a dispersion of spherical fine particle TiO_2 of various particle sizes and can effectively block a wide UV range including UV-B and

Table 3 Evaluation of Cosmeserve® dispersion (sunscreen milk formulation of W/O shake well type)

		Constituents	INCI	(wt%) WP-TTN (V)	WP-UF (V)	WP-LS
A	(1)	Silicone KF-6038	Lauryl PEG-9 Polydimethylsiloxyethyl dimethicone	1.50	1.50	1.50
	(2)	Cosmol 41 (V)	Polyglyceryl-2 isostearate	0.50	0.50	0.50
	(3)	O.D.O	Caprylic/capric triglyceride	2.00	2.00	2.00
	(4)	Cosmol 525	Neopentyl glycol diethylhexanoate	3.00	3.00	3.00
	(5)	Tocopherol	Tocopherol	0.05	0.05	0.05
	(6)	Propyl paraben	Propyl paraben	0.05	0.05	0.05
B	(7)	Cosmeserve WP-TTN (V)	(Dimethicone/Vinyl dimethicone) crosspolymer Polyglyceryl-2 Diisostearate Cyclopentasiloxane Stearic acid Titanium dioxide Aluminum hydroxide	25.00	—	—
	(8)	Cosmeserve WP-UF (V)	(Dimethicone/Vinyl dimethicone) crosspolymer Polyglyceryl-2 isostearate Cyclopentasiloxane Silica Stearic acid Titanium dioxide Aluminum hydroxide	—	21.60	—
	(9)	Cosmeserve WP-LS	PEG-9 dimethicone Cyclopentasiloxane Stearic acid Polyquaternium-61 Polyglyceryl-3 polydimethylsiloxyethyl Dimethicone Titanium dioxide Aluminum hydroxide	—	—	20.00
	(10)	Cosmeserve WPA-STD (V)	(Dimethicone/Vinyl dimethicone) crosspolymer Polyglyceryl-2 isostearate Dimethicone Phenyl trimethicone Zinc oxide	25.00	25.00	25.00
	(11)	Silicone DN-195	(Dimethicone/Vinyl dimethicone) crosspolymer Cyclopentasiloxane	1.50	1.50	1.50
	(12)	Silicone CY-5	Cyclopentasiloxane	17.50	20.90	22.50
C	(13)	Orgasol 2002 Exd. Nat. Cos.	Nylon-12	2.00	2.00	2.00
D	(14)	Purified water	Water	14.00	14.00	14.00
	(15)	1,3-butylene glycol	Butylene glycol	6.00	6.00	6.00
	(16)	Glycerin	Glycerin	1.00	1.00	1.00
	(17)	Methyl paraben	Methyl paraben	0.20	0.20	0.20
	(18)	Sodium chloride	Sodium chloride	0.70	0.70	0.70
			Total	100.00	100.00	100.00

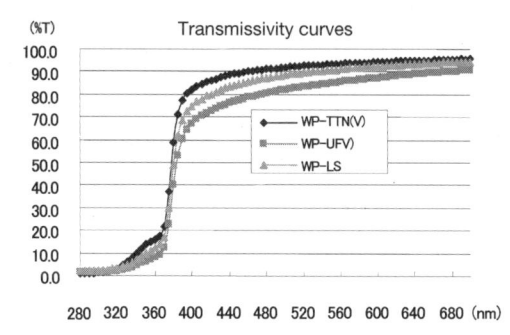

Figure 10 Comparison of light transmittance of the samples

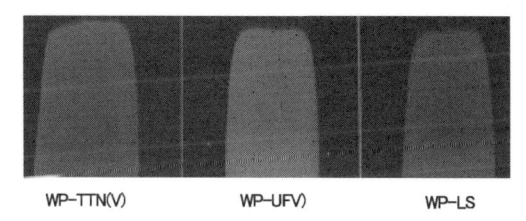

Measured using Spectrophotometer HITACHI U-3010 (Application thickness: 6 μm)

Figure 11 Transparency of the samples

UV-A. In Cosmeserve® WP-TTN (V), needle-shaped fine particles of TiO_2 are dispersed to reduce the blue color appearance of TiO_2 and increase transparency. Cosmeserve® WP-LS is a dispersion of fine TiO_2 in a form of spindles. The three Cosmeserve® products were evaluated by preparing W/O sunscreen milk specimens (amount of TiO_2 8%) (table 3). All specimens also used Cosmeserve® WP-STD (V) (amount of ZnO 15%), which is a dispersion of fine ZnO. The light transmittance and transparency of the specimens, which were prepared as shown in table 3, are shown in figure 10 and 11.

From the results, we found that Cosmeserve® WP-UF (V) excelled in blocking not only UV-B but also UV-A, and Cosmeserve® WP-TTN (V) and Cosmeserve® WP-LS excelled in transparency.

Cosmeserve® BP-60, in which fine particles of Fedoped TiO_2 are used, blocked not only the UV-B range but also the UV-A range. Its has a color similar to that of the skin, and thus adding Cosmeserve® BP-60 prevents resultant sunscreen cream and lotion from coloring the skin unnaturally[6].

From the above results, we understood that TiO_2 and ZnO can be coordinated in balance and be kept transparent on the skin and defensive against UV by improving the dispersibility as well as controlling particle diameter and shape.

8.5 Assignment and Future Development

As the influence of UV on the skin has been widely recognized, the demand for functional sunscreen products has increased. UV absorbents and UV scattering agents are used to develop sunscreen products. They are used combined in products of high UV defense effects. However, fine particle TiO_2 powder is easy to coagulate, and keeping the powder dispersed in a primary form has been an important topic.

On the other hand, because fine particles of ZnO are easy to dissolve in acid and alkali, it is effective to treat the surface of the particles with silicone oil and inorganic matters such as silica or the alumina. However, when the surface is not thoroughly coated, zinc ions[7] elute into water destabilizing the entire O/W system.

In Japan, the majority of sunscreen products are W/O products of the shake-well type aiming to be water resistant, pleasant to use and long lasting. They have a disadvantage of being difficult to remove with ordinary washing agent.

Development of O/W sunscreen products that are pleasant and easy to use, last long, and effective is awaited. A future topic will thus be development of fine particle ZnO ingredients that are functional and stable when used in such sunscreen products.

References

1. M. Ichihashi, Photo-aging of the Skin and Science of the Suncare, p.62, Fragrance Journal Ltd. (2000)
2. M. Sakamoto *et al. Fragrance Journal*, **24** (3), 72 (1996)
3. Y. Suzuki, *Fragrance Journal*, **24** (3), 64 (1996)
4. M. Abe, T. Suzuki, H. Fukui *et al.*, Latest products made in function wound of cosmetics, material development, applied technology, p.436, NTS Inc. (2007)
5. T. Yoshioka *et al. Fragrance Journal*, **27** (5). 64 to 66 (1999)
6. A. Sakai, *Fragrance Journal*, **31** (4), 87 (2003)
7. T. Okamoto *et al.*, Manufacturing Technology of Cosmetics, p.133, Fragrance Journal Ltd. (2001)

Chapter 9
Optical Analysis of the Skin and Development of the Multilayered Powder

Eiichiro Misaki

9.1 Trends of cosmetics development and backgrounds

In the development of base makeup cosmetics, investigations have focused on optical functionality, leading to rapid development of various analytical technologies and technologies for producing useful raw materials. On the market, cosmetics for sensitive skin have attracted attention, increasing the demand for developing cosmetics that are safe and do not irritate the skin.

With these backgrounds, today, development of cosmetic materials involves not only searches for new materials but also finding new combinations of materials for which safety has been confirmed to express new functions. Investigations are made on compositing two materials to result in both safety and functions that cannot be manifested alone. For example, a new optical function was obtained by coating pearl pigments with barium sulphate[1]. This involves extracting a new characteristic by compositing two common cosmetic materials: barium sulphate and pearl pigment. There have been reports on optical materials[2-4] prepared by compositing fine particles of titanium oxide on substrates of mica and talc using supercritical liquid of carbon dioxide. They are also combinations of materials that have been proven safe. Using materials that have been proven safe reduces the labor for checking safety, and thus can quickly develop products responding to demands of users and trends of the market. Composite materials account for a large percentage of newly developed cosmetic materials these years. This chapter describes development of composite powder for cosmetics, in which iron oxide, titanium oxide and mica are used to reproduce beautiful skin.

New attempts are also being made on methods for developing base makeup cosmetics. Users highly demand for more beautiful skin, having a beautiful look, and thus a younger appearance. Because these goals are difficult to quantitatively evaluate, cosmetics have been usually developed by designing formula using trial and error and evaluating the results by sensory evaluation. In this chapter, our new method of investigation is also described, which involves taking photographs of an ideal model from a number of viewpoints, extracting and quantifying factors constituting the beauty, developing makeup foundations using the factors as an index, and evaluating the resultant impression from a number of viewpoints. Visualization and digitalization of goals are effective for accelerating development. As represented by the development of computer technologies and advanced digital cameras, digital imaging technologies have rapidly advanced, enabling the method to be used in cosmetic development. The development of computer technologies has also enabled cosmetic materials to be designed using optical simulation analysis. The optical simulation method used for designing the structure of a pearl pigment is also described. The method has enabled us to predict the colors that can be expressed by pearl pigments, and we used the method to develop a pearl pigment that shows ideal color changes demanded for makeup foundations.

9. 2 Optical analysis of the skin and targets of development

9. 2. 1 Multiple-viewpoint image analyzer

We developed a multiple-viewpoint image analyzer to investigate the varied-angle properties of subjects[5–7]. An exterior view of the analyzer and photographs taken are shown in figures 1 and 2, respectively. The system is 1/4 of a sphere of a radius of 1.5 m and consists of 20 digital cameras and 40 sources of light, which are arranged on a spherical coordinate system. Thirteen cameras and 13 lights are arranged from 0 to 180 degrees at intervals of 15 degrees. Seven cameras are positioned so as to look down at the subject. Forty lights are arranged on a spherical coordinate system at intervals of 15 degrees. The system is designed so that the

Figure 1 Multiple-viewpoint image analyzer

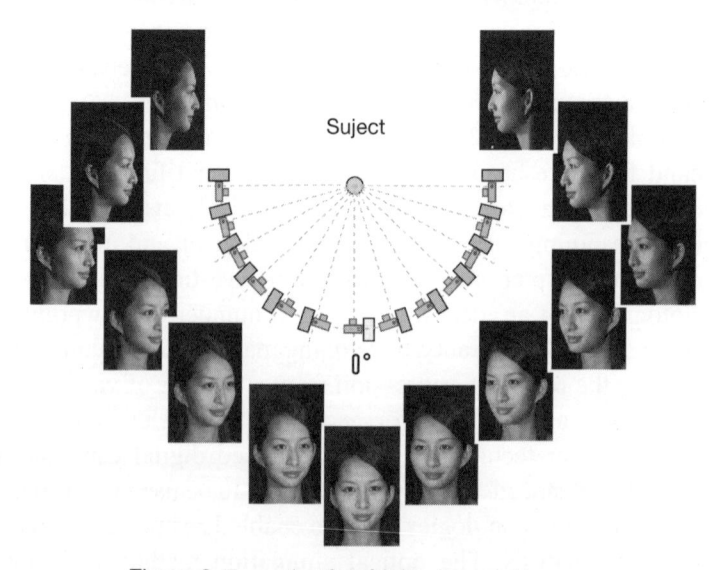

Figure 2 Example of multiple-viewpoint image

Chapter 9 Optical Analysis of the Skin and Development of the Multilayered Powder   115

subject is at center of all optical axes. All light sources are commercially available halogen lights (color temperature: 4700 k) and were adjusted for spectrum in advance by testing and controlling voltage because the uniformity in spectrum was not guaranteed. To simultaneously take 20 photographs, a system was constructed for controlling the cameras. The software program on a PC can control all cameras (for the entire series of tasks including setting and changing the conditions of taking photographs, taking photographs, and transmitting data). Because the photographs of the faces contain personal information, all data are controlled using a 1TB security server.

Conventional methods for measuring varied angles have mainly involved the use of goniospectrophotometers. The photometer can collect precise data of a spectrum level, but the area of measurement is limited and is not feasible for acquiring the general impression of a face of a person. On the other hand, our system can measure varied angels as in a goniospectrophotometer on an image base and can collect quantitative data on the impression of the entire face from a number of viewpoints by synchronously taking photographs. The system is highly beneficial because it enables makeup cosmetics to be evaluated in terms of total impression of the entire face.

9. 2. 2 Optical characteristic of beautiful and normal skins

Subjects who had beautiful and normal skins were measured using the multiple-viewpoint image analyzer to investigate the varied-angle characteristics of the beautiful and normal skins. The results are shown in figure 3. We found that the beautiful and normal skins differed in the optical property of chroma changes in the specular reflection range. The beautiful skin did not lose the chroma of the skin color in the specular reflection range and showed smooth color changes when the direction of view changed. Adding and improving the chroma of the skin color in the specular reflection range were decided to be the guideline for cosmetics development, and reproduction of the beautiful skin on the normal skin was attempted.

9. 3 Designing and developing optics materials (multilayered powder)

Pearl pigments have been long known as an optical material that develops color in the specular reflection range. The design of a pearl pigment structure that can be used as an optical material to reproduce the chroma-change profile of the beautiful skin was investigated. Theoretically, materials that have structural colors develop different colors depending on combination. Thus, a large number of trials must be made using trial and error for actual production. We first developed an optical simulation technology for predicting structural colors[8,9]. As a result, we

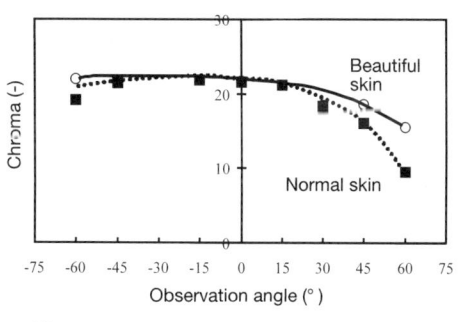

Figure 3 Changes in chroma of beautiful and normal skin

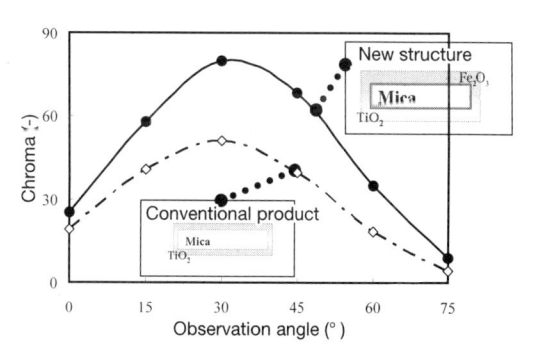

Figure 4 Simulation results

predicted that a pearl structure that contains iron oxide in its nanometer-thick inner layers can develop high-chroma skin color (gold) (figure 4).

Conventional methods for predicting the structural colors of pearl pigments could not produce results that agreed with actual measurement in spectrum level. Our investigations showed that the following points needed to be considered as well as those already considered in the conventional method:
1) Thickness distribution of substrate films,
2) Refractive index with absorption considered,
3) Orientation of board-shaped particles, and
4) All interference layers.

We added these points in our conventional model. The points are briefly described below.

9.3.1 Thickness distribution of substrate films

A SEM photograph of a cut pearl pigment is shown in the figure 5. Conventional design methods assumed the thickness of the substrate film to be infinite and calculate only the thickness of the surface films. As shown in figure 5, the thickness of the mica substrate is finite, showing that conventional calculations are not accurate. The film thickness seems to vary more than the grain size. Thus, the thickness of mica substrates was quantitatively measured under AFM. The measured thickness distribution is shown in figure 6. The thickness distribution of mica substrates was also considered for predicting the colors of pearl pigments. The estimated and

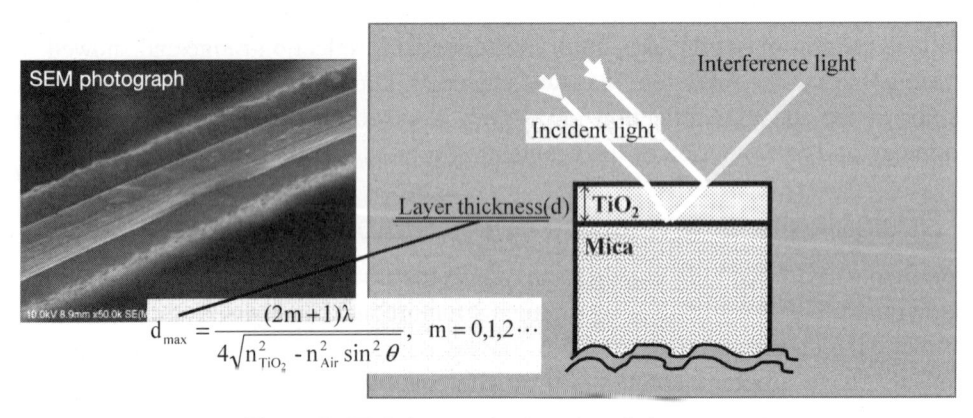

$$d_{max} = \frac{(2m+1)\lambda}{4\sqrt{n_{TiO_2}^2 - n_{Air}^2 \sin^2\theta}}, \quad m = 0,1,2\cdots$$

Figure 5 SEM photograph of a cut pearl pigment

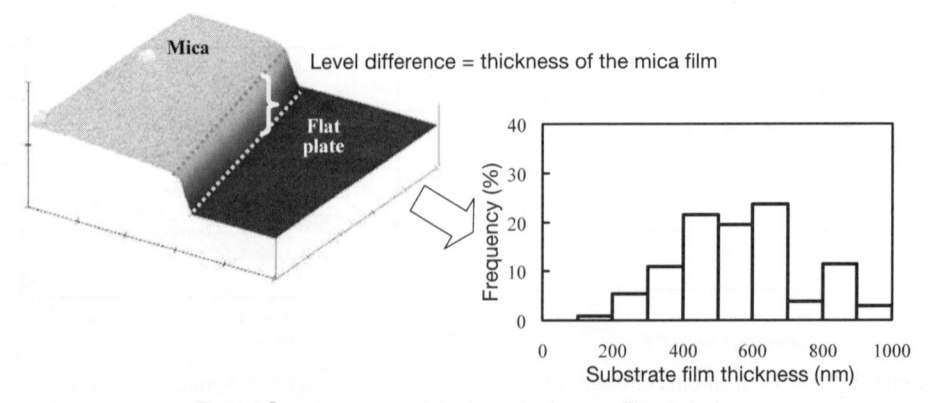

Figure 6 Thickness distribution of substrate film (mica)

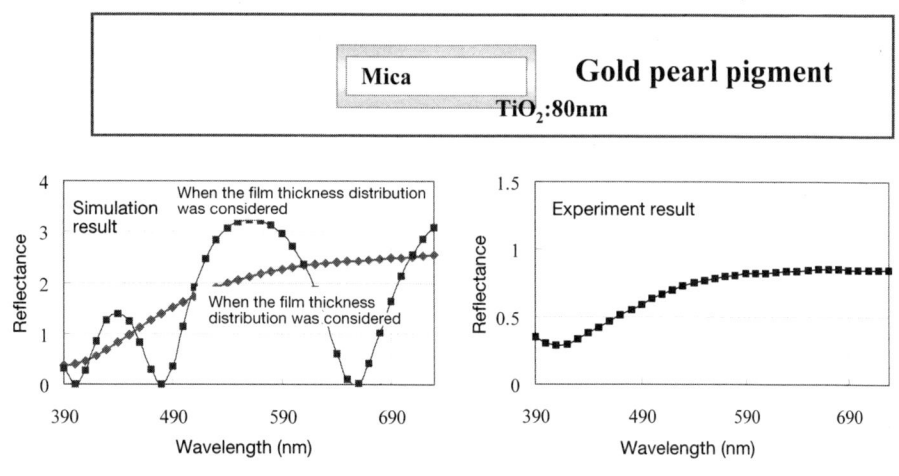

Figure 7 Influence of the film thickness distribution

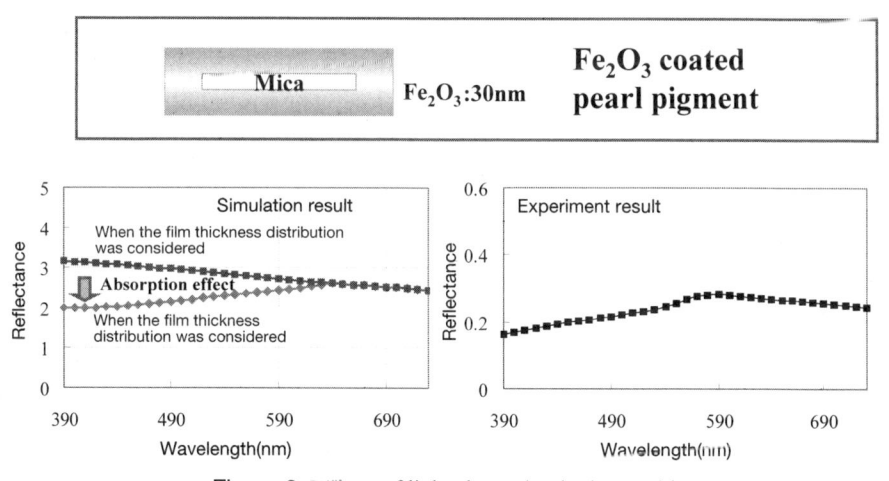

Figure 8 Effects of light absorption by iron oxide

measured results of an ordinary gold pearl pigment are shown as spectral reflectance curves in figure 7. The curves show that the consideration of the thickness distribution of the substrate increased the accuracy of prediction.

9.3.2 Refractive index with light absorption considered

The pearl pigments we designed contained colored materials, such as iron oxide. Thus the absorption of light needed to be also considered. Complex refractive index, in which light absorption is also considered, was used in simulation. Because iron oxide absorbs light of short wavelengths, the use of complex refractive index improved the accuracy of prediction (figure 8).

9.3.3 Orientation of board-shaped particles

Pearl pigments to be used in foundation cosmetics should develop good color when applied on the skin. In our study, we evaluated pearl pigments not by using the ordinary method that involves applying pigments on pieces of coating paper but by applying the pigments on

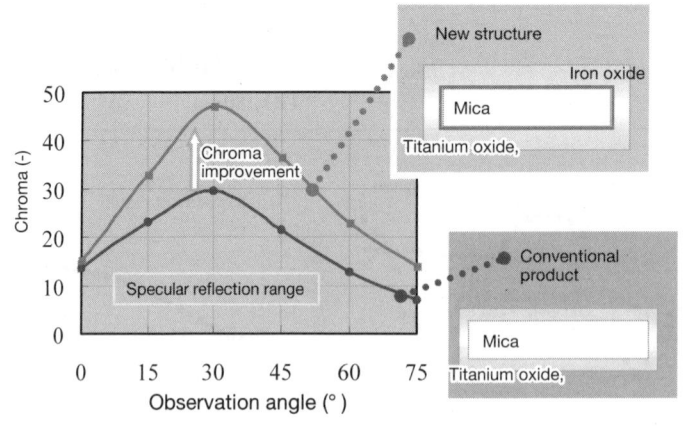

Figure 9 Actual measurement of the multilayered powder

artificial skin, which was black urethane and had unevenness similar to those of the skin, and measuring the samples using a goniospectrophotometer. Thus, the reflectance of the samples on which pearl pigments were applied on the rough surface needed to be known for the optical simulation analysis for predicting the colors. The orientation of the board-shaped particles was investigated using Torrance-Sparrow model. The mean inclination, which shows the coarseness of the surface, was determined so as to reproduce the varied-angle spectral reflectance measurements of the artificial skin with existing peal pigments applied.

9. 3. 4 All interference layers

In most conventional methods for predicting the color of pearl pigments, mica substrates are assumed to have an infinite thickness, and only the thicknesses of upper layers are calculated. In our study, we decided to consider all layers in our calculations because light is interfered and absorbed also on the back surface of a mica substrate.

An optimum structure was determined based on the results of the optical analysis, and a high-chroma pearl pigment was predicted to be produced (figure 9).

Multilayered pearl pigment of the design was prepared on a trial basis. The pigment was found to have chroma much larger than pigments on the market (figure 9). The multilayered pearl pigment is prepared from iron oxide, mica, and titanium oxide, which are common and have been confirmed to be safe, and shows high chroma by its structure.

9. 4 Development and inspection of base makeup cosmetics

We prepared powder foundation that contained 5% of the newly developed high-chroma pearl pigment. The powder foundation was applied on the skin of monitors of normal skins and the effects were evaluated using the multiple-viewpoint image analyzer. The foundation was found to reproduce the chroma changes of the beautiful skin (figure 10). Qualitatively, the product was evaluated by conducting a questionnaire survey on at least 50

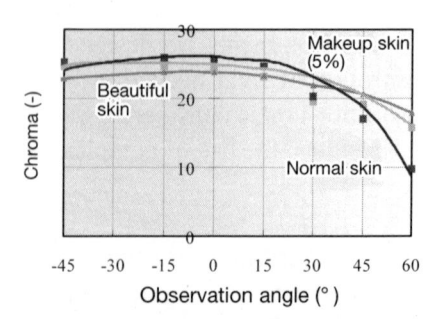

Figure 10 Inspection of 5% combination prescription (chrom changes)

monitors who used the foundation for 2 weeks, which showed that our aim of reproducing the beautiful skin was recognized by the subjects.

9.5 Conclusions

Setting the goals of developing base makeup cosmetics and examining the effects using the multiple-viewpoint image analyzer were briefly described. Rapid progress of digital imaging technologies and computer technologies will lead to development of similar methods. Pearl pigment that has high chroma was actually produced based on the design determined using an optical simulation method. Advance of nanotechnologies for measurement and production will enable powders that can develop structural colors like the feathers of peacocks and Morpho butterflies to be designed and produced. Nanotechnologies are expected to create safe and high functional artificial color pigments that are beautiful as those in the nature.

References

1. Yagi, K., *et al.*, "Optical rejuvenating Makeup Using An Innovative Shape-controlled Hybrid Powder", IFSCC Conference., Florence (2005)
2. N. Nojiri, *Fine Chemical*, **36**, 1, pp. 38-47 (2007)
3. H. Kubo, *et al.*, *Chemical engineering*, **1**, pp. 38-40 (2004)
4. H. Kubo, *et al.*, *Chemical Equipment*, 5, pp. 39-45 (2006)
5. E. Misaki, *et al.*, Substance of Imaging Conference Japan 2005 Fall Meeting, p. 45 (2005)
6. E. Misaki, *et al.*, *Journal of the Society of Powder Technology*, Japan, in print
7. E. Misaki, H. Shiomi *et al.*, IFSCC Osaka (2006)
8. H. Shiomi, E. Misaki *et al.*, FATIPEC Congress (2006)
9. H. Shiomi, E. Misaki *et al.*, *Journal of Coating Technology and Research*, in print (2007)

PART 3 Trends in Developments   [Emulsification]

Chapter 1
Polymer Micelles

Kunio Shimada

1. 1 Introduction

Nanotechnology studies started in January 2001 when the 42nd president of the United States, President Clinton, announced officially that they are a State strategy and gave them research funds of 2.70 hundred million dollars for that year and 4.95 hundred million dollars for 2002[1].

According to an investigation[2], cosmetics that will be noticed in 2004 include (1) anti-aging cosmetics for young people such as those in their 20's, (2) amino acids, which are recognized by customers as healthy food, (3) deep ocean water, which contains inorganic salts and other minerals in a good balance, such as water of off Muroto, off Okinawa and off Kochi, and (4) products of nanotechnology, which are expected to make sales of 95 hundred million yen in 2004. "Nanotechnology" is sometimes used for "fine" or "minute particles". We think nanotechnology, or production of minute particles, improves the quality and performances of cosmetic products and improves the affinity and penetration to the skin. Here, polymer nanotechnology cosmetics is briefly described, but my intention is not to follow a recent trend of calling new materials "nano-something" by adding "nano" to words as a prefix.

1. 2 Synthetic polymer micelles

Polymer micelles were noticed as soon as studies on liposomal technology started. Polymer micelles have a beneficial characteristic. They can be changed into hydrophobic monomers and into arbitrary hydrophilic monomer length, with which the time for releasing drugs can be controlled. Nippon Kayaku Co., Ltd. already started researching Adriamycin to develop carcinostatic medicine for clinical use using micelles of block polymers[3].

Nano is one over 10 hundred million (10^{-9}) meter in size. It is believed that emulsifier stability is better in smaller micelles particles, which results in better skin penetration. The creaming phenomenon, which is caused by separation of emulsion, occur when specific gravity differs between emulsifier particles and the medium. Emulsifier particles show Brownian movement, which is caused by collision of molecules in the medium. Creaming does not occur when the diffusion velocity caused by the Brownian movement exceeds the sedimentation velocity. When the medium is water, Brownian movement is effective for emulsifier particles smaller than 1 μm in size. Micro emulsion was shown to be more anti-inflammatory than normal emulsion when an edematization rat was treated either by micro emulsion of 87nm, which contained an oil phase of 0.4% Indomethacin, or emulsion of 10–100 μm.

Of radical polymers, a copolymer consisting of 2-metacryloyloxyethyl phosphorylcholine (MPC) and a long alkyl chain of strearyl methacrylate (SMA) is attracting attention. The copolymer spontaneously aggregates in water and forms 30–40 nm micelles although copolymers that contain over 30 mol% SMA are insoluble to water. The solubility of MPC/SMA=50/50(mol%) polymer was higher than that of other polymers[4]. During the drying process of nano dispersion, this copolymer spontaneously forms a lamella structure of 6 nm[5].

These polymer micelles can show nano dispersion without special emulsifier apparatus because they do not need shear stress energy to be dispersed. They are more stable against heat than liposome. They are safe since they are less cytotoxic than lysophospholipids, such as mono-chain acyl and polyglycerin fatty acid ester. Double-chain acyls, such as ceramide, are not cytotoxic, but their water holding capacity is lower than that of phospholipid polymers because the polarity level is low. Lamella rearrangement of the polymer micelles that enclose fat-soluble vitamins and other substances in a stable manner, which occurs when the micelles spread on a surface dries, is a promising characteristic for developing future nanotechnological material for cosmetics.

1. 3 Ingredients for cosmetics

On the other hand, 2-methacryloyloxyethylphospholylcholine (hereinafter referred to as MPC), which has a polar phospholipid group, is a biocompatible material designed to imitate the biological membrane. One of its derivatives, MPC polymer, is also as a biocompatible material and is studied for the application in artificial organs, such as artificial blood vessels and biosensors, since it has high protein adhesion-control ability and inter-cellular binding-control ability[6,7].

In the cosmetics field, MPC polymer (trade name: LIPIDURE) is widely used as a high functional material mainly in skin care products, for its excellent moisture-absorption/retention ability[8,9], skin-barrier function, skin-protection ability from irritant substances[5,6] and other advantageous functions.

Among the Lipidure series, only Lipidure-S (INCI: Polyquaternium-61) self assembles. The unique function of Lipidure-S is shown in figure 1; it disperses in water forming nanoparticles (average particle size: <50 nm) and forms a lamella structure on a surface when the nanoparticle dispersed solution is applied on the surface and dried.

In this article we introduce new Lipidure-S-based materials, which were developed for easy formulation of cosmetics, namely Lipidure-NR (polyalcohol solution of Lipidure-S) and Lipidure-NA (Lipidure-S's nanoparticle aqueous dispersion).

The largest disadvantage of liposomes is its instability against heat and surfactants. The polymer micelles are highly stable as experimentally shown that the transmission factor showed no changes when 1% aqueous solution of the copolymer (MPC-co-SMA) was stored at either room temperature or 40°C for six month (figure 3).

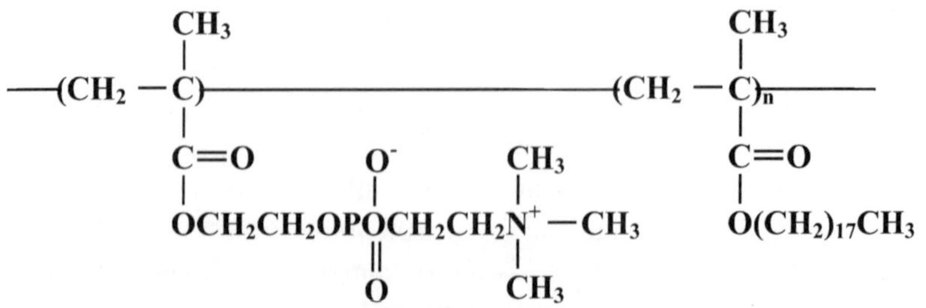

Figure 1 Chemical structure of Polyquaternium-61

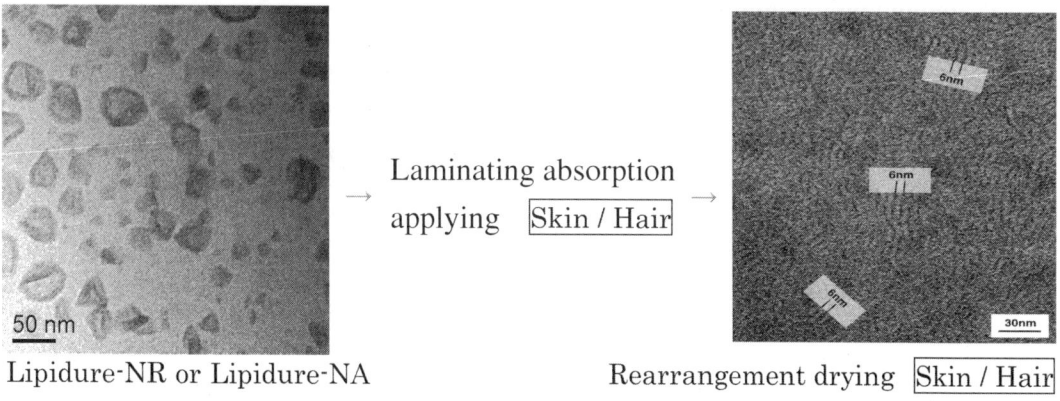

Chapter 1 Polymer Micelles 123

→ Laminating absorption applying Skin / Hair →

Lipidure-NR or Lipidure-NA aqueous dispersion

Rearrangement drying Skin / Hair
Formation of Lamella

Figure 2 Lamella layer formation on the skin and the hair (TEM image)

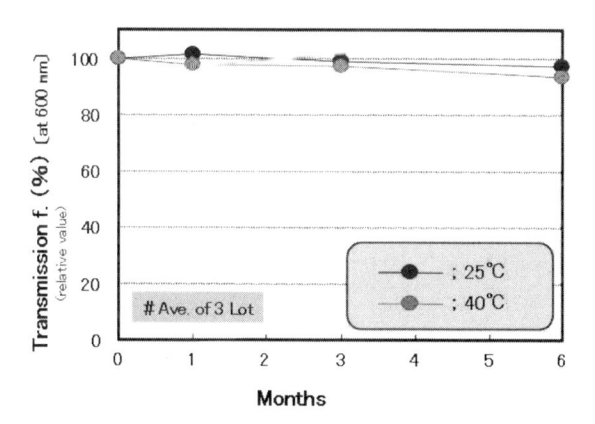

Figure 3 Stability for Polyquternium-61

1.4 Encapsulating active elements and maintaining effect

Lipidure-S is a hydrophobic polymer of MPC and a plant-derived (palm) stearylmethacrylate, which has an alkyl group. Its average molecular weight is approx. 0.1 million (figure 1). As described in figure 4, Lipidure-NR is polyalcohol added to Lipidure-S solution, and Lipidure-NA is aqueous dispersion of Lipidure-S, in which Lipidure-S forms nanoparticles.

As shown in figure 1, Lipidure-NA (aqueous dispersion of Lipidure-NR) has a unique characteristic that Lipidure-S forms a nanoscale lamella layer on its coated surface when it is dried. An X-ray structure analysis of a polymer film made by Lipidure-S showed that the lamella structure consists of regularly arranged Lipidure-S molecules at 5.7 nm intervals and regularly arranged stearyl groups at 0.42 nm intervals. The structure was also confirmed by transmission electron microscope observation[10,11].

Lipidure-NR and Lipidure-NA have the following characteristics: form nanoparticles in water, can encapsulate oil soluble materials in their nanoparticles, since the inside of a nanoparticle becomes hydrophobic when formed, and form lamella structures when dried. We now explain the encapsulation of functional materials in the nanoparticles and the advantages

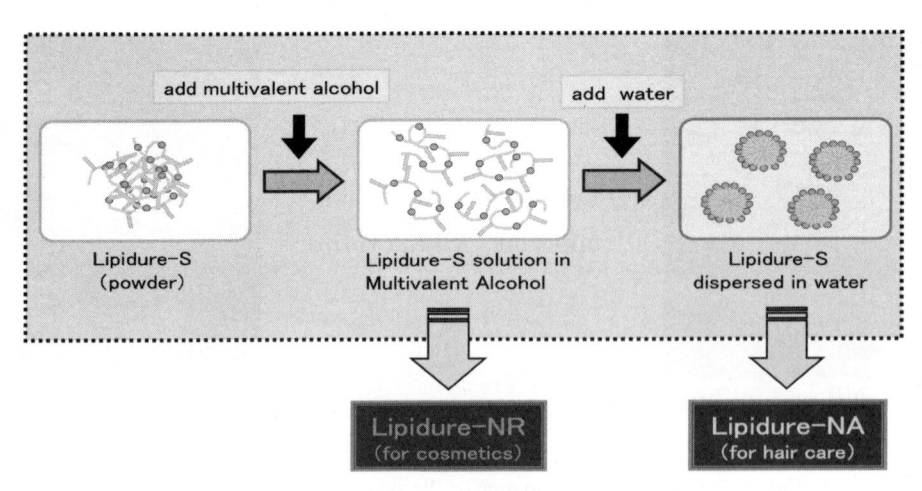

Figure 4 Lipidure-NR and Lipidure-NA

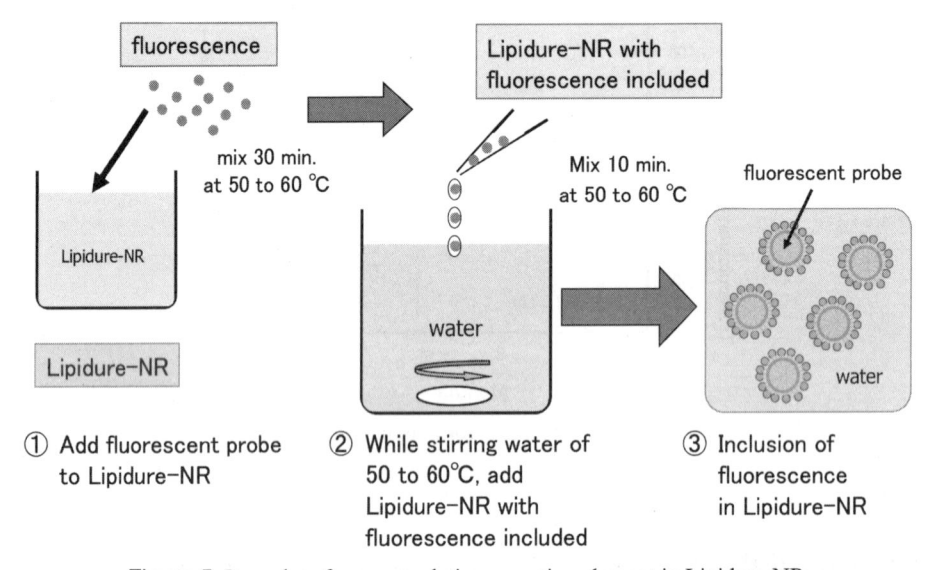

Figure 5 Procedure for encapsulating an active element in Lipidure-NR

of these materials in formulating skin care and hair care products.

Lipidure-NR can encapsulate oil soluble active elements, such as vitamins, without the use of homomixer or other emulsifying devices. We examined the encapsulation and the persistency on the coated surface using an oil soluble fluorescent material Nile red (NIR, E_x=540 nm) as a sample of active element.

We prepared nanoparticles that encapsulated NIR by adding NIR to Lipidure-NR, stirring for around 30 minutes in water bath at 50–60°C to solubilize the NIR, and dripping the solution into 50–60°C water under stirring (figure 5). Using the NIR-encapsulating Lipidure-NA (aqueous dispersion of Lipidure-NR) made this way, we evaluated the persistency of NIR on the surface of damaged hair.

The sample hair was damaged by soaking in a 1:1 mixture of 4.5% hydrogen peroxide water and 2.5% ammonia water for 20 minutes ten times. The damaged hair was then soaked in the NIR-encapsulating Lipidure-NA solution for one minute, rinsed in water bath for 10 minutes,

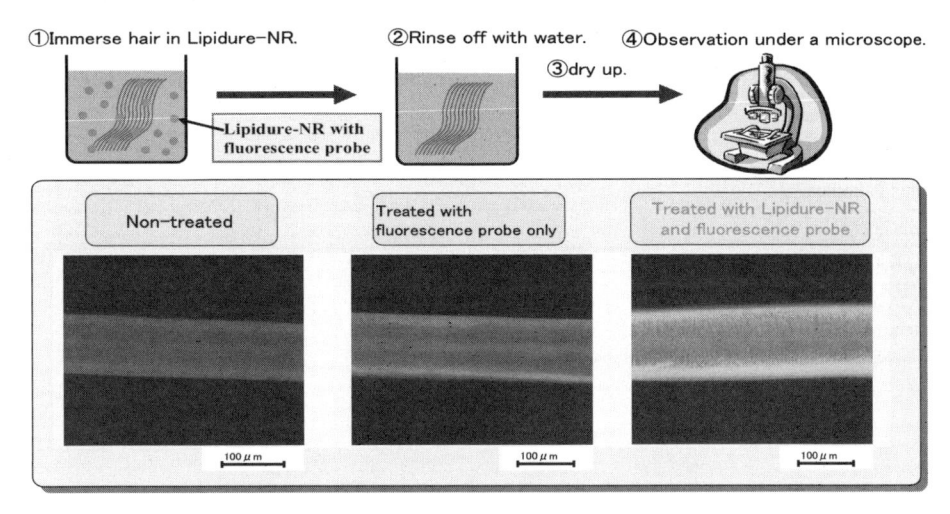

Figure 6 Fluorescent microscope pictures of hair

and dried. The treated hair was observed under an epi-fluorescent microscope (BHS-RFC, Olympus). Figure 6 shows the fluorescent microscope pictures of the non-treated damaged hair, hair treated with NIR surfactant-solubilized solution, and hair treated with the NIR encapsulating (solubilized) Lipidure-NA.

Strong fluorescent of NIR was observed only in the hair sample treated with the Lipidure-NA, showing that NIR was kept absorbed on the hair surface even after washing. This suggests that a lamella layer that contains active elements is formed on the hair surface when the hair is treated by Lipidure-NA that encapsulates active molecules.

Besides, oil soluble materials of high polarity, such as tocopherol acetate (vitamin E), should be easily encapsulated because glycerin and butylene glycol are used as the base of Lipidure-NR. On the contrary, oil soluble materials with lower polarity are less compatible with the base, and thus are more difficult to encapsulate and less stable after encapsulation. However, we found in experiments that use of PCA ethyl cocoyl alginate (CAE) and 1,2-hexanediol enables low polarity materials to be efficiently encapsulated.

The procedure of encapsulation is: add 1,2-hexanediol to the active element to encapsulate, dissolve it under stirring at 60°C for a certain period, add CAE and Lipidure-NR to the solution, and dissolve it under stirring at 60°C for an hour until it becomes uniform. Pouring this Lipidure-NR with the active element solubilized into 50–60°C water under stirring produces a dispersion of Lipidure-S nanoparticles.

Table 1 shows the results of encapsulation studies on various oil soluble active elements. It is very likely that the majority of oil soluble materials can be encapsulated since squalane, which is a material of low polarity, can be encapsulated stably. The stability after three weeks is shown in table 1. Ceramide, retinol palmitate, squalane and ascorbyl tetraisopalmitate were all included in polymers stably. Powder ingredients can be included by dissolving in squalane.

As shown in figure 7, inclusion of perfume in nano micelles enables resultant cosmetics to have aroma over a long period of time.

Polymer micelles also have disadvantages over liposomes. Liposomes are available in diverse production kits and empty capsules. Moreover, polymer micelles can only encapsulate small amounts of oily active ingredients. These disadvantages will be investigated and solved.

Table 1 Encapsulation of various oil soluble materials

materials (INCI)	NR	NR+CAE[*1]	NR+CAE+HG[*2]
Tocopherol Acetate	○	○	—
Synthetic Ceramide	×〜△	△〜○	○
Retinyl Palmitate	×〜△	△	○
Squalane	×〜△	△	○
Ascorbyl Tetraisopalmitate	×	×	○

*1: PCA ethyl cocoyl alginate, *2: 1,2-Hexanediol
<Criteria: eye observation at 1 week later>
○: Stable, △: Slightly unstable, ×: Unstable (precipitation or creaminess)

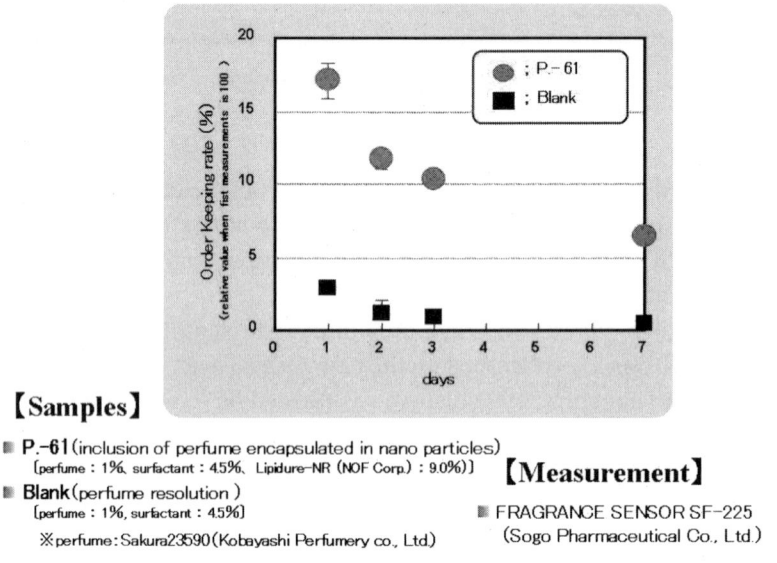

【Samples】
- P.-61 (inclusion of perfume encapsulated in nano particles)
  (perfume : 1%, surfactant : 4.5%, Lipidure-NR (NOF Corp.) : 9.0%)
- Blank (perfume resolution)
  (perfume : 1%, surfactant : 4.5%)
  ※perfume：Sakura23590 (Kobayashi Perfumery co., Ltd.)

【Measurement】
- FRAGRANCE SENSOR SF-225
  (Sogo Pharmaceutical Co., Ltd.)

Figure 7 Order keeping efficacy

1.5 Conclusions

In this article we have introduced a unique cosmetics ingredient that forms nanoparticles in water and a lamella layer once dried. Compared to the structures of phospholipids, Lipidure-S has a structure as if a number of natural phospholipids are bound together by covalent bonds. Phospholipid is the main component of the cell membrane, in which the size of the polar group is balanced with that of the hydrophobic group when the polar group is phospholylcholine, and is known to form a lamella layer similar to that formed by inter horny-cellular lipids. Since Lipidure-S has a structure similar to that of polymerized phospholipid and well-balanced hydrophilic and hydrophobic groups, Lipidure-S forms a lamella layer when it dries and becomes a film. In addition, usability evaluation tests showed that Lipidure-NR restored the skin that has lost its barrier functions by forming a pseudo-horny layer and restored damaged hair by producing a pseudo-F layer.

Lipidure-NR and Lipidure-NA are materials that fit to today's and future highlighted concepts, such as nanotechnology, lamella layers, barrier functions and encapsulation, and have functions to cope with the concepts. We believe that Lipidure-NR and NA will contribute much to the development of new functional cosmetics/hair products of these concepts.

Rererences

1. M. Shimomura, *Polymers*, **50**, 5, p.300 (2001)
2. Fuji-Keizai Tokoy Marketing Division, "Cos. Market. Develop. Strate. 2003"
3. K. Kataoka, *Nikkei Bio Business*, **9**, p.49 (2003)
4. N. Yamamoto, *et al.*, Polymer Preprints. Japan, **52**, 5, p.1107 (2003)
5. N. Yamamoto, *et al.*, Polymer Preprints. Japan, **53**, 1, p.1107 (2004)
6. K. Ishihara, N. Nakabayashi, K. Nishida, M. Sakakida and M.Shichiri, *CHEMTEC*, October 19 (1993)
7. K. Ishihara, *Journal of Japanese Society for Biomaterials*, **18** (2), 36 (1996)
8. T. Kii, K. Shimada, Y. Murata, K. Ishihara, N. Nakabayashi, *Journal of Oleo Science*, **48** (6), 35 (1999)
9. H. Kuroda, M. Shaku, A. Oba, *Fragrance Journal*, **24** (12), 12 (1996)
10. N. Yamamoto, T. Iwasa, H. Fukui, K. Shuto, Polymer Preprints, Japan, **52** (5), 1107 (2003)
11. N. Yamamoto, K. Shuto, Polymer Preprints, Japan, **53** (1), 2169 (2004)

Chapter 2
Lipid Capsules

Asako Mizoguchi, Hiroshi Oda, Yuka Shimoyama,
Makiko Fujii and Yoshiteru Watanabe

2.1 Introduction

Phospholipids, which are a main constituent of the biomembrane, have both hydrophilic and hydrophobic groups in their molecules and are known to form liposomes when dispersed in water, resembling closed endoplasmic reticulum. Liposomes are classified by their size and form, and there are unilamellar vesicles in which bilayers are connected into one and multi lamellar vesicles with many membranes (figure 1)[1]. By their configuration, liposomes can enclose water-soluble materials in their inner water phase and incorporate fat-soluble materials within their bimolecular membrane, acting as capsules. Liposomes are known to not only stabilize drugs but also improve the absorption of the drugs into living tissues by encapsulating the drugs, and are thus studied in various fields of medicine as carriers for drug delivery systems.

In the field of cosmetics, phospholipids have long been used as an emulsifier and moisturizing agent, and are recently attracting attention as ingredients of the living body origin. Liposomes made of phospholipids are of particular attention because they can encapsulate medicinal ingredient stable over a long period of time, promote percutaneous absorption, and are promising application of nanotechnology in cosmetics.

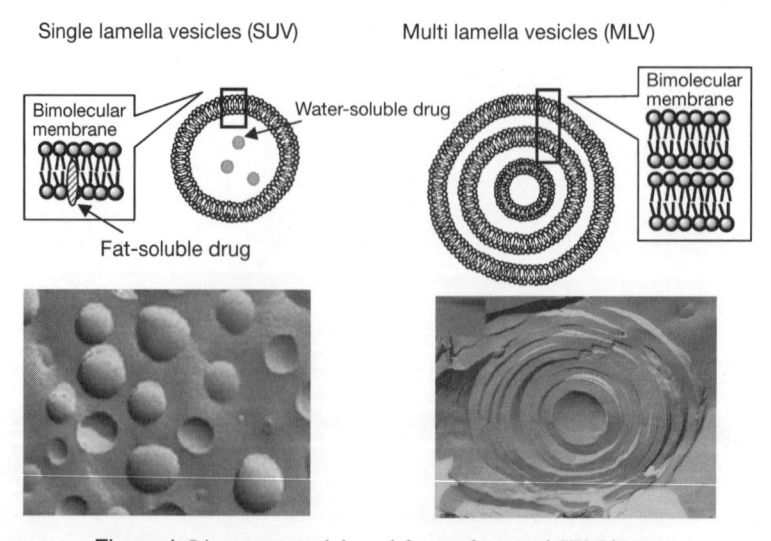

Figure 1 Liposome models and freeze-fractured SEM images

In this chapter, *in vitro* tests of skin penetration of liposomes are reported.

2.2 Penetration of liposomes into the skin

2.2.1 Penetration of water-soluble materials (*in vitro*)

Penetration of liposomes has been studied over a long time and reported[2-4], but its details have not been understood. We tested the skin penetration of unilamellar vesicles by encapsulating a water-soluble fluorescent substance (carboxy fluorescein: CF) as a model compound in liposomes and removing CF outside the liposomes by dialysis. The percutaneous absorption was examined using the open method, which involved applying liposomes that contained CF on the skin and drying.

First, a three-dimensional culture skin model, TESTSKIN (Toyobo), was used to examine the penetration of the CF liposomes. A cross section of the model skin is shown in figure 2 at 8 hours after the CF liposomes (200 nm) were applied on the skin at a concentration of 10 μL/cm^2. Fluorescence was observed in the epidermis, showing that the CF encapsulated in the liposomes penetrated into the skin model.

Next, the skin excised from Yucatan micropig (YMP) was used to examine the penetration of CF into the skin. Specimens were 1) CF solution (control), 2) CF encapsulated in liposomes (average particle size: 100 nm), and 3) CF encapsulated in liposomes (average particle size: 200 nm). The specimens were applied 10 μL/cm^2 on the skin. The amount of CF on the skin was quantified 24 hours after the application. The quantity of CF in the skin was determined by extracting CF from skin sections. The quantification involved excising skin sections, stripping out the skin sections into layers, extracting CF from each stripped layer, and determining the quantities of CF in the first and second layers as the surface and in the third to tenth layers as the stratum corneum. CF was also extracted from epidermis and dermis and quantified.

The quantification showed that a large amount of CF remained on the skin surface when the CF solution was applied. The amount of CF in the stratum corneum, which was used as the index of penetration, was the largest in liposome application (both (2) and (3) > (1) (CF solution)). The amount of CF in the epidermis was the largest in application of liposomes of 100 nm ((2) > (1)), showing significant differences in CF penetration by the use of liposomes (figure 3). Liposomes that contained CF penetrated significantly into the depths of the stratum corneum, and thus the distribution of CF in the epidermis was significantly higher than when CF was applied in solution. The penetration of CF was improved by encapsulating in liposomes. In addition, the particle size of liposomes was shown to affect the distribution in the skin.

To further investigate the penetrability of liposomes, we spread (1) and (2) on slide glasses,

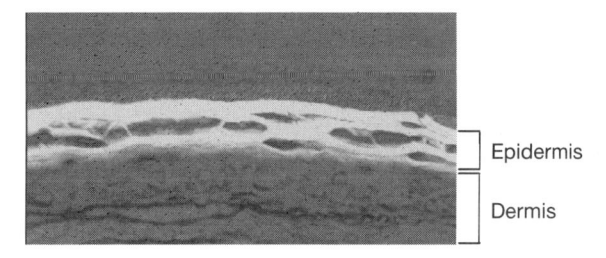

Figure 2 Penetration of CF containing liposomes (200 nm) into the skin
(Cross section of the skin 8 hours after the application of liposomes encapsulating CF on a three-dimensional culture skin model (TESTSKIN (Toyobo)) 10 μL/cm^2)

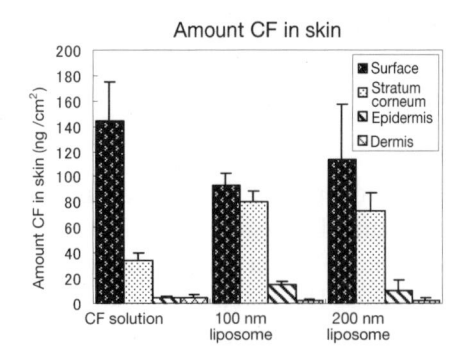

Figure 3 Amount of CF in each skin department by the application of CF in solution and in liposomes (Amount of CF on the surface, in the stratum corneum, in the epidermis and in the dermis)

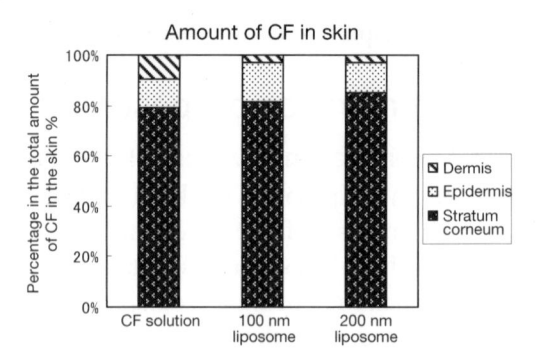

Figure 4 Distribution of CF in each skin department by the application of CF in solution and in liposomes (Percentages in the stratum corneum, epidermis and dermis)

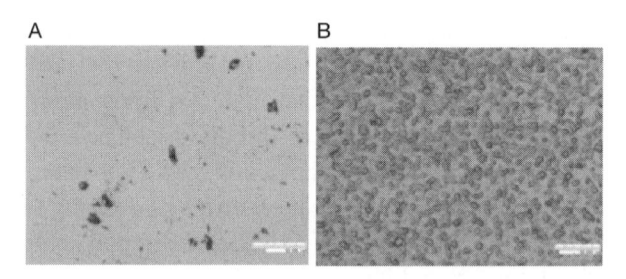

Figure 5 Microscopic images of the glass surfaces
(A) CF solution after drying, (B) CF encapsulating liposomes (100 nm) after drying

dried the surface, and observed the conditions of the surface under a microscope (figure 5). Crystals of CF were observed for the CF solution specimen (A), while no crystals were observed for the CF contained in liposomes (B). CF encapsulated in liposomes did not crystallize likely because the phospholipids that constituted the liposomes kept water molecules and prevented CF from crystallization. This property was a possible reason for the high penetrability of the liposome system into the skin.

2. 2. 2 Examining the water holding capacity and water absorption capacity (*in vivo*)

Phospholipids are known to have moisturizing effects because one phospholipid molecule can bind to 11 water molecules[5,6]. The moisturizing effects of liposomes have also been investigated, and have been reported to be higher than that of phospholipids[7]. In this report, the moisturizing effects of liposomes were compared with those of hyaluronic acid, which is a widely used moisturizing macromolecule.

The water content in the stratum corneum was measured to examine the water holding and absorption capacities. A 1 % liposome solution (1 % phospholipid water solution) and 1 % hyaluronic acid (HA) water solution were applied on the skin of the upper arm. As the control, no treatment was performed. The water content in the stratum corneum was measured at the time of application and after washing the arm. At the time of application, hyaluronic acid showed high water holding and absorption capacities, but the capacities dropped when the arm was washed to the levels of the non-treated section. On the other hand, the liposome solution

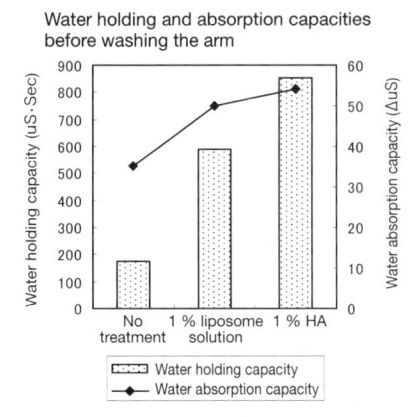

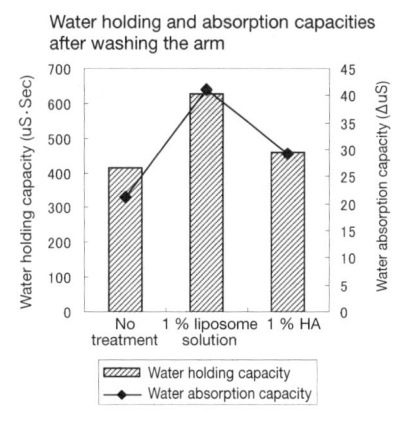

Figure 6 Water holding and absorption capacities of stratum corneum after application of 1 % liposome solution and 1 % hyaluronic acid (HA) water solution
Left: After application before washing, Right: After washing

showed water holding and absorption capacities slightly lower than those of hyaluronic acid but much larger than the control and maintained its effects even after washing the arm. These were likely because hyaluronic acid covered the skin surface showing moisturizing effects but lost the effects when it was washed away. On the other hand, liposomes kept the moisturizing effects even after washing likely because the liposomes penetrated into and stored in the inside of the skin.

2.3 Conclusions

Some functions of liposomes, which are capsules of phospholipids, were described. Phospholipids are of the living body origin and very attractive materials for manufacturing cosmetics as they have emulsifying and moisturizing effects and give a moist feeling when used. Moreover, encapsulating in liposomes stabilizes the encapsulated ingredients and improves transdermal absorption, and liposomes are highly promising tools in the fields of cosmetics and medical supplies.

References

1. H. Terada, T. Yoshimura, "Liposome in life science-an experiment manual" p.1 (1992)
2. M. Mezei et al., *Life Sci.*, **26**, 1473 (1980)
3. K. Egbaria and N. Weiner, *Cosmet. & Toilet.*, **106**, 79 (1991)
4. E. Touitou et al , *J. Pharm. Sci.*, **81** (2), 131 (1991)
5. S. Nojima, J. Sunamoto, K. Inoue, Liposome, p.82 (1988)
6. M. I. Ruocco and G. G. Shipley, *Biochem. Biophys. Acta*, **691**, 309 (1982)
7. T. Harada, *Hifu*, **36** (5), 697 (1994)

Chapter 3
Nanogel Engineering of Cholesterol-bearing Pullulan

Setsuko Yamane and Kazunari Akiyoshi

3. 1 Introduction

Nano-sized polymer hydrogels, which have characteristics of both nanoparticles and hydrogel, are recently called nanogels and are distinguished from conventional microgels. Nanogels are useful as sensing materials by surface modification and as nanocarrier in drug delivery systems to trap drugs and biological molecules (protein or nucleic acids) inside hydrogel nanomatrix. We have synthesized hydrophobized polymers such as polysaccharide or polyamino acid derivatives with various hydrophobic groups, and investigated their association behaviors[1-4]. Especially, cholesterol-bearing pullulan (CHP) formed monodispersive nanogels by self-assembly in water. CHP is already used in cosmetics[5]. This chapter describes various functions of CHP and hybrid materials with CHP.

3. 2 Formation of nanogel of cholesterol-bearing pullulan

Hydrophobized polysaccharide (CHP) was synthesized by introducing cholesteryl group (1–3 wt%) to pullulan (natural polysaccharide of molecular weight of 50 to 100 K) (figure 1). In dilute concentration, CHPs self-associated and formed uniform nanoparticles (ca. 30 nm) by ultrasonic or thermal treatment[3]. The number of association of CHP depended on the molecular weight of polysaccharide. We can estimate sugar density of one nanogel by using molecular weight and size of the nanoparticle. The one nanoparticle contained over 80% water. The rests of them were networks of polysaccharides. Four to five cholesterol groups of CHP associate and form hydrophobic domains in a nanoparticle[6]. The number of associating cholesterol groups is limited due to steric hindrance of the polysaccharide chain attached. The aggregate of CHP regards as a hydrogel nanoparticles (nanogel) in which the association regions of the cholesterol groups (hydrophobic domains) are cross-linking points of hydrogel. The size of nanogel particles, sugar density, hydrophobicity and the distribution of hydrophobic regions can be controlled by changing the structure and the degree of substitution of hydrophobic groups. Similar nanogel formation was also observed when other branching polysaccharides such as mannan and cluster dextrin were used instead of pullulan[7].

Functional CHPs are synthesized by introducing various functional units into a carbohydrate chain of CHP. These polysaccharide derivatives formed nanogels and also formed complex nanogels with other hydrophobized polymers. CHP nanogel derivatives with cellular affinitive carbohydrate chains, such as galactose, in CHP[8,9] and those with polyoxyethylene oxide side chain are useful as a drug delivery system[10]. Thermoresponsive nanogels were prepared by grafting PNIPAM polymer or pluronic polymer to CHP[11]. We also developed photoresponsive

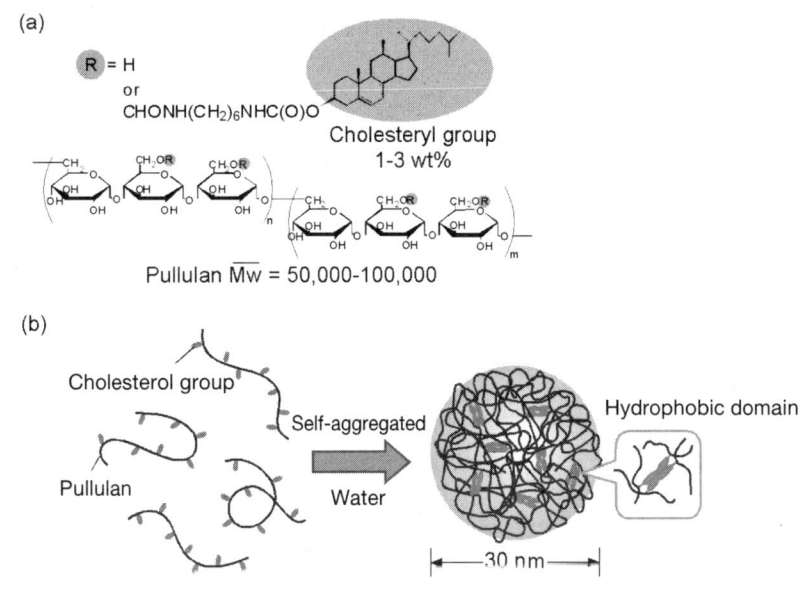

Figure 1 (a) Chemical structure of cholesterol-bearing pullulan,
(b) Schematic diagram of nanogel in water

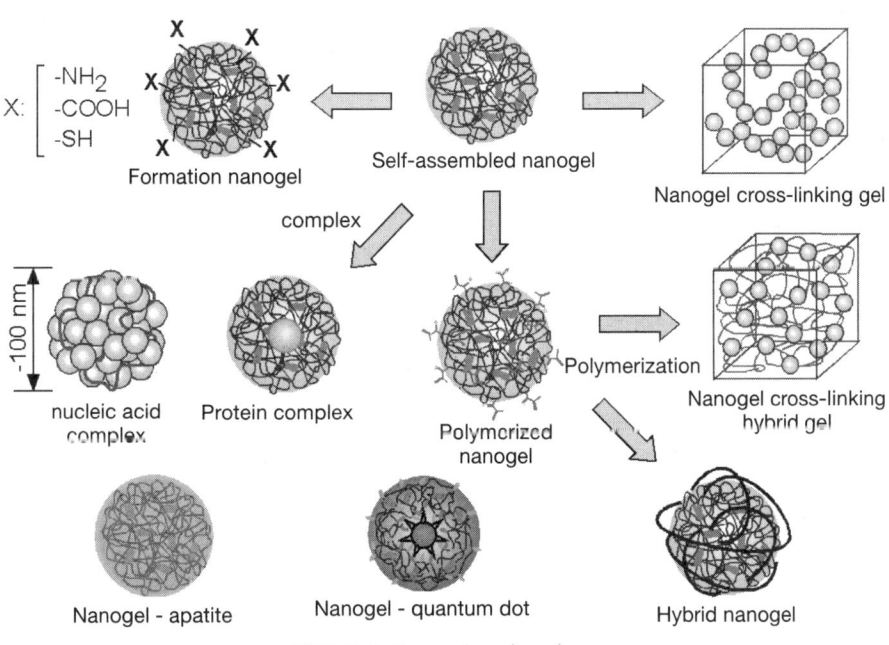

Figure 2 Nanogel engineering

nanogels which introduced spiropyrans group as a photoresponsive association unit instead of cholesterol[12].

3. 3 Formation of CHP nanogel-based hydrogel

3. 3. 1 Hydrogel by association of CHP nanogels

An aqueous dispersion of CHP nanogel was freeze-dried to prepare powder samples, which are relatively easy to suspend in water up to a high concentration. When the concentration of the nanogel increased over ~3 wt%, the nanogels entangled with each other, resulting in sudden increased in viscosity and finally in formation of a transparent macrogel. Pullulan that did not bear hydrophobic groups formed no macrogel even at high concentrations[13, 14]. A transmission electron microscope observation (freeze fracturing method) showed that CHP macrogel had a structure consisting of connected nano-sized spherical particles, which were observed in dilute solutions. The gel showed dynamic visco-elasticity behaviors similar to those of the Maxwell type that consists of a single relaxation process. The dynamics of the gel was controlled by the hydrophobicity of the hydrophobic groups.

3. 3. 2 Nanogel cross-linking hydrogel

Nanogel-based hydrogel was prepared by polymeraizable CHP nanogel with methacryloyl groups (CHPMA). For example, hybrid hydrogel with nanogel cross-linking domains was synthesized by copolymerizing of CHPMA, into which six methacryloyl groups were introduced per 100 simple sugars, and 2-methacryloyloxyethyl phosphorylcholine (MPC), which is a water-soluble polymer[15]. Hybrid nanoparticles were also synthesized by polymerization reaction in a dilute condition, in which CHP acted as seed nanoparticles[16, 17]. Hydrogel was also obtained by the reaction between CHP with acryloyl groups and the thiol derivatives of polyethylene glycol[18]. These techniques enabled novel hydrogel materials with a network structure controlled in nano-order. The nanomatrix gels was excellent scaffold with function of controlled release of cytokine in bone regeneration[19].

3. 4 Functions of CHP nanogel

3. 4. 1 Interaction of nanogels with hydrophobic molecules

CHP nanogels bind various hydrophobic molecules including anticancer agents such as adriamycin and chromophore of neocarzinostatin. The CHP nanogels were useful as a nanocarrier for cancer chemotherapy. Associating domains of cholesterol group in CHP nanogel act as hydrophobic binding sites. Bilirubin (BR) showed induced circular dichromatism under the presence of CHP nanogel, showing that there was chiral binding sites in the nanogel[20]. The hydrogen bond between the hydroxyl group of the carbohydrate chain and BR plays an important role in the induction of asymmetry

3. 4. 2 Interaction of nanogels with cyclodextrin

Cyclodextrin (CD) incorporates hydrophobic molecules into the hydrophobic cavity in water. The complex is water soluble. When β-CD was added to CHP nanogel, the nanogels dissociated due to the complexation of cholesterol groups of CHP with CD. When adamantane carboxylic acid, which is also hydrophobic and strongly interacts with β-CD, was added, CHP nanogels were regenerated because the complex of CD with cholesterol group of CHP was replaced to that of CD with adamantane carboxylic acid[23]. The association and dissociation of hydrophobized polymers can be reversely controlled using the molecular recognition of CD in this manner.

3. 4. 3 Interaction with proteins and molecular chaperone function

Nanogel effectively interacted with proteins since the amphiphilic gel matrix is a nano-meter size with comparable size of protein. CHP nanogel selectively interacted with various proteins depending on size and hydrophobicity of proteins[24–26]. The main driving force of binding of the protein with the nanogels is hydrophobic interaction. The experiments of size exclusion chromatography showed that CHP nanogel was stable even under the presence of an excessive amount of protein and acted as a host macromolecule of a protein (guest). The protein-nanogel complex is colloidally and thermally stable. This is a novel host-guest system that can be used as a new nanocarrier of proteins. In practice, nanogels are effective protein carriers for insulin[27] and for artificial vaccines by trapping with antigen proteins[28-30].

CHP nanogel inhibited the aggregation of proteins by selectively interacting with heat denatured proteins or refolding intermediate proteins like natural molecular chaperon. The incorporated proteins lost their functions within the nanogel but regained them spontaneously when they were released from the nanogel by addition of cyclodextrin[31]. It can also assist the regeneration of proteins form inclusion bodies. The nanogel cross-linked hydrogel, which is copolymerizing CHPMA nanogel and MPC (see section 3. 3. 2), showed a high chaperone-like activity as an artificial chaperone-immobilization column[15] and offered a new technology of protein functional analysis in the post-genomics.

3. 4. 4 Nanogel-calcium phosphate hybrid nanoparticles: nanogel template mineralization

Calcium phosphate-based hybrid nanoparticles are promising candidates for use in drug delivery systems (DDS) due to the excellent biocompatibility of calcium phosphate. We reported nanogel-template mineralization, a novel method for preparing nanohybrid materials of calcium phosphate. CHP nanogel was added to a dilute solution of hydroxylapatite (HAp) of a Ca^{2+} concentration of 0.8 mM, and the pH of the solution was raised from 5 to 8 by evaporating carbon dioxide dissolved in advance. The procedure resulted in dispersion of nano particles of a size of about 30 nm (figure 3)[32]. CHP nanogel acted as nucleus on which calcium phosphate deposited and the nanogel structure contributed to the dispersion of the particles. The nanoparticles formed under the presence of CHP nanogel were suggested to be amorphous calcium phosphate (ACP) from their energy dispersive X-ray spectra and electron diffraction patterns. The ACP nanoparticles were colloidally sable for several months. When a protein-nanogel complex was used as a template, complex nanoparticles containing protein were also obtained. Calcium phosphate-based hybrid nanogels containing drugs are promising candidates for controlled-release DDS due to the improved stability of nanogels upon hybridization with inorganic compounds and also their pH-dependent dissolution properties in calcium phosphate. Nanogel-coated liposomes acted as templates for mineralization of calcium phosphate and ACP-coated liposomes were obatined.

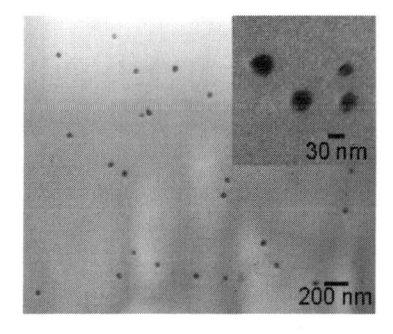

Figure 3 TEM image of nanogel - ACP hybrid nano particles

3.5 Interaction between CHP nanogel and colloids

3.5.1 Interaction with surfactants

CHP nanogel formed macrogel in a semi-dilute solution (-5 wt %) by connecting nanogels. The macrogel dissolved when surfactant such as SDS was added and the gel became a viscous liquid. When SDS was added to CHP nanogel of various concentrations, the viscosity suddenly jumped for dilute CHP nanogel solutions (-1 wt %), and the whole solution gelated for those of high CHP concentrations (-2 wt %). Further addition of the surfactant resulted in drops in viscosity and finally the gel was dissolved[33, 34]. Mix micelles between the cholesterol groups of CHP and the surfactants was formed and these became new networks between polysaccharide chains of CHPs. As a result, the viscosity increased (figure 4 (a)). In the presence of excess amount of the surfactants, almost all the hydrophobic groups of the polysaccharide were dissolved in the surfactant micelles, collapsing the polymer network and decreasing the viscosity. Sol to gel transition and the dynamic viscosity of nanogel suspension were controlled by the addition of surfactants.

3.5.2 Interaction between hydrophobized polymers

Various hydrophobized polymers that are hydrophobically modified water-soluble polymers have been reported. Mixing CHP and other hydrophobized polymer that has different polymer chains results in hybrid nanogel in which the two kinds of polymers are associated in nanometer scale by the self-assembly of hydrophobic groups linked to two hydrophobized polymer in water. For example, thermoresponsive nanogels were prepared by mixing of cholesterol-bearing pullulan (CHP) and hydrophobized poly-N-isopropylacrylamide (a copolymer of N-isopropylacrylamide (NIPAM) and N-[4-(1-pyrenyl)butyl]-N-n-octadecylacrylamide) (PNIPAM-C18Py)). After ultrasonication of a mixture of CHP and PNIPAM-C18Py (5:1 by weight) at 25 °C, monodisperse nanoparticles (Dh=45 nm) were obtained, consisting of self-assembly of the two polymers associated via their hydrophobic moieties. Evidence from fluorescence and

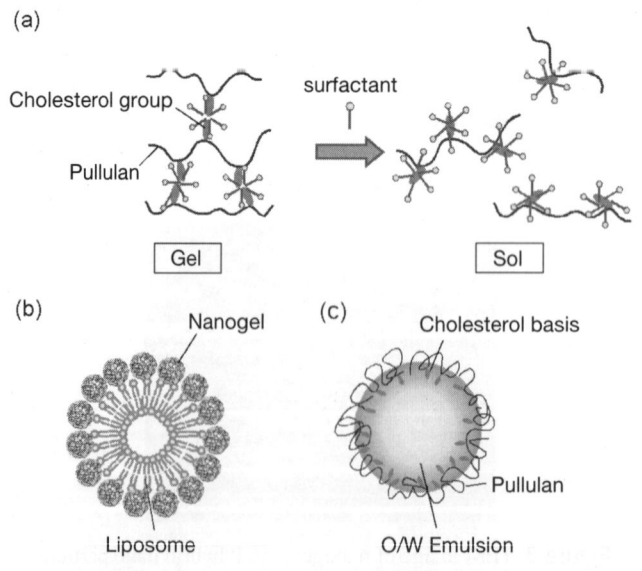

Figure 4 (a) Interaction between CHP nanogel and surfactant, (b) Nanogel-coated liposome, (c) CHP-coated O/W emulsion

dynamic light scattering demonstrated that, above 32 °C, the lower critical solution temperature (LCST) of PNIPAM-C18Py, the colloidal mixed nanoparticles increase in diameter (from 47 to 160 nm), but no macroscopic aggregation could be detected. This phenomenon was thermoreversible: upon cooling the particles recovered their original diameter[35]. This mixing of two polymers via association of hydrophobic groups represents a new preparation method of stable functional nanogels.

3. 5. 3 Interaction with liposomes
CHP nanogels interacted with liposomes and adsorbed on the surface of the liposomes without disturbing but enhancing the structural stability of the liposome membrane (figure 4 (b)). Under a confocal laser fluorescent microscope[36], it was observed visually that fluorescent-labeled CHP nanogel existed only at the most outer layer of a giant liposome. The adsorption behavior of the liposome can be analyzed using the adsorption isotherm equation of Langmuir, which showed that CHP nanogel was adsorbed as a single layer on the liposomal surface[37]. CHP nanogel interacted efficiently with a black film[38] and a monolayer film[39]. CHP nanogel also coated and stabilized plant protoplasts[40].

3. 5. 4 Stabilization of O/W emulsions
O/W emulsions are thermodynamically unstable colloidal particles and various surfactants have been used for their stabilization. Water-soluble polymers adsorbed onto the surface of emulsions and stabilized the emulsions by steric effect. A stable emulsion was prepared by adding CHP nanogel to an O/W emulsion consisting of triglyceride[41]. The particle size of the resultant emulsion changed depending on the concentration of CHP nanogel added. An addition of CHP nanogel at a weight ratio of at least 0.02 resulted in the formation of a relatively stable emulsion. The cholesterol group of the CHP nanogel served as anchors of polymer chains to facilitate efficient coating on the emulsion surface, resulting in a stable emulsion (figure 4 (c)). On the other hand, under the presence of pullulan with non-modified cholesterol groups, stable emulsions were not formed. CHP-coated O/W triglyceride emulsions were stable even in the blood and were found to be an effective drug delivery system[42, 43].

Hererences

1. K. Akiyoshi, *Expected Material for the Future*, **2**, 36 (2004)
2. K. Akiyoshi, *Kagaku (Chemistry)*, **60**, 29 (2005)
3. K. Akiyoshi, S. Deguchi, N. Moriguchi, S. Yamaguchi and J. Sunamoto, *Macromolecules*, **26**, 3062 (1993)
4. K. Akiyoshi, A. Ueminami, S. Kurumada and Y. Nomura, *Macromolecules*, **33**, 6752 (2000)
5. K. Inomota, K. Shimada, A. Hayashi, K. Akiyoshi and J. Sunamoto, *FRAGRANCE JOURNAL*, **7**, 74 (2002)
6. K. Akiyoshi, S. Deguchi, T. Tajima, T. Nishikawa and J. Sunamoto, *Macromolecules*, **30**, 857 (1997)
7. E. Akiyama, N. Morimoto, P. Kujawa, Y. Ozawa, F. M. Winnik and K. Akiyoshi, *Biomacromolecules*, **8**, 2366 (2007)
8. I. Taniguchi, K. Akiyoshi and J. Sunamoto, *Macromol. Chem, Phys.*, **200**, 1386 (1999)
9. I. Taniguchi, K. Akiyoshi and J. Sunamoto, *Macromol. Chem, Phys.*, **200**, 1804 (2004)
10. K. Akiyoshi and J. Sunamoto, *Supramolecular Science*, **3**, 157 (1996)
11. S. Deguchi, K. Akiyoshi and J. Sunamoto, *Macromol. Rapid Commun.*, **15**, 705 (1994)

12. T. Hirakura, Y. Nomura, Y. Aoyama and K. Akiyoshi, *Biomacromolecules*, **5**, 1804 (2004)
13. K. Akiyoshi, K. Kuroda and J. Sunamoto, *Kobunshi Ronbunshu*, **55**, 781 (1998)
14. K. Kuroda, K. Fujimoto, J. Sunamoto and K. Akiyoshi, *Langmuir*, **18**, 3780-3786 (2002)
15. N. Morimoto, T. Endo, Y. Iwasaki and K. Akiyoshi, *Biomacromolecules*, **6**, 1829 (2005)
16. N. Morimoto, T. Endo, M. Ohtomo, Y. Iwasaki and K. Akiyoshi, *Macromol. Biosci.*, **5**, 710 (2005)
17. N. Morimoto, F.M. Winnik and K. Akiyoshi, *Langmuir*, **23**,217 (2007)
18. U. Hasegawa and K. Akiyoshi, *Cell Technology*, **26**, 679 (2007)
19. N. Kato, U. Hasegawa, N. Morimoto, Y. Saita, K. Nakashima, Y. Ezura, H. Kurosawa, K. Akiyoshi and M. Noda, *J. Cell. Biochem.*, **101**, 1063 (2007)
20. K. Akiyoshi, S. Deguchi, H. Tajima, T. Nishikawa and J. Sunamoto, *Proc. Japan Acad.*, **71**, (B), 15 (1995)
21. K. Akiyoshi, I. Taniguchi, H. Fukui and J. Sunamoto, *Eur. J. Pharm. Biopharm.*, **42**, 286 (1996)
22. I. Taniguchi, M. Fujiwara, K. Akiyoshi and J. Sunamoto, *Bull. Chem. Soc. Jpn.*, **71**, 2681 (1998)
23. K. Akiyoshi, Y. Sasaki, K. Kuroda and J. Sunamoto, *Chem. Lett.*, **27**, 93 (1998)
24. K. Akiyoshi, Y. Sasaki and J. Sunamoto, *Bioconjug. Chem.*, **10**, 321 (1999)
25. Y. Nomura, M. Ikeda, N. Yamaguchi, Y. Aoyama and K. Akiyoshi, *FEBS Lett.*, **553**, 271 (2003)
26. Y. Nomura, Y. Sasaki, M. Takagi, T. Narita, Y. Aoyama and K. Akiyoshi, *Biomacromolecules*, **6**, 447 (2005)
27. K. Akiyoshi, S. Kobayashi, S. shichibe, D. Mix, M. Baudys, S. W. Kim and J. Sunamoto, *J. Control. Release*, **54**, 313 (1998)
28. X-G. Gu, M. Schmitt, A. Hiasa, Y. Nagata, H. Ikeda, Y. Sasaki, K. Akiyoshi, J. Sunamoto, H. Nakamura, K. Kuribayashi and H. Shiku, *Cancer Res.*, **58**, 3385(1998)
29. Y. Ikuta, N. Katayama, L. Wang, T. Okurawa, Y. Takahashi, M. Schmitt, X-G. Gu, M. Watanabe, K. Akiyoshi, H. Nakamura, K. Kuribayashi, J. Sunamoto and H. Shiku, *Blood*, **99**, 3717 (2002)
30. Kitano S, Kageyama S, Nagata Y, Miyahara Y, Hiasa A, Naota H, Okumura S, Imai H, Shiraishi T, Nasuya M, Nishikawa M, Sunamoto J, Akiyoshi K, Kanematsu T, Scott AM, Murphy R, Hoffiman EW, Oil LJ, Shiku H, *Clin. Cancer Res.*, **12**, 7397 (2006)
31. K. Akiyoshi, *Chemistry Today*, **428**, 30 (2006)
32. A. Sugawara, S. Yamane and K. Akiyoshi, Macromol, *Rapid Commun.*, **27**, 441 (2006)
33. S. Deguchi, K. Akiyoshi, B. Lindman and J. Sunamoto, *Macromol. Symp.*, **109**, 1 (1996)
34. S. Deguchi, K. Kuroda, K. Akiyoshi, B. Lindmana and J. Sunamoto, *Colloids and Surfaces A: Physicochem. Eng. Aspects*, **147**, 203 (1999)
35. K. Akiyoshi, E-C. Kang, S. Kurumada, J Sunamoto, T. Principi and F.M. Winnik, *Macromolecules*, **33**, 3244 (2000)
36. S. Ueda, J. Lee, Y. Nakatani, G. Ourisson and J. Sunamoto, *Chem. Lett.*, **27**,14 (1997)
37. E-C. Kang, K. Akiyoshi and J. Sunamoto, *J. Bioact. Compat. Polym*, **12**, 14 (1997)
38. J. Moellerfeld, W. Prass, H. Ringsdrof, H. Hamazaki and J. Sunamoto, *Biochim. Biophys. Acta.*, **857**, 265 (1986)
39. A. Baszkin, V. Rosilio, G. Albrecht and J. Sunamoto, *J. Colloid. Interface Sci.*, **145**, 502 (1991)
40. Z. Guo, S. Kallus, K. Akiyoshi, J. Sunamoto, *Chem. Lett.*, **24**, 415 (1995)
41. H. Fukui, K. Akiyoshi and J. Sunamoto, *Bull. Chem. Soc., Jpn.*, **69**, 3659, (1996)
42. H. Fukui, K. Akiyoshi, T. Sato and J. Sunamoto, *J. Bioact. Compat. Polym.*, **8**, 305 (1993)
43. H. Fukui, K. Akiyoshi and J. Sunamoto, *J. Biomater. Sci. Polym. Ed.*, **7**, 829 (1996)

Chapter 4
Technology for Using Cyclodextrins in the Field of Cosmetics

Keiji Terao

4.1 Introduction

Cyclodextrins (CD) was first discovered in 1891, and its industrial production in Japan started in the 1970's. However, wide industrial use of cyclodextrins started only very recently. After 1980, investigations on the use of cyclodextrins in various fields have been increasingly active in Japan and other countries. The numbers of patents and papers on cyclodextrins increased remarkably from 1980 to 2000. The same tendency was observed also in the field of cosmetics, and early applications have been summarized in several reviews and references[1-3]. Although the cosmetics industry in Japan once led the world in development of cyclodextrin applications, cyclodextrins is still not a common ingredient for cosmetics unlike in Europe and the US. The biggest reason is because still many researchers think that only β-cyclodextrin is economically feasible.

In the 1980's, the progress of biotechnology led to the development of a high selective and economic manufacturing process for three kinds of α-cyclodextrin, β-cyclodextrin and γ-cyclodextrin[4]. All these cyclodextrins are produced today by Wacker Chemical Corporation in the US using an economic manufacturing method. The price of γ-cyclodextrin, which was very expensive, was reduced to 1/100, sharply increasing the use in the field of cosmetics. Although α-cyclodextrin and γ-cyclodextrin have been little investigated for applications, their water solubilities at room temperature are 8 and 13 times larger than that of β-cyclodextrin, respectively, suggesting they can be highly useful in cosmetics.

Today, hydroxypropylated cyclodextrins, in which the hydroxyl group is chemically modified to improve water solubility, and maltoxylated cyclodextrins (also called a branch-type CD), which are enzymatically modified, are industrially produced and used in cosmetics[5]. This chapter gives an outline of the trend of using cyclodextrins for cosmetics in Japan, Europe and the US.

4.2 What is cyclodextrin?

Cyclodextrins are cyclic oligosaccharides in which D-(+)-glucopyranose units are linked by α (1→4) glycosidic bonds. There are three common cyclodextrins, which differ in the size of the ring and solubility: α-cyclodextrin, β-cyclodextrin and γ-cyclodextrin. They consist of 6, 7 and 8 glucose units, respectively, and have solubility to water of 14, 2 and 23 g/100mL. The solubility of β-cyclodextrin is extremely low (figure 1).

Cyclodextrins are natural compounds produced from starch, which is a polymer of glucose. In nature, starch is enzymatically decomposed by specific microorganisms into cyclodextrins

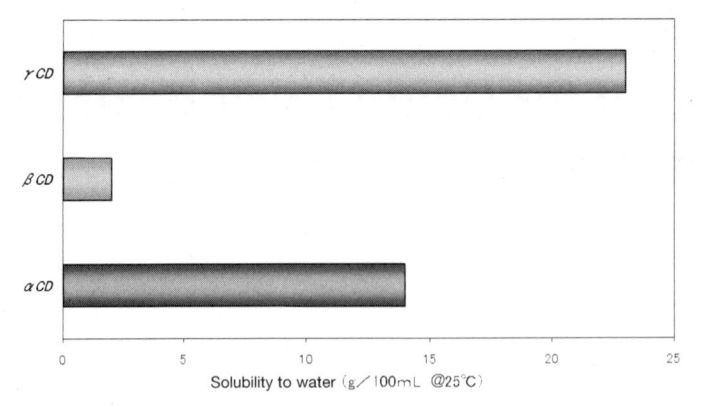

Figure 1 Solubility of cyclodextrins to water (g/100mL)

and saccharide chains. Highly active bacteria that can selectively produce oligosaccharide rings have been selected by combining modern and refined techniques[4]. The enzyme for producing oligosaccharide rings has a formal name of 1,4-α-D-glucan4-α-D-(1,4-α-D-glucano)-transferase (cyclizing) [12. 4. 1. 19], and it is also called cyclodextrin glucosyltransferase or merely CDTase. A research group in Freigburg succeeded in crystallizing the enzyme in 1991 and reported the crystal structure of CGTase[6].

Cyclodextrins have a toroid molecular structure, and the inner wall has a high electron density due to the arrangement of hydroxyl groups and is hydrophobic. On the other hand, the outer wall is hydrophilic. The secondary hydroxyl groups are located at the larger opening of the toroid. The primary hydroxyl groups, which are located at the smaller opening, are rotatable in a manner that they can change the size the opening. These hydroxyl groups determine the water solubility of cyclodextrin. Cyclodextrins are slightly sweet white powder, are highly more stable than starch and sucrose, and can be stored over years under appropriate conditions.

4. 3 Purposes of using cyclodextrins in cosmetics

Because the inside is hydrophobic and the outer surface is hydrophilic, cyclodextrin can take in various hydrophobes in its cavity and forms inclusion complexes. For example, α-cyclodextrin forms inclusion complexes with hydrocarbon chains and gases such as carbon dioxide and propane. β-cyclodextrin forms inclusion complexes with monocyclic aromatic complexes and low molecular weight terpenoids. γ-cyclodextrin can form inclusion complexes with Vitamin D_3 and large cyclic complexes of a large three-dimensional size[7]. Even when the entire guest molecule is not included in cyclodextrin, the molecule is possibly stabilized and solubilized by the complexation of a part of guest molecule with cyclodextrins.

Irrespective of whether the guest molecule is in gaseous, liquid or solid state, the resulted inclusion complexes is always solid power. Stable powder is much easier to handle than unstable volatile substances, such as aroma oil. Converting volatile and unstable active substances into inclusion complexes facilitates adding precise amounts of the substances and storing them stably until they are used for cosmetics production.

Today, cyclodextrins are used in cosmetics for the following purposes:
- Stabilization (of active ingredients, emulsions, and volatile substances against oxidation, photo reaction, and thermal decomposition)
- Reduction (of unpleasant smell, irritation to mucous membranes, irritation to the skin, surfactant content, *etc.*)

- Controlled release (of perfumes, active plant extracts, *etc.*)
- Improvement of the bioavailability (for vitamins, medicines, active plant extracts, *etc.*)
- Solubilizing (hydrophobic materials in water phase systems)

Because the inclusion in cyclodextrins is reversible, the included active ingredients can be released gradually or at a certain condition. Thus, active ingredients can be released in a controlled manner. Recently manufactured cosmetics using this property of cyclodextrins are exemplified below:
- Hair care products (permanent wave, shampoo, hair conditioner, hair lotion *etc.*)
- Skin care products (ointment, powder cosmetics, bath salt, medical cream, pack cosmetics (for washing), pack cosmetics (for moisturizing), conditioner cream, self-turning cream, lotion *etc.*)
- Products for nails (enamel remover)

4. 4 Stabilization effect

Cosmetics are stored at homes possibly over a long period under various bad conditions such as temperature changes, oxidation and UV irradiation and are prone to deterioration. Therefore, stabilization of products is very important. Cyclodextrins are effective not only for stabilizing the ingredients included but also for stabilizing products.

Use of cyclodextrins for stabilizing emulsions is effective for reducing the amount of surfactants or totally eliminating the use of surfactants. However, cyclodextrins cannot be used for all emulsion systems because the resultant behaviors vary by the ingredients contained.

On the other hand, the use of cyclodextrins is always effective for stabilizing volatile substances and other unstable active ingredients. Inclusion complexes are highly more stable than their guest active substances against oxidation and UV. Stable formulations can be obtained from ingredients of mutually bad affinity by forming inclusion complexes with one of the ingredients. For example, the affinity between vitamin C and anthocyanin is bad, but they can be mixed in a product by complexing anthocyanin in γ-cyclodextrin[8].

4. 4. 1 Vegetable oil containing unsaturated fatty acid triglyceride

Vegetable oils of a higher unsaturated fatty acid triglyceride content, such as evening primrose oil and borage oil, are prone to oxidation, and thus their storage stability is low. Regiert *et al.* stabilized these vegetable oils by using γ-cyclodextrin[9] (figure 2). Stable emulsions can also be formed by mixing γ-cyclodextrin : vegetable oil: water at 1:2:7 and using no emulsifying agent.

4. 4. 2 Vitamin A (Retinol)

Retinol is known as an effective anti-aging material and mitigates wrinkles and scars on the skin. However, retinol is even more unstable against oxidation than unsaturated fatty acid and decomposes easily at the room temperature, disturbing the use in cosmetics. There are several articles that contain retinol, which is stabilized by esterification with acetic acid or palmitic acid. However, the anti-aging effect of the esterified retinol is only one-tenth of that of intact retinol. Microencapsulation using liposomes and combined use with an antioxidant, such as butyl hydroxyanisole (BHA) and butylated hydroxytoluene (BHT) have been tested, but the resultant stability has not been satisfactory. Regiert et al. succeeded in full stabilization of retinol by using γ-cyclodextrin as in stabilization of unsaturated fatty acid[10, 11] (figure 3).

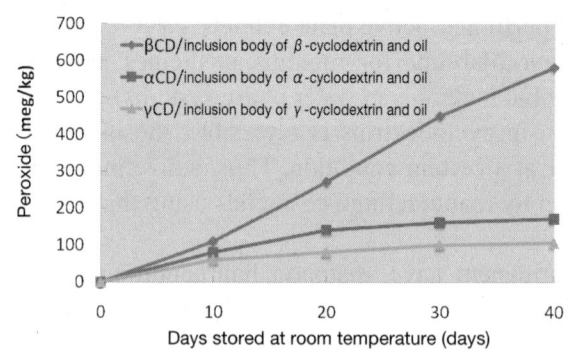

Figure 2 Storage stability of an inclusion compound of cyclodextrin and evening primrose oil under sunlight exposure

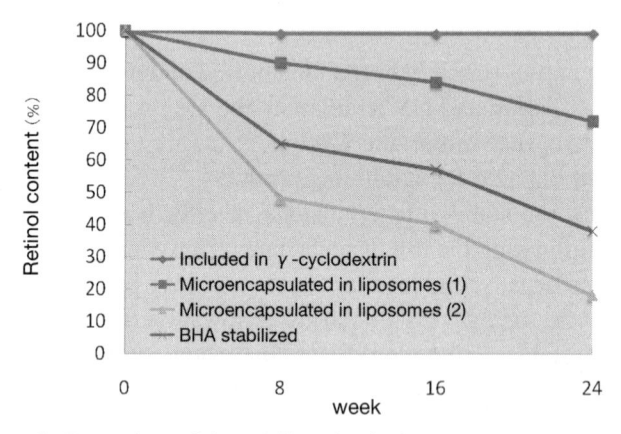

Figure 3 Comparison of the stability of retinol cream at room temperature

4. 4. 3 Phthalimide caproic acid peroxide (PIOC)

PIOC is an active ingredient for whitening, sterilizing and deodorizing the skin developed as a personal care product by Italian Ausimont Corporation and is soft to the skin. However, its storage stability is low disturbing development of new products. Recently, a product with increased stability was developed by including PIOC in β-cyclodextrin, and is sold with a brand name of EURECO®HCW7[12].

4. 4. 4 Linoleic acid (Vitamin F)

Linoleic acid is an essential fatty acid and is attracting attention these days for its ability of restraining melanin production (reducing stains and whitening) has been confirmed. However, linoleic acid is an unsaturated fatty acid, unstable against oxidation, and prone to deterioration. As described above, γ-cyclodextrin is highly effective for stabilizing triglyceride, but α-cyclodextrins has been found most effective for stabilizing free unsaturated fatty acids. The results of a stability test of linoleic acid and an inclusion complex of α-cyclodextrin and linoleic acid, which were mixed in various kinds of color cosmetics, are shown in Figure 4[13]. The inclusion complex of α-cyclodextrin and linoleic acid won the best personal care material prize in Europe in 2006.

Chapter 4 Technology for Using Cyclodextrins in the Field of Cosmetics 143

Stability of color cosmetics containing 1% linoleic acid 1% (45°C)

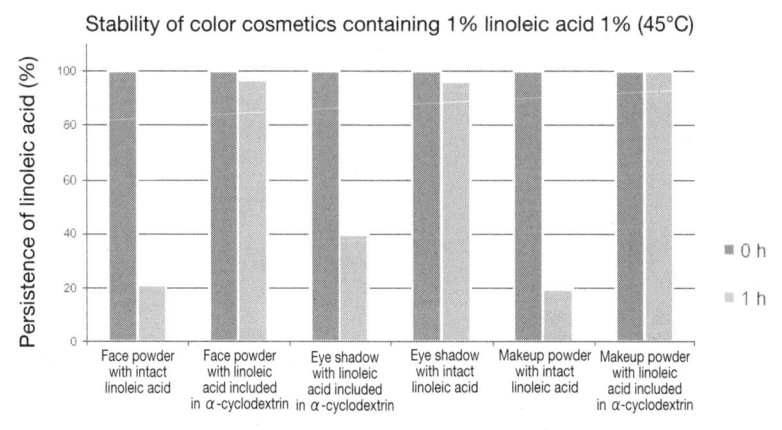

Figure 4 Stabilization of linoleic acid against UV-A/B by α-cyclodextrin

4. 5 Reduction effect

4. 5. 1 Reducing unpleasant smell (deodorizing effect)

Because cyclodextrins easily form inclusion complex with various smelling molecules, empty cyclodextrins are deodorants by themselves. For example, a self-tanning cream, which colors the skin slightly brown, has been popular among young women in Europe and the US. The cream contains γ-cyclodextrin for neutralizing the smell that accompanies the use. "Odor-eliminating powder", which is effective against the odor of foot, which is attributable to lipids, also uses the deodorizing effect of cyclodextrins. An inclusion compound of β-cyclodextrin and aroma molecule can be added to the "odor-eliminating powder" to produce a deodorant that gives fragrance long. A formulation of the odor-eliminating powder is shown in table 1.

4. 5. 1. 1 Iodine

Nippoh Chemicals Co., Ltd., which is an iodine manufacturer, and Wacker Chemical Corporation jointly developed cyclodextrin-iodine inclusion complex (CD-I) as a nature antibacterial deodorant. The use of CD-I is studied now in various fields including the field of cosmetics[14, 15]. Inclusion in cyclodextrin stabilizes iodine, which is otherwise sublimating, and keeps its antibacterial and deodorizing effects. The deodorization of ammonia smell by CD-I is shown in figure 5 as an example.

Uses of CD-I, which is the safest antibacterial deodorant, are being investigated such as for feed additives, water treatment agents, and soil improvement agents. For example, cyclodextrin is included in breath care products to deodorize the smell of fish and garlic, and even higher effect can be achieved by using CD-I[16]. CD-I has been also reported to prevent the odor of the body when included in bath salts and deodorant and antiperspiration sprays. CD-I is likely to

Table 1 Formulation of odor-eliminating powder (Example)

Formulation of smell absorption powder	
1) Fine-particle talc	70.0%
2) Starch	20.0
3) Perfume included in β-cyclodextrin	6.0
4) γ CD	4.0
	100.0

Mixing method: The four kinds of powder were mixed until the entire system becomes uniform.

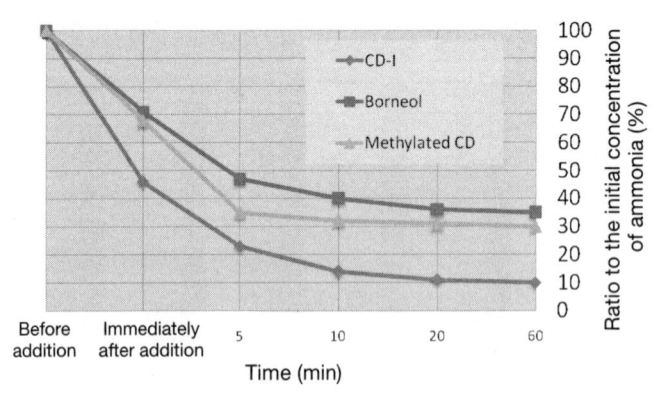

Figure 5 Ammonia deodorizing effects of various deodorants
(Test using gas detecting tube)

control odors produced by lipid decomposing bacteria (bacteriostasis) on the skin by the effects of iodine as well as by the cyclodextrin's original effects of removing odor by complexation.

4.5.2 Reducing irritation
Because rose oil and/or lemon oil is added in shampoos to give fragrance, shampoos may irritate the eyes. Cyclodextrins can suppress the irritation by enclosing the aroma oils. γ-cyclodextrin causes no irritation and is safe to human eyes even the powder itself gets into the eyes.

4.5.2.1 Salicylic acid
Salicylic acid is used in preservatives and medicines for treating skin diseases. Irritation to the skin and stability against sublimation can be improved by the complexation with cyclodextrins, and use of cyclodextrins also facilitates diffusion. The use of the inclusion compounds in skin care products for treating acne is being investigated in Europe and the US.

4.6 Controlling release
Water molecules mediate the complexation of guest molecules into cyclodextrins. Water molecules also mediate the dissociation of the guest molecules from cyclodextrins. The skin surface is always moist due to sweating, a condition at which dissociation is easy to occur. Cyclodextrins can be used to release perfume, which is volatile, and extracts of plants, which are unstable against oxygen and light, only when dissociation is wanted, for example when the person sweats and starts to smell. The controlled release enables the effects of ingredients, such as perfumes and medically active ingredients, to be maintained long.

4.6.1 Menthol
Menthol is known as sweet smelling, refreshing, local analgesic and antiphlogistic compound that is insoluble to water and soluble to organic solvents. It easily forms an inclusion complex with β-cyclodextrin and γ-cyclodextrin. Because the stabilization coefficient with γ-cyclodextrin is smaller than that with β-cyclodextrin, the fixation into γ-cyclodextrin is looser and the release is faster (figure 6). Therefore γ-cyclodextrin and β-cyclodextrin are used depending on purpose. For example, γ-cyclodextrin is used in medical chewing gum for controlling the inflammation of the gums.

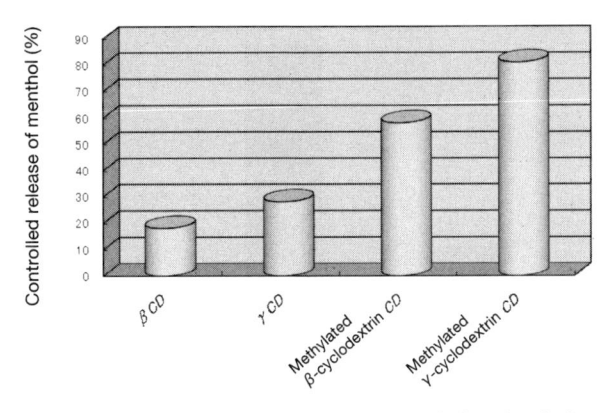

Figure 6 Control of menthol release by various cyclodextrins during storage at a high temperature (100°C, 30 days)

4.6.2 Tea tree oil

Tea tree oil is attracting attention as a natural antibacterial substance. It is used in skin care products for treating acne. Because the oil has a strong peculiar smell, the use has been limited. Inclusion in α-cyclodextrin and β-cyclodextrin has enabled to control release and smell. Effects on prolonging the antibacterial activity and improving the stability have also been shown.

4.7 Improving bioavailability (Vitamin E (tocopherol) and coenzyme Q10, as examples)

Zanotti et al. have made detailed investigation on improvement of activities by including natural tocopherol and δ-tocopherol in cyclodextrins[17]. δ-Tocopherol is known to eliminate free radicals, which triggers a series of skin damage reactions. Including δ-tocopherol in cyclodextrins was found to increase and prolong the radical elimination activity. The results show that cyclodextrins are useful not only for skin care products but also for makeups, which are stored long.

In Japan, the use of coenzyme Q10 (CoQ10) in cosmetics has been permitted since 2004. However, CoQ10 is soluble to lipids, almost insoluble to water, photolytic, unstable in air, and thus difficult to process into cosmetics. Its use is restricted to only 0.03 % the maximum, and methods for effectively using the antioxidation activity of CoQ10 are demanded. Cyclochem Co., Ltd. has examined the effects of including CoQ10 in γ-cyclodextrin on its antioxidation activity using the 1,1-diphenyl-2-picryhydrazyl (DPPH) radical elimination activity test method. Forming inclusion complex with γ-cyclodextrin was found to increase the radical elimination activity by 4.3 times as shown in figure 7[18].

4.8 Conclusion

When industrial production of cyclodextrins started, manufacturers of cyclodextrins concentrated on production and sales of cyclodextrins and left the development of inclusion complexes to those in the application fields. Today, application users increasingly use inclusion complexes developed by manufacturers of cyclodextrins. Cyclodextrin manufacturers have accumulated knowledge on methods for including various kinds of active ingredients to respond to demands

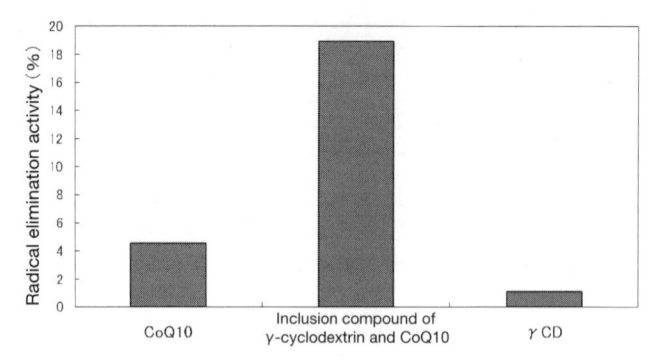

Figure 7 Improved antioxidation effect of CoQ10 by formation of inclusion compound with γ-cyclodextrin

Table 2 CD inclusion compounds for personal care products

Functional ingredients	CD Kind of cyclodextrin	Functional ingredients	CD Kind of cyclodextrin
Retinol	γ CD	Phthalimide caproic acid peroxide	β CD
CoenzymeQ10	γ CD	Squalane	γ CD
α-Lipoic acid	γ CD	Tocopherol (vitamin E)	γ CD
Astaxanthin	γ CD	Parsol 1789	β CD
Menthol	γ CD	Ferulic acid	γ CD
Linoleic acid	α CD	Orange oil	γ CD
Farnesol	β CD	Hinokitiol	γ CD
Tea tree oil	β CD	Citronellal	β CD
Vegetable oil (evening primrose oil etc.)	γ CD	Lavender oil	β CD
Iodine	β CD		

of various application fields. Sometimes, special and original skills and means are needed as well as ordinary mixing and dissolving methods. As mentioned in Introduction, not only β-cyclodextrin but also α- and γ-cyclodextrins are now available from Wacker Chemical Corporation. As technologies for using cyclodextrins have progressed, it is the time to retest items that were once tested using β-cyclodextrin, which was "magic powder" at that time, and ended unsuccessful. Using α-cyclodextrin, γ-cyclodextrin and modified cyclodextrins may produce a successful result. Finally, recently developed cyclodextrin inclusion compounds for personal care products are listed in table 2 as a reference for those engaged in development of cosmetics.

References

1. D. Duchene, D. Wouessidjewe and M-C Poelman, Dermal uses of cyclodextrins and derivatives, in New Trends in Cyclodextrins and Derivatives, D. Duchemes, *ed*, Paris; Edition de Sante, pp 449-481 (1991)
2. C. Vaution, M. Hutin and D. Duchene, The use of cyclodextrins in various industries, in Cyclodextrins and their Industrial Uses, Paris: Edition de Sante, pp 297-350 (1987)
3. J. Szejtli, Cyclodextrins in food, cosmetics and toiletries, *Starch/Staerke*, **34**, 379 (1982)

4. G. Schmid, O. S. Huber and H. J. Eberle, Selective complexing agents for the production of γ-cyclodextrin, Proc. 4th International Symposium Cyclodextrins, 87-92 (1988)
5. SHISEIDO Co., Ltd., Publication of (unexamined) patent applications 3-58906
6. C. Klein and G. Schulz, Structure of Cyclodextrin glucosyl transferase refined at 2A resolution, *J. Mol. Biol.*, **217**, 737-750 (1991)
7. M. Amann and G. Dressnandt, Solving problems with cyclodextrins in cosmetics, *Cosmetics & Toiletries*, **108**, 90-95, November (1993)
8. M. do Carmo Guedes, INF/COL-III, USA, 19-22 April (1998)
9. M. Regiert, T. Wimmer and J.-P. Moldenhauer, Application of γ-cyclodextrin of vegetable oils containing triglycerides of polyunsaturated acids, 8th International Cyclodextrin Symposium, 575-578 (1996)
10. T. Wimmer, M. Regiert, J. P. Moldenhauer, 9th International Cyclodextrin Symposium, Santiago de Compostela, Spain, May 31-June 3 (1998)
11. M. Regiert, J. P. Moldenhauer, DE 19847633
12. U.P. Bianchi, P. Iengo, 10th International Cyclodextrin Symposium, Am Arbor, Michigan USA, May 21-24 (2000)
13. M. Regiert, *Cosmetics and Toiletries magazine*, **121**, No. 4, P.43-50, April (2006)
14. K. Terao, I. Tachi, H. Suzuki M. Kunishima, S. Tani, Proceeding of 28th annual conference of the Society for Antibacterial and Antifungal Agent, Japan, PS36 (2001. 5. 24.)
15. Publication of (unexamined) patent applications, Syowa 51-88625, Syowa 51-100892, Syowa 51-1011123, Syowa 51-101124, Syowa 51-110074, Syowa 51-112538, Syowa 51-112552, Syowa 51-118643, Syowa 51-118859, Syowa 51-140964, Syowa 52-28966, Syowa 52-15806 (TOYO INK MGF Co., Ltd.)
16. K. Terao, D. Nakata, I. Tachi, S. Hagiwara, H. Suzuki, M. Kunishima, S. Tani, (Cyclochem Co., Ltd., Nippoh Chemicals Co., Ltd, Kobe Gakuin Unv.), Proceedings of the 19th Cyclodextrins Symposium, p44-45 (Kyoto 2001)
17. F. Zanotti, I complessi di ciciclodestrina in cosmetic, Cosmetics & Toiletrion, P124-125 (1999)
18. J. Morita, D. Nakata, K. Terao, M. Kunishima, I. Tachi, S. Tani, (Cyclochem Co., Ltd., Kobe Gakuin Unv.), Proceedings of the 23th Cyclodextrins Symposium, P124-125 (Nishinomiya 2005)

Chapter 5
Possibility of Antioxidation Vitamin Derivative Capsules

Shinobu Ito

5. 1 Introduction

Vitamins are classified into two general groups of water-soluble vitamins and lipid-soluble vitamins (table 1). On the other hand, derivatives of vitamins, which are recently produced, are "amphipathic" or soluble to both water and oil. However, ascorbyl palmitate, which is one of amphipathic vitamin derivatives, has been used only as an additive for antioxidizing lipids in food, feed, and cosmetics because its solubility to oil is not as prominent as that to water. A purpose of modifying vitamin derivatives with hydrophilic groups and lipid-base ester is to improve the stability in and solubility to solvents. However, few studies have been conducted on using the amphipathy and surface activity of modified vitamin derivatives compared to those on modification of amino acids and polysaccharides with lipid-base ester. DDS nano capsule for ascorbyl 2-phosphate 6-palmitate (APP), sodium tocopheryl phosphate (TPNa) (figure 1), and ascorbyl tocopheryl maleate (CME) were recently developed by ITO Co., Ltd. These derivatives of vitamins have relatively large surface activities and are accelerating studies on clathrates that use amphipathic vitamin antioxidants.

APP is a water-soluble derivative of ascorbic acid that is little soluble to oil but has a surface activity and can be used as an emulsifier and nanocapsule clathrate. When a water solution

Table 1 Major vitamin C and derivatives registered in the list of ingredients for cosmetics in Japan.

(lipid-soluble derivatives)	(Water-soluble derivatives)
Ascorbyl 6-stearate	Disodium L-ascorbic acid sulfate
Ascorbyl 6-palmitate	Sodium ascorbate
Ascorbyl 2,6-dipalmitate	Magnesium L-ascorbyl 2-phosphate (quasi drug ingredient)
Ascorbyl 2,3,5,6-tetraisopalmitate	Sodium L-ascorbyl 2-phosphate (quasi drug ingredient)
Potassium ascorbyl tocopheryl phosphate	Ascorbic acid 2-glucoside (quasi drug ingredient)

Ascorbyl 2-phosphate 6-palmitate (APP)   Sodium tocopheryl phosphate (TPNa)

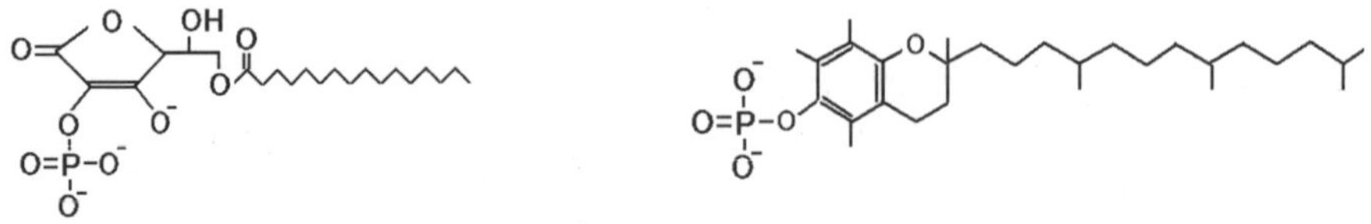

Figure 1

Chapter 5 Possibility of Antioxidation Vitamin Derivative Capsules   149

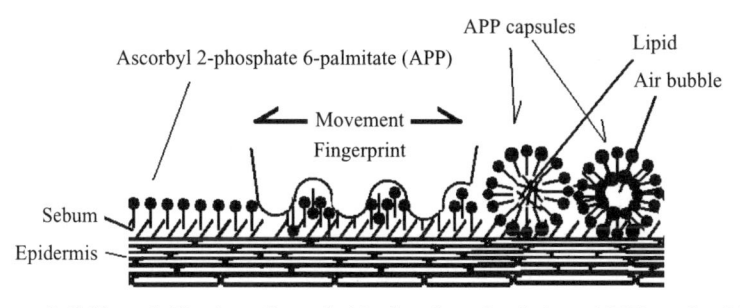

Figure 2 Self-emulsification of ascorbyl 2-phosphate 6-palmitate (APP) on the skin

containing 1% APP is applied on the skin and rubbed in, the transparent solution becomes cloudy and turns into cream-like gel on the skin (self-emulsification). When further rubbed in, it becomes transparent again and agrees to the skin like ordinary creams. This is likely because the APP solution incorporates squalane, cholesterol and other fatty materials present on the skin and forms micelles by self emulsification, trapping in air and forming minute bubbles (figure 2). APP solution becomes "emulsified cream" by mechanical action between the sulcus cutis of the finger and the skin.

5. 2 Derivative of vitamin C with surface activity

Some emulsion products produced using the surface activity of APP are shown in figure 3. Surfactants are used in many articles we daily use, including cosmetics, foods, medical supplies and paint, but chemical surfactants are prone to irritating the skin. The skin is known to have a structure consisting of piled up keratinized cells with layers of lipids and water. Because the cell membrane consists of two layers of phospholipids, amphipathic materials are easy to penetrate into the skin. Amphipathic materials that are toxic to the cell (destroy the cell membrane, *etc.*) have a tendency of skin irritation. Because many consumers believe "chemical surfactants are irritating", they have a trend to use natural surfactants as much as possible. APP is an amino acid system that transforms itself entirely into vitamin C, phosphoric acid and palmitic acid

Figure 3 Examples of articles manufactured using APP as surfactant on a trial basis
From the left top, soap bubbles in a glass (bubbles), soft ice cream, cosmetics (whitening cream), bath salt, cream on the cake, and toothpaste

within the body. In our laboratory, we experimentally produced a skin-care cream, bath salt, toothpaste, shampoo, hair mousse, kitchen detergent, laundry detergent, cake, ice cream and soap bubbles using the surface activity of APP (Some are shown in figure 3). These products are likely to be welcomed by consumers who care about the safety of surfactants.

Many of the products, in which APP was used, received good scores on the feeling of use, such as "soft" and "mild". The reason for being "mild" is not clear, but it was possibly because the active oxygen elimination activity of the vitamin C surfactant may have influenced the taste and tactual sense. Factors that trigger the generation of active oxygen are believed to exist in quantity in daily life, such as in tap water, solar radiation, photocatalysis, medical supplies, pesticides, building materials, paint and adhesives. Vitamin C surfactants improve the feeling of use possibly because their strong oxygen elimination activity reduces active oxygen in the living environment. We have also developed applications other than those shown in figure 3, such as for dissolution aids of medical supplies, paint, photographic paper, fiber, gauze and surgical thread, and started joint studies for some products. Although APP is likely promising in various fields, the use is legally accepted only for cosmetics, and legal registration is needed for commercial use of APP in foods, feeds, and medicines.

5. 3 Derivative of water-soluble vitamin E with an ability to form transparent gel: TPNa

Sodium tocopheryl phosphate (TPNa) is water-soluble white crystals and forms gel when dissolved in water at a concentration of 1 % to 2 %. The gel gives a stretching feeling like polysaccharides, forms a strong water-holding barrier on the skin, and becomes a transparent viscous fluid of a good and non-sticky touch. Furthermore, TPNa is attracting attention for it controls hyperoxidation of lipids, prevents roughness of the skin caused surfactants (SDS), kills acne bacteria, controls the expression of the annexin gene under the exposure to UV-B, mitigates photo aging, and reduces the toxic effects of drugs. In the field of cosmetics in Japan, it is registered as an effective ingredient for preventing roughness of the skin and is used in quasi-drugs category (functional cosmetics). Like APP, TPNa can be used as an emulsion film agent, and it is used for coating capsules, which highly increases the stability of lipids against oxidation because TPNa has a vitamin E structure in itself. It is also effective for controlling natural pigments from fading, such as astaxanthin.

5. 4 Ascorbyl-2-phosphate (AP)

APP is a derivative of ascorbyl-2-phosphate with an additional fatty acid at its sixth place. It is easily converted into ascorbyl-2-phosphate (AP) in the living body by the activity of esterase. APP can be used as a supplier of ascorbyl phosphate, which is widely used today and has been reported to express the majority of the efficacies of AP[1]. Studies on AP are briefly described below[2,3].

AP is more stable than ascorbic acid. AP is hydrolyzed into ascorbic acid by the activity of alkaline phosphatase, which exists in the serum and near the stratum basale epidermidis. Because AP is gradually converted into ascorbic acid in the stratum basale epidermidis, it does not cause sudden increases in the concentration of free ascorbic acid on and in the skin. At high concentrations, free ascorbic acid is toxic because it turns into ascorbate radicals by UV-B. AP controls the cytotoxicity of ultraviolet ray B to human keratinocytes (HaCaT) but has been reported to not control free ascorbic acid[4]. AP has been used as a so-called skin lightning ingredient for preventing stains and freckles caused by sunburn since the 1970's[5], and has been reported to prevent and improve chromatosis such as chloasma[4,5], control disorders caused by UV irradiation, and promote collagen synthesis[7]. AP is produced in stable salts of Mg and Na rather than in a free form. Its Na salt is an anion in water solutions and penetrates into the depth of the skin easily by iontophoresis, and is particularly effective against chloasma[10] and chromatosis[8,9].

The effects of AP on the skin tissue include inhibition of melanin synthesis by inhibition of the tyrosinase activity[10] and those against UV irradiation disorders, such as canceling activated oxygen[11], controlling oxidation of lipids[12] and preventing apoptosis in the skin[11]. Other physiological effects have also been reported: promoting collagen synthesis in skin fibroblast[14], promoting basal membrane formation and cytodifferentiation[15], controlling the shortening of telomere[16], inducing the formation of tissues and organelles[17], activating the synthesis of collagen types I, III, and IV, promoting the construction of extracellular matrics such as sulfated proteoglycan and glycosaminoglycan[18], and preventing damages to DNAs[19]. Recently, it has been reported to improve acnes[20], reduce sebum secretion, and reduce the size of pores[21]. Many applicational studies have also been conducted in the fields of feed, food and medical supplies.

The anti-oxidants like AP, TPNa and Fullerene were effective as an inhibiter of these abnormal behaviors which the limited assembly of stimulant and opioid like methamphetamines was induced. It is interesting that the extreme increase of AA* that was thought iron protein relation was also found in striatum tissue where oxidation is renewed by administration of stimulant drug.

5. 5 High concentration of the ascorbic acid in the cell

APP is characterized by high absorbency into the living body and expression of high ascorbic acid activity. The principal effects of APP are described below. Kato *et al.* compared the incorporation of APP into normal human skin keratinized cells HaCaT with that of AP, which is a conventional derivative of ascorbic acid, by culturing the cells in media containing various concentrations of APP and AP [26]. The method involved transferring a certain number of cultured HaCaT cells to a medium containing 0.1 mM of either AP or APP, collecting the cells at 3, 6, and 18 hours of culture, rinsing and crushing the cells, and measuring the content of ascorbic acid. The measurement showed that the amount of ascorbic acid accumulated in the cells was larger when cultured with APP than with AP and the addition of the hydrophobic group increased the intake markedly. Miwa *et al.* reported in detail that AP and APP were both very effective provitamin C to be applied on the skin[27,28]. They showed in an APP incorporation experiment using the actual human skin tissue, that the incorporation of APP into the skin tissue, particularly into the dermis, was especially high among various derivatives of ascorbic acid. These and other many reports have suggested that APP added to a culture medium at a low concentration is efficiently incorporated into cells and releases ascorbic acid within the cells and shows antioxidation activities.

5. 6 Promotion of collagen synthesis and inhibition of tyrosinase activity by APP

Kato *et al.* reported that APP highly promoted collagen synthesis and caused large accumulation of vitamin C, which inhibits tyrosinase activity, in the cells and thus had a skin lightning effect[20]. They reported that human fibroblast cells cultured in media containing various concentrations of APP or AP ranging 1–10 μM for a period of 72 hours showed significant promotion of collagen synthesis even at low APP concentrations. APP has been reported to inhibit the activity of tyrosinase in a concentration-depending manner in human melanoma, mouse melanoma, and mushroom cells.

5.7 Production of micro and nano capsules (ITO-Nano DDS) using APP

APP is amphipathic and forms a liquid crystal structure when it is used as a surfactant together with lipids, enabling capsules consisting of APP film and containing lipids to be prepared. APP gel in a liquid crystal state has a multilayer structure and thus can incorporate water-soluble materials in its aqueous layer (figure 4).

5.7.1 Multilayered liquid crystal structure of lipid-containing APP capsules of the self-emulsification type (ITO-NanoDDS Capsule)

The APP capsules prepared as described above transform into self-emulsifying gel, with which minute emulsion dispersions can be prepared just by adding water. The resultant suspension of self-emulsified dispersions is anionic because APP forms the capsule film; and the suspension shows electrophoresis when voltage is applied.

5.7.2 Electrophoresis of APP capsules

APP capsules (ITO-Nano DDS capsule) are negatively charged and minute and thus shows electrophoresis when voltage is applied. figure 5 (1) shows microscope image of ITO-Nano DDS capsule which was dissolved in between micro electrodes before voltage application. When voltage was applied, the ITO-Nano DDS capsule particles moved to the direction of the arrow direction away from the (−) pole (2) with the black zone stretching from the left to right. When the current was reversed, the ITO-Nano DDS capsule particles moved back, causing the black zone to shrink (3). The semicircle objects at the right and left ends are the microelectrodes.

Using this phenomenon, it is possible to introduce non-electrically charged materials, such as lipids, and water-soluble materials, such as peptides, carbohydrates and antibodies using iontophoresis. We have already succeeded in producing self-emulsifying and lipid-containing APP capsules that enclosed CoQ_{10}, astaxanthine, VCIP (ascorbyl tetraisopalmitate), and fullerene (table 2). These APP capsules showed improved stability of APP from that of APP

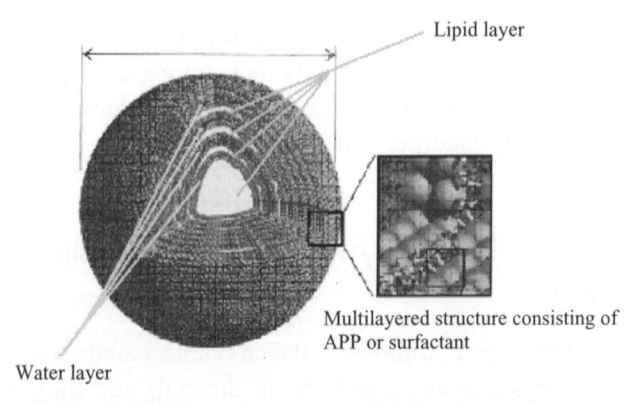

Figure 4 Multilayered structure of APP liquid crystal

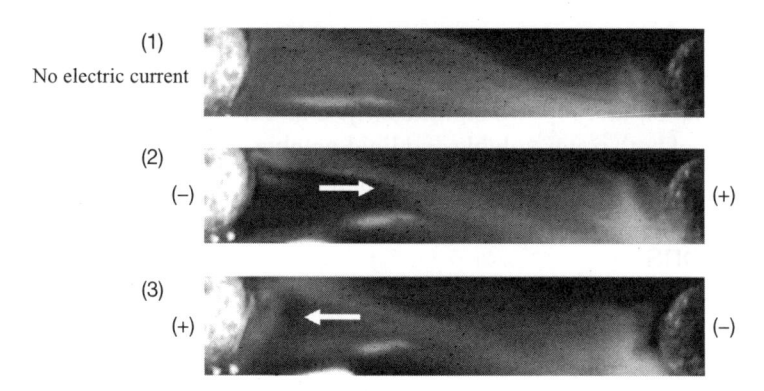

Figure 5 Electrophoresis of APP capsules

Table 2 Kinds and uses of nanomics

Brand name	Main active ingredients	Target active oxygen	Target
Nanomic C	APP	Hydroxy radical	Whitening, acne
Nanomic C + E	APP, Water-soluble vitamine E	Hydroxy radical	Antiaging
		Superoxide	Hair loss, menopause
Nanomic Q	Coenzyme Q_{10}	Superoxide	Dry skin
	Biotin		Atopic dermatitis
Nanomic A	Astaxanthine	Singlet oxygen	Wrinkle
Nanomic F	Fullerene	Various kinds of active oxygen	Antiaging

solutions and improved stability of lipids contained in the capsules. Particularly the APP capsules that contained various kinds of water-soluble and lipid-soluble antioxidants showed especially high improvements in the stability of the APP and the antioxidants. We estimate that the stabilization of both the APP film and antioxidants by the ITO-Nano DDS capsule was attributable to the artificial reproduction of the redox balance system of antioxidants in living cells.

5.8 Synergy with fullerene

Fullerene is frequently described as a "nanocapsule" because it has a shape of a "bird cage". Enclosure of hydrogen and other molecules in fullerene has recently been introduced by the media and enhanced the image of being a nanocapsule. However, cosmetic articles containing fullerene on the market are unrelated to nanotechnology. Today, fullerene is added to cosmetics as an antioxidant because it can stably absorb a large quantity of activated oxygen and has received a nickname of being a "radical sponge". In most cosmetics, fullerene is added as an antioxidant like vitamins C and E, and the use has no relationship with nanotechnology. However, we obtained an interesting result recently when we examined the ITO-Nano DDS capsule properties of fullerene. When ITO-Nano DDS capsule that contained an ascorbic acid derivative at a high concentration was exposed to a radical environment, such as UV, radicals of ascorbic acid were produced. However, when fullerene was added to the nanocapsules, the generation of ascorbic acid radicals was reduced significantly. Our experiment showed that fullerene controlled the generation of ascorbic acid radicals by UV under the presence of ascorbic acid at high concentrations. Fullerene was also found to inhibit the decomposition of carotenoids, such as β carotene and astaxanthine, under a hyperoxidation condition of lipids. We also found that fullerene inhibited the generation of superoxide radicals in the skin by infrared laser irradiation at 1440 nm significantly. Our experiments suggested that fullerene inhibits oxidization of both water-soluble and lipid-soluble antioxidants, the property of which possibly acts synergistically with amphipathic vitamins such as APP.

5.9 Conclusions

Many derivatives of ascorbic acid are used today in various fields, of which APP is of special interest because it also has properties of vitamin C unlike the other derivatives. The amphipathic properties of APP are likely promising tools for developing anti-aging systems with an effective redox balance, in which water-soluble and lipid-soluble redox molecules resonate with each other, and for applications in diverse industrial fields, including cosmetics and medical supplies.

References

1. N. Miwa, Provitamin C as bio-antioxidizer, FRAGRANCE JOURNAL Ltd. (1999)
2. S. Ito, *Science and industry Japanese*, **6**, 52-56 (1999)
3. S. Ito, *Fragrance Journal*, **29**, 68-72 (2001)
4. S. Yasuda et al., *In Vitro Cell Dev Biol Anim.*, **40** (3-4), 71-3 (2004)
5. H. Takashima et al., *Ame. Perfumer and cosmetics*, **86**, 29-36 (1971)
6. Ichikawa, *Jpn. J. Clinical Derm.*, **23**, 372-331 (1969)
7. T. Imai et al. *JAP. J. Pharmacol.*, **17**, 317-324 (1967)
8. Kihara, *Skin Surgery*, **2**, 20-24 (1993)
9. Suzuki, *Nishibi Gaiho*, **20**, 8-29(1998)
10. K. Kameyama et al., *J. Am. A cad Dermatol.*, **34** (1), 29-33 (1996)
11. Kobayashi, et al., *Photochem-photobiol.*, **67**, 6, (1998)
12. S. Kobayashi, et al., *Photobio.*, **64** (1), 224-228 (1996)
13. Kanatate et al., *Cell Mol Biol Res.*, **41** (6), 561-567 (1995)
14. R. Hata et al., *J. Cell Physol.*, **138** (1), 8-16 (1989)
15. Kitagawa et al., *Vitamins*, **66** (8), 431-436 (1992)
16. Furumoto et al., *Life Sci.*, **63** (11), 935-948 (1998)
17. Hata et al., *Vitamins*, **69**, 12 (1995)
18. Skikawa et al., *J. Dermatol Sci.*, **8** (3), 203-207 (1994)
19. Ano et al., *Agric. Biol. Chemi.*, **55** (12), 2967-2970 (1991)
20. Ikeno, Ohmori, *Cosmetic Dermatol.*, **16** (10), 29-33 (2003)
21. Ustugi et al., *Fragrance Journal*, **30**, 54-58 (2002)
22. S. Ito, T. Mori, T. Sawaguchi, *Behav. Pharmacol*, **17**, 691-701 (2006)
23. S. Ito, T. Mori, M. Namiki, T. Suzuki,T.Sawaguchi, *J. Pharmacol Sci.*, **105**,326-333 (2007)
24. S. Ito, T. Mori, T. Sawaguchi, *Behav. Pharmacol*, **19**, 113-119 (2008)
25. S. Ito, T. Mori, H. Kanazawa, T. Sawaguchi, *Toxicology*, **240**, 96-110 (2007)
26. Kato, Tsuzuki, Miwa., *Fragrance Journal*, **32**, 55-60 (2004)
27. Miwa, Unknown works of vitamin C, MARUZEN Co., Ltd. (1992)
28. Miwa ed., Beautiful skin, skin defense and biotechniques", CMC Publishing Co., Ltd. (2003)

Chapter 6
Biosurfactants as Cosmetic Ingredients

Tomohiro Imura, Tokuma Fukuoka, Tomotake Morita,
Masaru Kitakawa, Atsushi Sogabe and Dai Kitamoto

6. 1 Introduction

Recently, along with advancement of studies on the structure and physiology of the skin, products with high functionality and utility are increasingly demanded in the area of cosmetics. Phospholipids of the living body origin are widely used as functional materials. Phospholipids moisturize the skin for they have a lamellar structure and also serve as a drug delivery system (DDS) because they form liposomes (capsules of bimolecular film). As an ingredient of cosmetics, liposomes have superior functions, such as facilitating stable inclusion of active ingredients, increasing penetration and keeping the ingredients long on the skin, have attracted attention in the US, Europe and Japan, and have been used in cosmetics[1].

Materials that form lamellar structures in water, such as phospholipids, are known to have superior moisturizing effects by preventing water loss from the skin. Particularly, natural ceramide, which forms a lamellar structure in the space among horny cells of the epidermis, is a representative ingredient that has superior moisturizing effect and has been used in special care cosmetics. However, its separation and refinement is complicated because it is produced from plants, and high-purity ceramides are as expensive as ten thousand to million yen/kg. Ceramides and ceramide-like chemicals have been produced using chemical techniques, but they are also very expensive because the production involves complicated reactions.

Therefore, "new type" moisturizing ingredients that form a lamellar structure or liposomes (or vesicles), have moisturizing effects and are cheaper than ceramide are demanded. Cosmetic ingredients as a whole are increasingly nature-oriented from the viewpoints of safety and environmental protection, and demands for materials of plant and microbe origins are increasing.

With such a background, National Institute of Advanced Industrial Science and Technology (AIST) has performed studies on "biosurfactants", which are effective ingredients of the microbe origin. A recent joint study with TOYOBO. revealed that bio surfactants are a new moisture ingredient useful for D skin care cosmetics. This report gives an outline of the structure, material properties, biochemical and surface chemical characteristic of biosurfactants and describes their potential as a cosmetic ingredient.

6. 2 What is biosurfactant

Biosurfactant is a general term for interface-active substances of the living body origin in a broad sense. In the field of research and development, it denotes amphipathic lipids extracellularly produced by microorganisms (therefore, lecithin and saponin of the plant origin and bile acid, casein and other materials of the animal origin are not included in the category of biosurfactant)[2].

Recently, this kind of new materials of the microbe origin are increasing attracting attention due to developments of biotechnologies, and resultant drops in cost of the down stream process, increases in the production of plant resources, and the progress of separation technologies.

Biosurfactants were first recognized in the studies on hydrocarbon fermentation (fermentation of petroleum), which started in the 1960's. It was known at that time that amphipathic lipids were produced in the medium when a certain kind of microorganisms was cultured in a medium containing hydrocarbon. Because the amount of production was considerably large, the amiphipathic lipids or biosurfactants started to attract attention.

Most of the first studies on biosurfactants focused on the use as environment-friendly surfactants, which are biodegradable and safe. The focus has shifted these years as nanotechnology and life science approaches revealed that biosurfactants have molecular aggregation (self aggregation) and physiological activities (cell activation) much higher than those of the other surfactants and lipids[3,4]. Today, applications of biosurfactants for high-performance products with added values are actively investigated so as to use these and other specific characteristics of biosurfactants.

6. 3 Structure of biosurfactants

Biosurfactants are classified into the (1) sugar type, (2) amino acid type, (3) organic acid type and (4) polymers type by the structure of the hydrophilic base. Today, several tens of kinds are known. Typical hydrophilic bases are saccharides and amino acids, and typical hydrophobic bases are medium-length and long chains of fatty acids (saturated, unsaturated, branched, hydroxyl, *etc.*)[5].

The industrial use of BS is still limited, but it has various advantages to surfactants of the plant and animal origins. For example, biosurfactants are easy to expand the structure and functions, and their production can be high in production and separation efficiencies and little dependent on raw materials. Application studies are being actively conducted in various technical fields. The structures of representative biosurfactants and the organisms that produced them are shown in figure 1.

Compared with synthetic surfactants, biosurfactants are marked by the following structural characteristics: They have (1) two or more functional groups (hydroxyl groups, carboxyl groups and amino groups) or asymmetrical carbons, (2) complicated structures and large volumes, and (3) structures that are easy to biodegrade. Their functional characteristics include (1) high interface activity at low concentrations, (2) slow and long-lasting activity, (3) high molecular aggregation and liquid crystal formation activities, and (4) various physiological activities.

These properties and functions are attributable to their neatly arranged structures and ingenious combinations. Biosurfactants are synthesized by enzymatic reactions of microorganisms, which involve two-dimensional and three-dimensional positional selections. Thus, the resultant molecules are uniform in shape and orientation, enabling efficient molecular aggregation and orientation at the interface and resulting in high activity at low concentration. This characteristic is a very big advantage in designing the formulations of cosmetics and other products.

The combination of hydrophilic and hydrophobic bases is optimized during the long history of evolution of the organisms. Therefore, biosurfactants can perform "feats" that cannot be done by conventional surfactants.

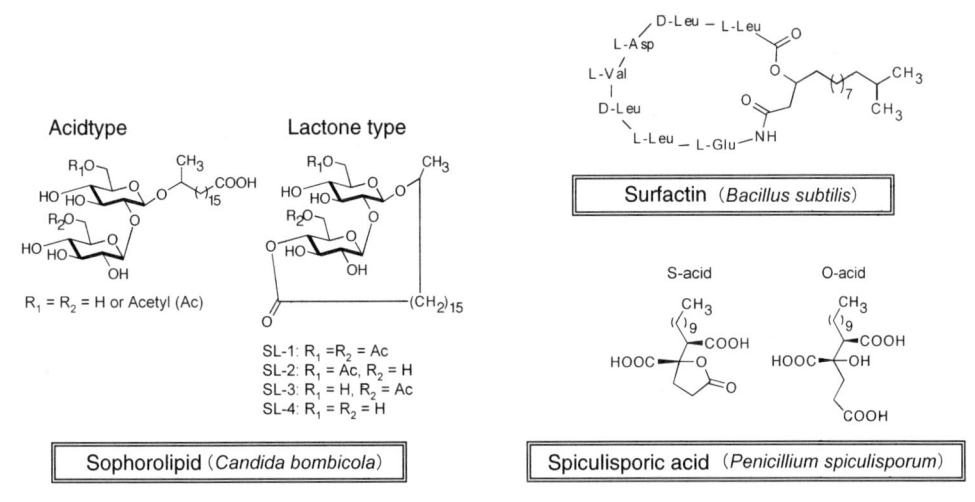

Figure 1 Structures of representative biosurfactants and productive microorganisms

6.4 Examples of biosurfactant use

Some biosurfactants are already in industrial use. Those were achieved by Japanese companies, and Japan is leading other countries in technologies for practical implementation of biosurfactants.

Sophorolipid (figure 1) is saccharide-type biosurfactant produced by yeast, and is produced in mass using a medium containing vegetable oil or glucose. It was first used in cosmetics as a skin care ingredient (moisturizing ingredient)[6]. It is also used in commercialized detergent for tableware washing machines by making use of its superior detergency and biodegradability (SARAYA Co., Ltd.).

Sophorolipid has a washing performance equivalent to that of synthetic surfactants, such as dodecyl-β-D-maltoside, linear alkylbenzen sulfonete and sodium lauryl sulfate (SLS), but produces less foam, making it suitable for machine washing. The sophorolipid content is only about 1 %, and it is an example of being highly effective even at low concentrations[7].

Surfactin (SF figure 1) is an amino acid type biosurfactant produced by *Bacillus subtilis*. It was first discovered as an antibacterial (in 1968), but today, it is used as a skin care ingredient (SHOWA DENKO) by making use of its strong emulsification ability and superior skin-care characteristics. Surfactin easily forms big stick-shaped micelles with the peptide part taking a high-order β-sheet-like structure and shows a superior interface activity[8]. For example, the critical micelle concentration (cmc) is 2.4×10^{-5} M, the surface tension is 27 mN/m and the surface tension of water/*n*-hexadecane is 1 mN/m. In case of sodium salt surfactin, the cmc is 3×10^{-6} M, and it shows high emulsification stability and dispersibility even at very low concentrations. It also has foaming ability and bubble stability. In addition, it is characterized by causing much less skin irritation than conventional amino acid system surfactants[9].

Since it is easy to prepare transparent gels of various oil kinds by using surfactin, it is useful in the formulation of cosmetics. The demand is spreading as new cosmetics material for washing foam, cleansing agent and milky lotions of various kinds.

6.5 Microbe production of biosurfactants

Vegetable oil and fat, such as soybean oil and rapeseed oil, are mainly used as the raw materials.

Usually, there are no large differences in productivity by vegetable source, and a wide range of vegetable sources can be used. Some productive microorganisms can produce biosurfactants from other lipid-based and sugar-based raw materials, such as fatty acids, alcohol, ester and glucose. Widely used microorganisms are ordinary yeast and *Bacillus subtilis* var. Of various kinds of biosurfactants, sugar-type biosurfactants show the highest productivity, are advantageous in terms of both raw materials (as sugar-type biomass can be used) and function (as reactions are specific to the organisms), and have been most investigated[10]. Production of biosurfactant, which is investigated at AIST, is described below.

For mass production of a specific biosurfactant, microorganism that produces the biosurfactant must be efficiently collected first. A common method involves screening from species in the nature using a selective medium. In a screening from microorganisms living in the soil, flowers and trees, which was conducted using a medium that contained only soybean oil, a yeast species living on fruit skin, etc. (*Pseudozyma antarctica*) was found to produce a biosurfactant that has a structure different from known biosurfactants. The new biosurfactant has mannose and erythritol in its hydrophilic group and 2 mol of middle chain fatty acids in its hydrophobic group, and was named mannosylerythritol lipid (MEL, figure 2)[2,3]. Subsequent investigations showed that *Pseudozyma* species other than *Pseudozyma antarctica* can also

MEL-A: $R^1 = R^2$ = Acetyl (Ac)
MEL-B: R^1 = Ac, R^2 = H
MEL-C: R^1 = H, R^2 = Ac
(n = 6 to 10)

Mannosylerythritol lipid
(*Pseudozyma antarctica*)

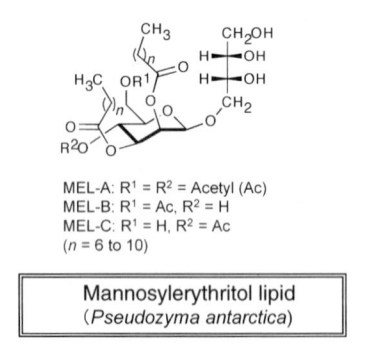

Figure 2 Structure of mannoerythritollipid

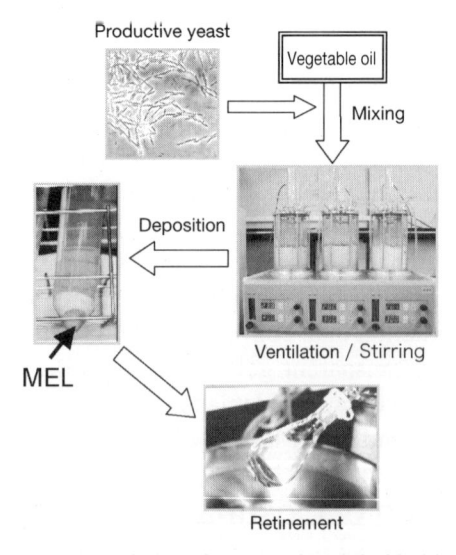

Figure 3 Productive scheme of mannosylerythritol lipid using yeast

produce MEL[11, 12]. Interestingly, the structure of MEL produced differs by the species used, enabling various kinds of MELs, which have mutually different properties and functions, to be produced[13].

When appropriate culture conditions are used, tri[14] and monoester[15] type MEL of different hydrophile-lipophile balances (HLB) can also be produced.

These yeast species can produce MEL using both the fermentative method (culture by adding nutritive salts for growth) and the resting cell method (culture by adding no nutritive salts) because MEL production is not linked to the proliferative phase. For example, MEL is easily produced extracellularly just by mixing the fungus body (resting cells), which was subcultured in glucose or other saccharides, with water-insoluble carbon sources (figure 3).

In a laboratory-scale culture of the resting cell method, stable MEL production can be achieved just by continuously adding soybean oil into the reactor. The final production exceeds 100 g/L of the culture medium[16].

6. 6 Surface chemical characteristics of biosurfactants

MEL shows a high interface activity even at a very small critical micelle concentration (cmc= 2.7×10^{-6} M) although its chain of hydrophobic bases (fatty acid) is short (C_8–C_{12}). The surface tension at the cmc is 27 mN/m, and the interface tension between water and n-Hexadecane is reduced to as low as less than 2 mN/m[17]. MEL has an emulsification ability for soybean oil and hydrocarbons at least twice as large as that of sucrose fatty acid ester and polyoxyethylene sorbitan fatty acid ester, which are synthetic surfactants[16].

MEL shows a very unique self-assembly characteristic and a self-organization structure in water solutions. It forms a oil-drop-like structure of a diameter of 1–2 μm when hydrating a film of MEL-A(1 mM). Transmission electron microscope observation and investigation using the small angle X-ray scattering method confirmed that the structure is in the sponge phase (L_3 phase: A three-dimensional network of randomly connected bimolecular membranes)[18]. On the other hand, MEL-B, which has one less acetyl radical than MEL-A, does not show the sponge phase but forms huge liposomes of diameters of 10–20 μm (lamellar phase, L_α

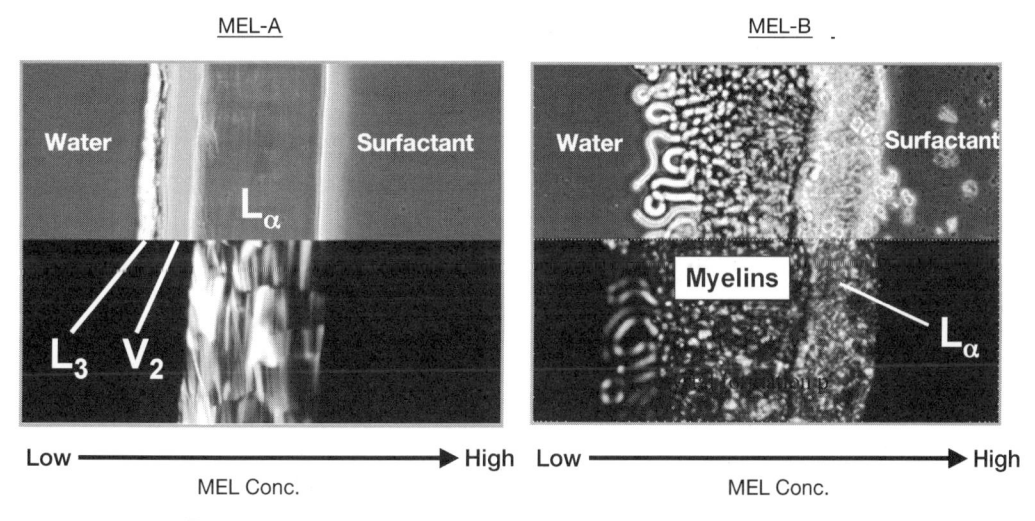

Figure 4 Liquid crystal formation patterns of mannosylerythritol lipid
(the water immersion method)

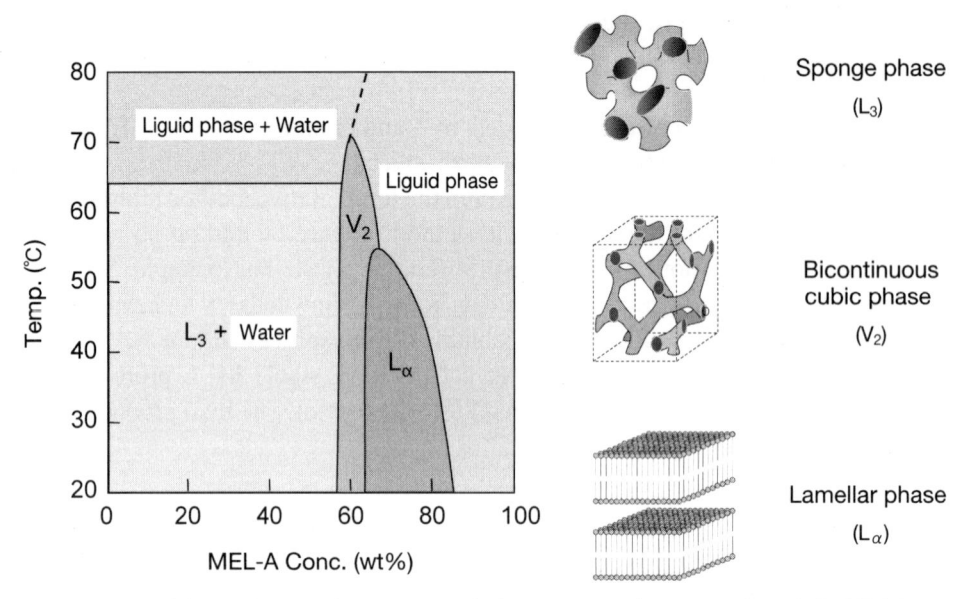

Figure 5 Phase diagram in a water solution system of mannosylerythritol lipid

phase)[19]. The presence of one hydroxyl group on the mannose determines the direction of self-organization and induces a dramatic change of the structure from the sponge phase (random structure) to lamellar phase (structure in order).

General surfactants and amphiphilic lipids self assemble in water solutions and easily form micelles. Only limited substances can forms huge liposomes with a bimolecular film structure. It is widely known that phosphatide (lecithin) forms liposomes, but there are very few glucolipid systems that form vesicles by themselves.

An evaluation of the liquid crystal formation ability of MEL performed using the water immersion method is shown in figure 4[17]. It shows that small differences in molecular structure influence the pattern of liquid crystal formation greatly.

In figure 5, a phase diagram (formation of lyotropic liquid crystals) of MEL-A in a water solution is shown. As shown in the diagram, MEL can be in any liquid crystal phase (sponge phase, bicontinuous cubic phase, lamellar phase, *etc.*) at a wide range of concentration and temperature, and has properties different from those of the existing surfactants and lipids[20].

6.7 Biochemical characteristics of biosurfactants

MEL has superior antibacterial activity and inhibits the growth of Gram negative bacteria, such as *Bacillus subtilis* and *Staphylococcus aureus*, even at low concentrations[2]. The action density is only 1/100 to 1/300 of that of a synthetic surfactant of the saccharide type described above.

MEL also inhibits proliferation and induce differentiation of leukemia cells of various types, including the human acute promyelocytic leukemia cells (HL60 cells), at concentrations of 5–10 μM[2]. MEL promotes extension of neurites and induces differentiation of rat pheochromocytoma derivative cells (PC12), and inhibits proliferation and induces apoptosis of mouse melanoma cells (B16), which cause malignant tumors[5].

As described above, MEL shows superior lamellar formation ability, which is effective for developing high-performance liposomes by combining with phospholipids and other ingredients and drug delivery systems for medicines and cosmetics[21]. Actually, cationic liposomes that contain MEL have found to stabilize DNA and have extremely high efficiency in introducing genes (transfection) into animal cells (HeLa, NIH3T3 *etc.*)[22,23].

Interestingly, these biochemical characteristics of MEL are similar to those of gangliosides (GMI, GM3, *etc.*), which are gylolipids in the cortex of animal cells. Because gangliosides are found in very small amounts and are complicated to separate, refine and synthesize, the use as a functional material is difficult. On the other hand, MEL has similar characteristics and can be produced in mass, and it is a promising tool in the fields of life science and medicine.

6. 8 Application of biosurfactants for skin care cosmetics

As described above, AIST has studied biosurfactants as a project for developing advanced and environment-friendly functional materials and has elaborated their production technologies. It has also analyzed the detailed material properties and functions of biosurfactants and has conducted application studies using their properties. Meanwhile, TOYOBO has investigated functional materials and developed their usage in its new project for developing high-functional biomaterials using its biotechnological knowledge accumulated. In 2006, a joint study started, and they developed mass production methods of biosurfactants and a new type of skin care material using biosurfactant.

The material developed this time is based on MEL. Like natural ceramide used for moisturizing the skin, it has a high moisture retaining effect and can be used for cosmetics and ointments. It has a simple structure consisting of fatty acids and sugar connected with each other and is superior in biocompatibility (safety) and environmental adaptability (biodegradability, living body influence *etc.*).

Natural ceramide shows a superior moisturizing effect and is an important skin care material equaling hyaluronic acid today, but is as expensive as hundred thousands to million yen/kg because it is produced by animals and plants. Therefore, cheap and safe high-functional materials that have equivalent activities have been demanded. The demand can be met by the aforementioned bioprocess that uses specific yeast species and relatively cheap vegetable oil.

The unique structure of MEL (containing sugar, sugar alcohol and fatty acid at the same time) was noticed when investigating its uses in cosmetics. A molecular modeling evaluation suggested that its structure was similar to that of natural ceramide, which is an intercellular lipid. Because there were few efficient and precise techniques for evaluating skin moisturizing effect, it was difficult to verify the effect. Utilization of the three-dimensional culture skin model (Test Skin) of TOYOBO accelerated the verification of the superior humidity retention effect of MEL, and MEL was shown to be applicable to cosmetics and ointments.

Some results of the evaluation using Test Skin are shown in figure 6. First, "a rough skin condition" was induced on cultured skin cells (two layers of epidermic cells: keratinization cells on the upper layer and fibroblasts in the lower layer) by applying a surfactant, such as SDS. Then, a test ingredient was added on the upper layer. The survival rate of the cells was measured, and the recovery rate of the cells from the rough skin state was determined. For example, in olive oil, almost all cells perished, and recovery was not seen. In natural ceramide (an animal origin), a superior recovery rate exceeding 90 % was seen. In MEL, the cell survival was equivalent to that of ceramide, showing the superior roughness improvement effect of MEL.

Based on the results of investigations in the past, the mechanisms of the moisture retaining effect of MEL are likely to involve (1) easy incorporation of MEL into the intercellular spaces

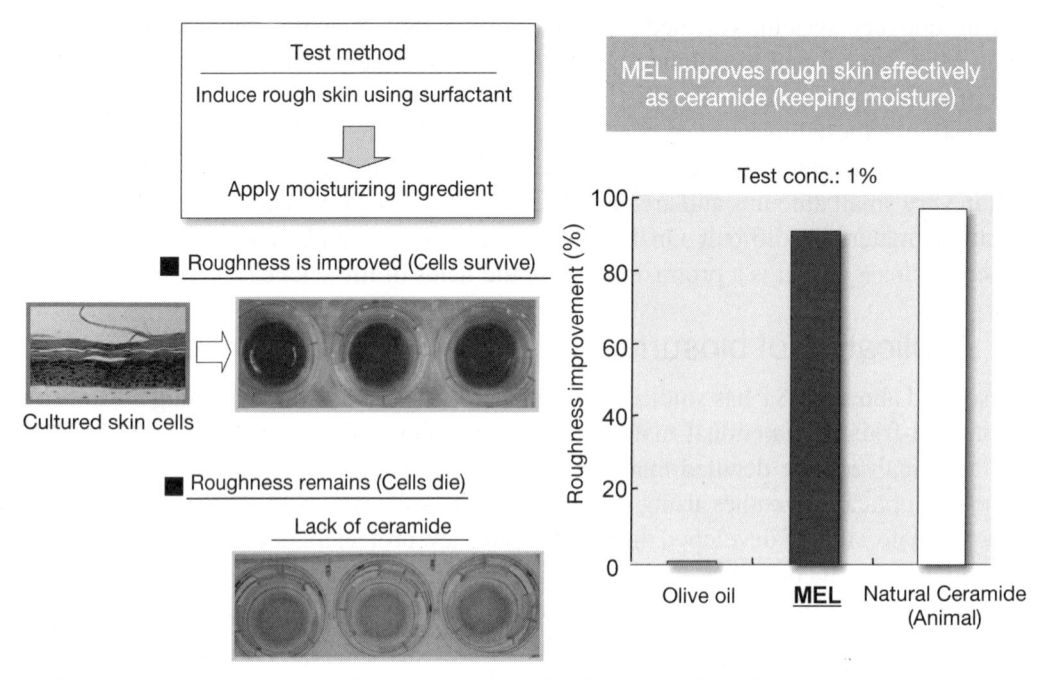

Figure 6 Moisturizing effect of mannosylerythritol lipid in a three-dimensional culture skin model

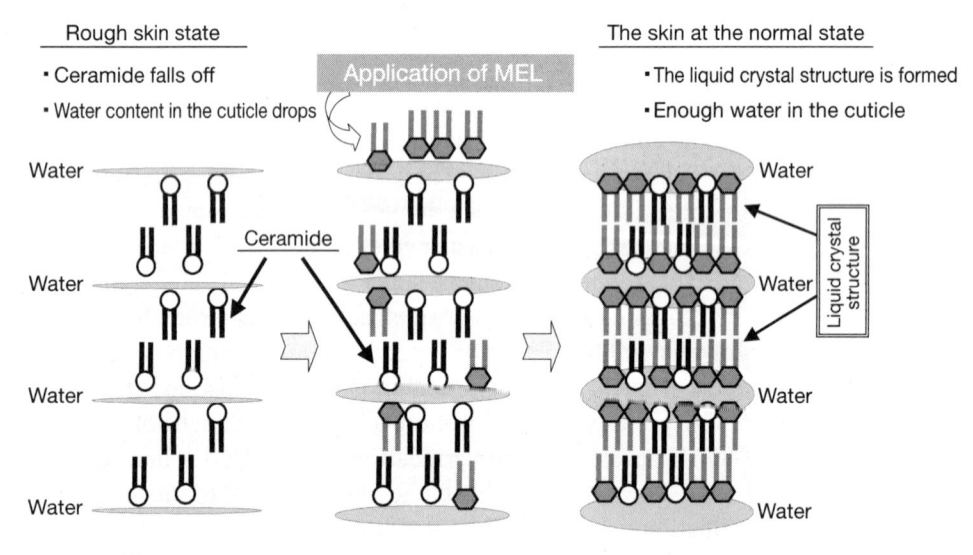

Figure 7 Effect of mannosylerythritol lipid in preventing roughness (prospects)

of the horny layer because its three-dimensional structure resembles that of ceramide, and (2) its liquid crystal structure effective in keeping water in the intercellular spaces (figure 7).

MEL can also be used to stabilize cosmetics ingredients (protection of the ingredients by encapsulating) and improve the penetration of the ingredients into the skin (the skin affinity effect of the capsules) because liquid crystals of various kinds, such as liposomes, can be formed at a wide range of concentration.

The production costs can be reduced to one-fifth to one-tenth of those (hundred thousands to several million yen/kg) of the conventional methods, which involve extracting natural ceramides from plants and chemically synthesizing pseudo-ceramide, because the yeast fermentation process produces biosurfactants at high degrees of purity. The process uses no materials of the petroleum origin and can produce biosurfactants in one step, which involves fermentation of vegetable oil, which is a biomass resource. It imposes less load to the environment than chemical synthesis, and is thus resource saving and environment protecting.

Functional biomaterials that can be used in skin care products are attracting attention[24], and AIST continues searching for biosurfactants of new structures and properties and developing methods (for expanding the line-up) using microbe biotechnologies. On the other hand, TOYOBO places the functional biomaterials developed using its biotechnologies as a main shaft of new business and started providing samples to manufacturers (inquiries: Bio research group, Toyobo Research Center, TOYOBO).

6.9 Conclusion

Although biosurfactants have peculiar structures, they can be produced in mass from biomass resources of various kinds without depending on organic synthesis. The cost of microorganism production is always an important technological issue for practical application of biosurfactants in a wide range of fields. Recent rapid innovation in biotechnology and progress in related technologies will enable early breakthrough. Practical use of sophorolipid in kitchen detergent is a good example.

The unique material properties and functions of biosurfactants briefly described here are based on their "neatly arranged three-dimensional structures" and "ingenious self-assemblies" in principle. Biosurfactants are cosmetics materials of a new concept. Development of new materials, such as surfactants, accompanies the need of developing safety evaluation technologies. The safety of new materials is evaluated as much as possible, but methods should be continuously revised using the state-of-the-art knowledge. We want to spread the use of biosurfactants in the international society by closely cooperating with the partner company.

References

1. N. Naito et al., "Development of new liposome applications", p.644-650, NTS Inc. (2005)
2. D. Kitamoto, *Oleoscience*, **3**, 663-672 (2003)
3. D. Kitamoto, *KAGAKU TO SEIBUTU*, **41**, 410 416 (2003)
4. D. Kitamoto et al., "Glycolipid-based bionanomaterials", Handbook of Nanostructured Biomaterials and their Applications in Nanotechnology, Vol.1, p.239-271, American Scientific Publishers (2005)
5. D. Kitamoto et al., *J. Bioeng. Biosci.*, **94**, 187-207 (2002)
6. Y. Kimura, *Fragrance Journal*, **20**, 22 (1992)
7. D. Kitamoto, "Interfaces and Surfactants", p.61-69, Japan Oil Chemists Society(2005)
8. S. S. Cameotra and R. S. Makker, *Curr. Opin. Microbiology*, **7**, 262-266(2004)
9. T. Yoneda et al., *Fragrance Journal*, **29**, 12 (2001)
10. D. Kitamoto, *Fragrance Journal*, **30**, 5 (2002)
11. T. Morita et al., *Appl. Microbiol. Biotechnol.*, **73**, 305-313 (2006)
12. T. Morita et al., *FEMS Yeast Res.*, **7**, 286-292 (2007)
13. M. Konishi et al., *Appl. Microbiol. Biotechnol.*, **75**, 521-531 (2007)
14. T. Fukuoka et al., *Biotechnol. Lett.*, **29**, 1111-1118 (2007)
15. T. Fukuoka et al., *Appl. Microbiol. Biotechnol.*, **76**, 4 (2007)
16. D. Kitamoto et al., *Biotechnol. Lett.*, **23**, 1709-1714 (2001)

17. T. Imura *et al.*, *Chem, E. J.*, **12**, 2434-2440 (2006)
18. T. Imura *et al.*, *J. Am. Chem. Soc.*, **126**, 10804-10805 (2000)
19. D. Kitamoto *et al.*, *Chem. Commun.*, **2000**, 860-861 (2000)
20. T. Imura *et al.*, *Langmuir*, **23**, 1659-1663 (2007)
21. T. Imura *et al.*, *Colloid Surf. B*, **43**, 115-121 (2005)
22. K. Inoh *et al.*, *Biochem. Biophys. Res. Commun.*, **289**, 57-61 (2001)
23. K. Inoh *et al.*, *J. Control. Release.*, **94**, 423-431 (2004)
24. "Horizons in the future, research and development of functional material letting a market have a foreboding in the next generation" p.125, Fuji-Keizai (2007)

Chapter 7
Application of PLGA Nanoparticles to Skin Care Cosmetics

Hiroyuki Tsujimoto and Nobuhiko Miwa

7.1 Introduction

Drug delivery systems (DDS) aim to (1) deliver a drug to the target site (targeting), (2) express the drug at the required quantity at the required speed (control release), and (3) ensure safe absorption (absorption improvement). DDS has been used in diverse application fields. Recently, functional skin care cosmetics are being actively developed by using DDS to deliver effective ingredients through the barrier of the stratum corneum to the deep parts of the skin and express their efficacies long. In the majority of these cosmetics, the penetrability of effective ingredients into the skin is enhanced by preparing the ingredients into nano-size particles or enclosing them in nano-size carriers, such as liposomes and nanospheres. The carriers should not only promote the penetration but must also be safe to the physiological functions of the skin even when they are applied over a long period of time. This chapter describes the functions of poly(lactic-*co*-glycolic acid) (PLGA), which is a copolymer of lactate and glycolitic acid developed as a state-of-the-art nano-size carrier that satisfies these two criteria, and exemplifies uses of the copolymer in hair restorer and essence lotions.

7.2 Safety and application of PLGA nanoparticles to skin care cosmetics

7.2.1 Safety of PLGA nanoparticles

PLGA nanoparticles have been proven safe in all tests required by GLP: (1) acute toxicity test (rat LD_{50} > 2000 mg/kg), (2) primary skin irritation test (rabbit), (3) continuous irritation test (rabbit), (4) maximized sensitization test (guinea pig), (5) phototoxicity test (guinea pig), (6) acute eye irritation test (rabbit), (7) mutagenicity test, (8) human patch test, and (9) skin photosensitization test (guinea pig, no ultraviolet rays absorption reply in PLGA). In Japan, the use as an additive for quasi-drug was permitted in 2007. In the skin, PLGA is decomposed into lactic acid and glycolic acid by hydrolysis and is finally discharged from the body in forms of water and carbon dioxide. Some nanoparticles have recently been criticized for persisting inside the bodies, but PLGA does not persist in the bodies and is thus safe. From a viewpoint of including into cosmetics that is applied on the skin on daily basis, the effects of PLGA nanoparticles on metastasis and invasion into the basal membrane of melanoma (severe metastatic skin cancer) have been particularly investigated, and safety has been confirmed[1,2].

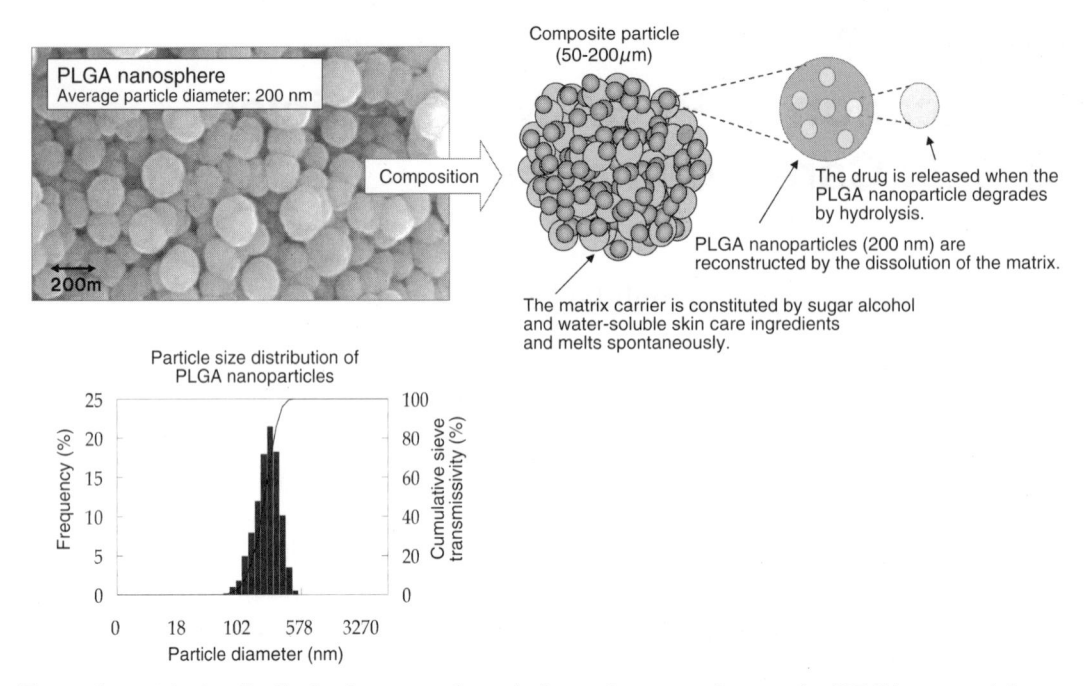

Figure 1 Particle size distribution in water and atomic force microscope photograph of PLGA nanoparticles and a schematic view of the composite nanoparticle structure

7.2.2 Stability of PLGA nanoparticles

PLGA, which has a low glass transition temperature at around 45 °C and is decomposed by hydrolysis, is unstable during storage because it is prone to coagulation by heat and decomposition by moisture, disturbing actual application. The authors found that the properties related to storage stability of drug (effective ingredients) carriers can be improved by preparing nanoparticles of a composite structure (apparent size of several tens of micrometers) that consists of a matrix of diluent and/or surface modifier and a dispersion phase consisting of PLGA nanoparticles of a mean particle diameter of 200 nm, which enclose skin care ingredients, using an original method (figure 1) When the composite particles are mixed with and used together with skin lotion or essence lotion, the matrix dissolves immediately, leading to the reconstruction of nanoparticle, and the original functions of the nanoparticles are manifested. When PLGA composite nanoparticles are dispersed in water, PLGA nanoparticles are also reconstructed immediately, and the particle size distribution and their photograph are shown in figure 1.

7.3 Functions of PLGA nanoparticles and application to skin care technology

In experiments using excised human skin specimens, PLGA nanoparticles have been shown to be an effective carrier of skin care ingredients into the epidermis and to release the enclosing skin care ingredients over a long period of time while the particles are hydrolyzed. PLGA nanoparticles have been used in skin care for whitening, reducing stain, improving and reducing wrinkles and scalp care, aiming to promote hair growth.

7.3.1 Skin penetration of PLGA nanoparticles

7.3.1.1 Evaluation of penetrability of PLGA nanoparticles and dermis deliverabilty of drugs enclosed in the nanoparticles (using excised human skin specimens and modified Bronaugh diffusion chamber) [3]

Figure 2 shows the results of an evaluation using excised human skin specimens. When fluorescent coumarin was applied on the specimens in a water dispersion solution, almost no pigment penetrated into the skin even when a large amount was applied on the skin. When the pigment was enclosed into PLGA nanoparticles and a water dispersion of the nanoparticles was applied on the skin, a large amount of coumarin penetrated into the skin and remained there for at least 4 hours. Enclosing a vitamin C derivative (VC-IP: (INCI name) ascorbyl tetraisopalmitate) in PLGA nanoparticles of the same size sharply increased the amount of reduced-form vitamin C in the skin from that when an oil-in-water dispersion of VC-IP was applied, which is known to be highly lipid soluble and high penetrable into the skin. The increase continued for 48 hours. This was caused by the PLGA nanoparticles having penetrated through the stratum corneum into the skin, been gradually hydrolyzed, and released VC-IP slowly, which was deesterified and converted into ascorbic acid.

The ability of the nanoparticles to supply reduced-form vitamin C into the dermis tested using excised human skin specimens is shown in figure 3. VC-IP enclosed in PLGA nanoparticles started to show reduced-form vitamin C in 2 hours and the cumulative amount in 4 hours was at least 10 times larger than that when VC-IP was applied as a simple dispersion. This suggests that a sufficient amount of reduced-form vitamin C is present in the skin at the noon, or 4 hours

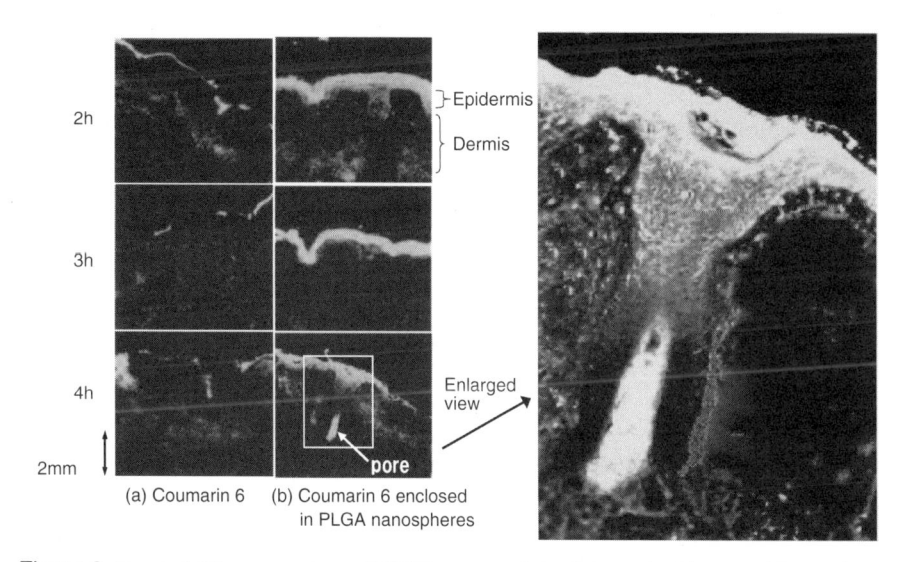

Figure 2 Penetrability comparison of PLGA nanoparticles into excised human skin specimens
(a) Water dispersion solution of coumarin 6 (conc. 10wt%). Surfactant: Pluronic F68 0.5wt%. Quantity of application: 0.3 mL.
(b) Water dispersion solution of PLGA nanoparticles enclosing coumarin 6. Concentration of the PLGA nanospheres: 0.2 wt%.
   Enclosure rate of coumarin 6 in PLGA: 0.05 wt%. Concentration of coumarin 6 in water: 0.00001 wt% (1/100 000 of (a)). Quantity of application: 0.3 mL.
Skin tissue excised from the side of a 35-years old woman and divided into 6 small pieces
Note: Stratum corneum (Skin surface >0.01 mm), epidermis (>0.1 mm), and dermis (>1.7 mm)
   (Photograph & evaluation: Prefectural University of Hiroshima, Faculty of Life and Environmental Sciences, Miwa laboratory)

168   Nanotechnology for Producing Novel Cosmetics in Japan

after applying cosmetics in the morning, when the ultraviolet rays are strong, serves as a front defender by removing active oxygen produced by ultraviolet rays, and protects the skin.

7.3.1.2 Evaluation of skin penetrability of drugs enclosed in PLGA nanoparticles (using rats and a horizontal diffusion cell)

The penetrability of PLGA nanoparticles that contained Ketoprofen (molecular weight: 254.28g/mol, antiphlogistic agent) into rat skin in 12 hours was measured using a horizontal diffusion cell (area: 0.95 cm^2). The skin specimens were excised from the right and left parts of the abdomen of a hairless rat after shaving (no stripping). Enclosing the drug in PLGA nanoparticles (S-1 to S-3) increased the penetrability into the skin by 2–8 times from a suspension of the drug, which was used as the control (phosphate bugger). Large increases were observed when the surface of the nanoparticles was modified with chitosan to give

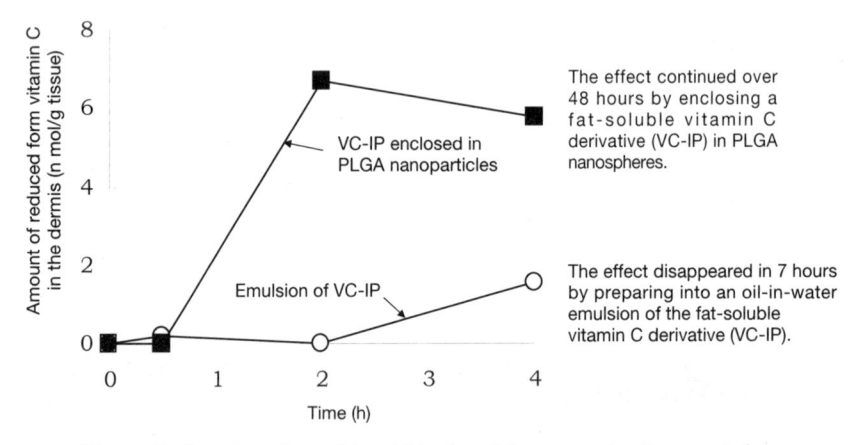

Figure 3 Quantity of ascorbic acid (reduced form vitamin C) converted from VC-IP in the dermis
(Excised human skin specimens: 52 years old woman, the upper eyelid)

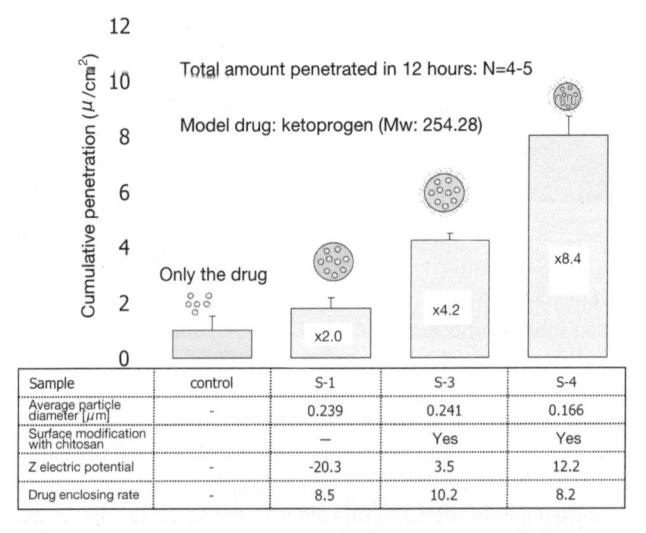

Sample	control	S-1	S-3	S-4
Average particle diameter [μm]	-	0.239	0.241	0.166
Surface modification with chitosan	—	—	Yes	Yes
Z electric potential	-	-20.3	3.5	12.2
Drug enclosing rate	-	8.5	10.2	8.2

Figure 4 Increases in skin penetrability and permeability of a drug (keoprofen) by enclosure in PLGA nanoparticles (measured using a horizontal diffusion cell)

viscosity (S-2 and S-3) and the size of the particles was reduced (S-3). Although the effects of electric charges on the particle surface are not clear, but zeta electric potential is an index of the dispersion stability of nanosphere in solution. Nanoparticles modified with cationic chitosan were localized in the stratum corneum, the outermost layer of the skin, increasing the concentrations of the nanoparticles and drug, and diffused into the dermis and through the skin membrane.

7.3.1.3 Evaluation of the skin penetrability of solution components (using a three-dimensional artificial skin and a Franz diffusion cell)

Increases of the skin penetrability of enclosed components were described above. In cosmetics, nanoparticles are dispersed in a liquid, such as skin lotion and essence lotion. The penetrability of the components in the liquid that interact with PLGA nanoparticles also increases by the skin penetrability of the nanoparticles.

To test the increase, a water-soluble drug model (tranexamic acid, molecular weight: 157.21g/mol) dissolvable to the dispersion phase of the PLGA nanoparticles (water or skin care lotion) was selected. The amino group of this medicine and the carboxyl group of the PLGA nanoparticle interact ionically. The effects of adding the PLGA nanoparticles on the permeability of a water solution of the model drug was evaluated by comparing the permeability of a 2.5% water solution of the drug to which PLGA nanoparticles were added so as to make 0.5 wt% and that of a 2.5 wt% water solution to which PLGA nanoparticles (0.5 wt%), ascorbyl phosphate (3 wt%) and mannitol (0.5 wt%) so as to reproduce cosmetics into a three-dimensional human skin culture model (TESTSKIN^{TM} (LSE, TOYOBO CO., LTD.)) using a Franz diffusion cell. The results are shown in figure 5.

The skin permeability of the drug increased by 2.5 times by the addition of the PLGA nanoparticles. A 4-times increase was observed when magnesium ascorbyl phosphate and mannitol were added together with PLGA nanoparticles because the penetration pressure of the donor solution increased. Enclosing hydrophobic substances in PLGA nanoparticles is doubtless effective, adding PLGA nanoparticles to water solutions of water-soluble ingredients was found to be also effective as they interact with each other.

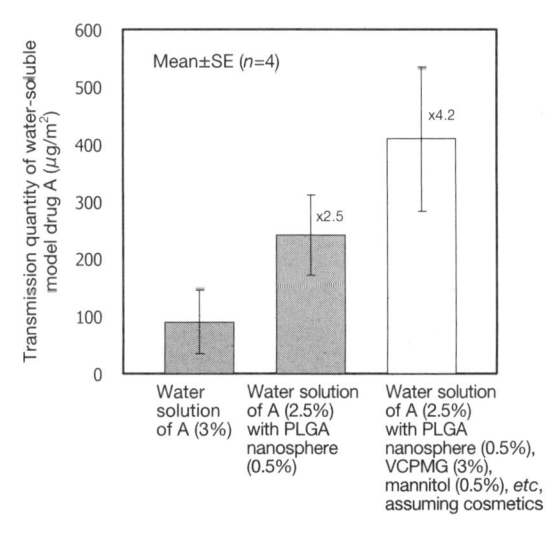

Figure 5 Increases in skin penetrability and permeability of a water-soluble drug by addition of PLGA nanoparticles (measured using a Franz diffusion cell)

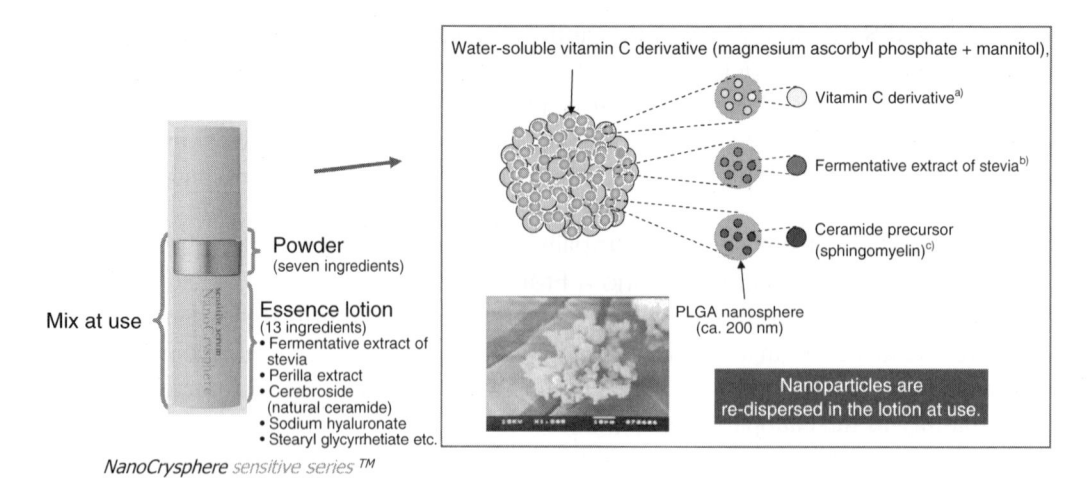

Figure 6 Functional cosmetics NanoCrysphere Prime Serum containing biocompatible composite nanoparticles and the structure of the functional PLGA composite nanoparticle (>100μm) and an electron micrograph

a) (Ascorbyl tetraisopalmitate) ⇨ VC, Whitening (stain and somberness), reducing wrinkles, increasing immunity (improving the abnormal immunity of atopic dermatitis), b) Relieving itchiness, antihistamine action, c) Restoring the barrier in a cellular level from the lipids within the stratum corneum lipids, Barrier layer restoration from a cell level (Ceramide is synthesized from sphingomyelin by sphingomyelinase)

7. 3. 2 Example of using composite nanoparticles in cosmetics

Functional nanoparticles (bulk) of various properties can be prepared by enclosing functional ingredients of the aimed characteristics and used in cosmetics. PLGA composite nanoparticles available from HOSOKAWA can be broadly classified into those for (1) whitening and anti-aging, (2) sensitive skin (atopic dermatitis), and (3) hair growth (either for men or women). Derivatives of vitamins C, E and A are enclosed in (1); derivatives of vitamin C, fermentative extract of stevia, and sphingomyelin (a ceramide precursor) are enclosed in (2); and crude drug extracts (sophora extract, cinchona extract, hinokitiol), derivatives of vitamin E and glycyrrhizin acid are enclosed in (3).

These bulks are used in NanoCrysphere Prime Series (whitening, anti-aging), Sensitive Series (for sensitive skin (figure 6)) and Nano Impact Series (for scalp care) from HOSOKAWA MICRON COSMETICS CO., LTD.

7. 3. 3 Skin care and scalp care technologies of PLGA nanoparticles
7. 3. 3. 1 Skin care technology

As described in 7. 3. 1. 1, PLGA nanoparticles that enclose VC-IP can efficiently deliver the included vitamins to the basal layer of the skin and have been proven to have whitening effect (by a melanogenisis restraining test using a human melanogenic cell culture HMV II and monitor tests) and improve and prevent stains and wrinkles (evaluation by the high-speed artificial wrinkle formation system)[4].

The whitening effect of skin lotion that contained VC-I in PLGA nanoparticles (Nano Crysphere Prime Series) tested using a human melanogenic cell culture HMV II is shown in figure 7. HMV II cells produce melanin and become brown when exposed to theophylline, which is a caffeine-like substance contained in tea leaves. This is an *in vitro* reproduction of the skin becoming tanned in several days after exposure to ultraviolet rays (UV). When the cells

[Examination of the whitening effect]

No theophylline treatment (a state before sunburn) (Theophylline activates melanocytes.)

Electron micrographs of human melanogenic cell HMV-II (×1000)

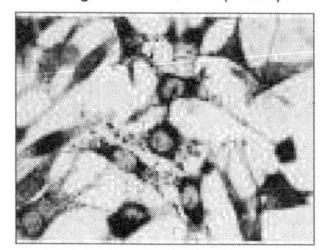

3 days after theophylline treatment (Reproducing a state 3 days after exposure to ultraviolet rays
⇨ Melanin accumulated in the cytoplasm around the nucleus.)

(Concentration of the essence lotion in the cell culture medium: 0.01%)

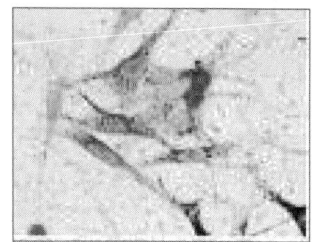

3 days after theophylline treatment after application of Nano Crysphere application
⇨ The antioxidation activity of the lotion controlled active oxygen and resultant melanogenisis.

Figure 7 Examining the whitening effects of a functional essence lotion (Nano Crysphere Prime Serum™) (Photograph & evaluation: Prefectural University of Hiroshima, Faculty of Life and Environmental Sciences, Miwa laboratory)

were treated with the PLGA nanoparticles 4 hours before exposing them to theophylline, the production of melanin was suppressed to a level almost equal to the level before theophylline treatment. Thus, the product was shown to suppress melanin production even 4 hours after application because the vitamins were gradually released from the PLGA nanospheres.

Using the high-speed artificial wrinkle formation system[3], wrinkles were formed on excised human skin specimens by intermittently irradiating UV of a critical dose that did not kill the cells, and the effects of the product in suppressing wrinkle formation were tested. That shows replicates of the skin surface of the human skin specimens, which were prepared by excising the skin around an ear of a person (55-years old female) and dividing into pieces of equal sizes, after UV-A irradiation of UV-A of 3 J/cm^2 twice a day for five days in a modified Bronaugh diffusion chamber. Irradiation of UV-A caused formation of deep lines and narrow but sharp-edge wrinkles. A line histogram evaluation also showed irregular unevenness of sulcus cutis. On the other hand, when this product was applied before UV irradiation, almost no lines and no wrinkles were formed. The histogram also showed smooth and regular sulcus cutis formation (fine). This was likely because provitamins enclosed in the PLGA reached the dermis and eliminated the active oxygen produced by UV irradiation.

7.3.3.2 Example of functional cosmetics containing PLGA nanoparticles (for sensitive skin)

Widely used methods for treating relatively slight atopic dermatitis involve applying cream that contains ceramide, which accounts for 50% of intercellular lipids, to restore the barrier function of the stratum corneum. Applying ceramide for several weeks increases the water content of the skin and improves its conditions. However, the moisture content sharply drops in several days after ceasing application.

The skin care product that contains PLGA nanoparticles for sensitive skin (Figure 6) also contains ceramide in its substrate to supply ceramide from the surface of the stratum corneum. It also contains three kinds of PLGA nanoparticles to supply a vitamin C derivative, fermentative extract of stevia and sphingomyelin (a precursor of ceramide) to the deep parts of the skin. Vitamin C increases the immunity strength (and improves the abnormal immunity of atopic dermatitis), and the fermentative extract of stevia activates the epidermis and dermis. Sphingomyelin is a raw material of intercellular lipids of the stratum corneum (ceramide:

glue among cells), promotes production of ceramide in deep parts of the skin, and leads to restoration of the healthy stratum corneum. Thus, the skin can keep moisture over a long period of time. In this manner, PLGA nanoparticles enable continued care of the skin.

7. 3. 3. 3 Scalp care technology

Test application of PLGA nanoparticles to scalp care technology has shown that the nanoparticles can deliver 3–5 times larger amounts of hair growth ingredients to the papilla pili deep in pores and keep releasing the ingredients long[5]. Using this technology of adding PLGA nanoparticles, hair growth lotions that need to be applied only once a day at night and keep being effective throughout the day as if a small amount of the lotion is constantly supplied can be developed.

Hair growth is promoted by cell activation, blood stream promotion, male hormone restraining, and inflammation control, sterilization and moisture retention at pores. Ingredients that have these functions should be combined and should be absorbed efficiently by scalp cells. As shown in figure 8, some plants extracts have been known to stimulate hair growth from old times and have been included in hair restorers on the market. We developed a hair growth technology, which involves enclosing these extracts in PLGA nanoparticles. The figure shows images of cross sections at pores of skin specimens excised from the head of a healthy 40-years old woman 4 hours after application of a water solution of hinokitiol (control) and a water dispersion of LPGA nanoparticles enclosing hinokitiol and intensities of the luminescence of hinokitiol measured from the surface of the scalp to the depth direction. As in coumarin described above, enclosing in LPGA nanoparticles sharply promoted the penetration of hinokitiol into the scalp particularly near the pore. The distribution measurements of luminescence (area values) along the depth direction showed that the PLGA nanoparticles increased the hinokitiol

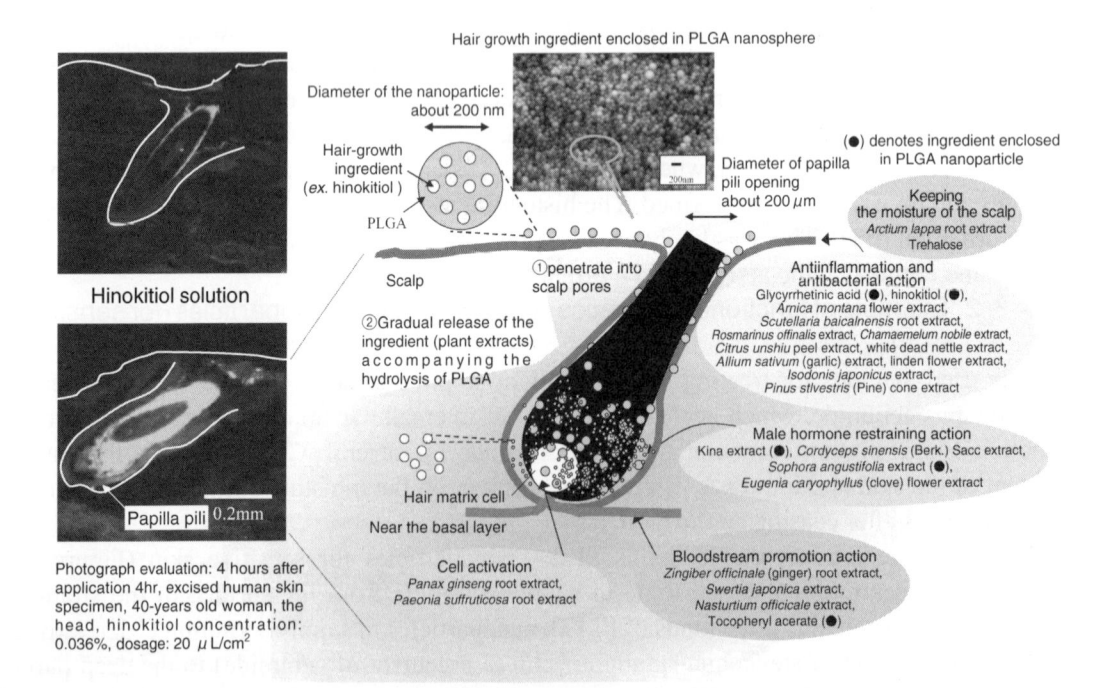

Figure 8 Examining the penetration of ingredients enclosed in PLGA nanoparticles into scalp pores and a schematic diagram of a system for delivering drugs to hair follicles (Photograph & evaluation: Prefectural University of Hiroshima, Faculty of Life and Environmental Sciences, Miwa laboratory)

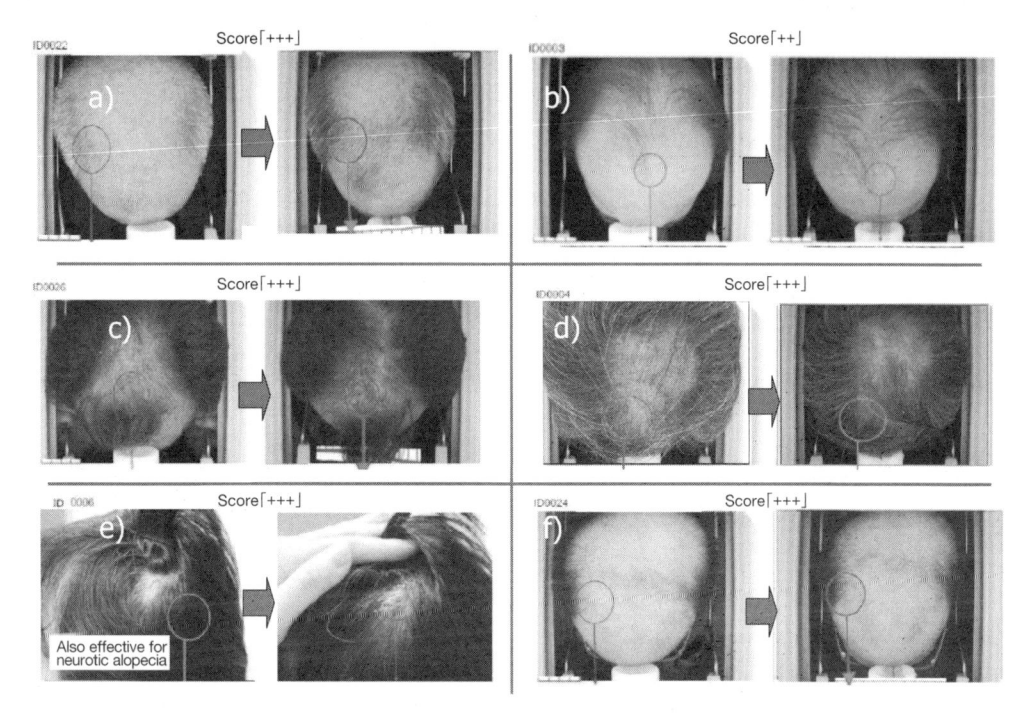

Figure 9 Examples of scalp care using a crude drug lotion containing PLGA nanoparticles (in 3-4 months of use)

penetration by 3 folds at the hair follicle, as shown in the photographs, and the increase was likely to be even larger at the hair root. PLGA nanoparticles were shown to be effective carriers of plant extracts to hair roots. When PLGA reaches inside a pore, it degrades gradually by hydrolysis, slowly releasing the ingredients enclosed and enabling their maximum efficacies to be expressed.

An *in vivo* test of the PLGA nanoparticles was conducted using C3H mouse, which is widely used for hair growth tests because, once the hair is shaved, all the hair on the shaved section starts to grow synchronously after a certain period of time. The hair on the lower back of a C3H mouse was cut, and was shaved on the next day. Plant extracts were applied on the skin once a day either enclosed in PLGA nanoparticles or alone, and their effects on switching the hair follicles to start growing hair were investigated. The hair growth was found to be more notable when the extracts were enclosed in PLGA nanoparticles than when they were used alone, and the switching from resting to growing periods was also shown[5,6]. A monitor test was also conducted on about 70 men who had sparse hair (figure 9). A general hair-growth and a scalp improvement effect were shown, showing the effectiveness of the scalp care technology.

7.4 Conclusions

PLGA composite nanoparticle for cosmetics, which uses safe and effective PLGA nanoparticles, has enabled functional cosmetics of high skin penetrability to be produced, which were deemed difficult. The bulks have been used in lotions for whitening, anti-aging, improving sensitive skin, and stimulating hair growth.

References

1. H. Tsujimoto, K. Hara, H. Mimura, N. Miwa, *J. Soc. Powder Technol. Jpn*, **43**, 2, 86-91 (2006)
2. K. Hara, H. Tsujimoto, C.C. Huang, Y. Kawashima, H. Mimura, N. Miwa, *Oncology reports*, **16**,1215-1220 (2006)
3. N. Miwa *ed.*, "Rejuvenation and skin protection, and biotechnology", CMC Publishing CO.,LTD, (2003)
4. H. Tsujimoto, K. Hara, C.C. Huang, T. Yokoyama, H. Yamamoto, H. Takenouchi, Y. Kawashima, N. Akagi, N. Miwa, *J. Soc. Powder Technol. Jpn*, **41**,867-875 (2004)
5. H. Tsujimoto, *Drug Delivery System*, **7**, 405-415 (2006)
6. H. Tsujimoto, K. Hara, Y. Tsukada, C.C. Huang, Y. Kawashima, M. Arakaki, H. Okayasu, H. Mimura, N. Miwa, *Bioorganic & medicinal chemistry letters*, **17**, 4771-4777 (2007)

Chapter 8
NANOEGG® and NANOCUBE®

Yoko Yamaguchi and Rie Igarashi

8.1 Introduction

Drug delivery systems (DDS) are systems for delivering the necessary amount of necessary medicine at necessary timing to the body parts of disorder, and are an indispensable technology for enhancing the desired effects and reducing the side effects of drugs. Biotechnology and nanotechnology have attracted attention these years, leading to DDS studies in diverse fields. Also, the importance of regenerative medicine has increased accompanying shifts of clinical and social needs to improving the quality of life (QOL) in the aging society and to gene therapies, which are a fruit of the genome decoding. Development of new DDS technologies is highly awaited to answer these and other various needs.

Endermic absorption systems are ideal because they impose little burdens to patients. However, they have a big disadvantage of low absorbency and permeability. To overcome the issue, new technologies are being energetically developed aiming for practical use of endermic absorption systems (such as iontophoresis, electroporation and microneedles)[1]. We have developed a number of new DDS technologies and succeeded in developing new transdermal delivery systems (TDS) called NANOEGG® and NANOCUBE®, in which nanotechnology is used to improve the absorption from the skin. These technologies can also sharply improve the dispersibility of fat-soluble drugs to water and thus enable drugs that could have been administered only orally to be used as injections (for example, all-trans-retinoic acid (atRA)). In this chapter, NANOEGG® is described first, which is a new DDS technology that we developed.

The skin is the organ that serves as the interface between the living body and environment and is in charge of homeostatic regulation, or preventing events caused by changes in the environment from adversely affecting the living tissues. The stratum corneum plays the main role and acts as a living barrier. The stratum corneum consists of flat cells piling up like bricks. The cells are mutually connected by corneodesmosomes, and the spaces are filled by intercellular lipids[2]. Intercellular lipids are secreted from lamellated granules of granule cells right under the stratum corneum and consist of ceramide, cholesterol and fatty acids at a molecular ratio of about 1:1:1. The constituents form a multilayered lamellar structure parallel to the cell membranes and serve as a barrier blocking microbes, toxic substances and even small molecules, such as water. The lamellar structure is composed of hydrophilic base layers and hydrophobic base layers alternatively lying top of the other, which contain water and oil, respectively, and form a multilayered moisture retaining membrane. Each layer is about 5–15 nm thick.

Regeneration of the skin occurs when the skin is damaged and inflamed. Multiplication and differentiation of epidermic cells, or the turnover of the skin, are temporarily accelerated, and acanthosis (thickening) of the skin occurs. These reactions are attributable to the homeostasis of a living body and occur accompanying various symptoms such as atopic xerosis, atopic dermatitis, contact dermatitis, drug eruption, and sunburn by ultraviolet rays, all of which involve decline of the barrier functions of the stratum corneum[2]. These and many other skin diseases are believed to be related with the decline in the functions of the stratum corneum. The epidermis restores the barrier quickly when it is damaged or destroyed[2]. A series of restoration reactions is triggered not only by visible damages to the skin, such as cuts and scratches, but also by a damage of a microscopic level.

We have studied a new system called NANOCUBE®, which promotes the regeneration of the skin by causing a temporary drop in the barrier function of the stratum corneum to trigger the natural homeostatic reactions of the skin. NANOCUBE® has also been suggested to effectively raise the absorption of drugs into the skin, and studies are being conducted on transdermal absorption. This chapter also describes the mesoscopic internal structure of intercellular lipids in the stratum corneum, which is a soft matter, and a new skin regeneration process NANOCUBE®, which involves structural transition.

8. 2 New DDS technology —birth of NANOEGG—[3]

All-trans-retinoic acid (atRA), which is the physiologically active part of Vitamin A, is known to accelerate the turnover[4] of the skin in which the cells have declined in division potential due to aging (by inducing differentiation and multiplication of keratinocytes) and has been used to treat stains and wrinkles (because it induces production of hyaluronic acid[5,6] in the epidermis). In the US, ointment containing retinoic acid has been used as a medical supply for more than ten years, but in Japan, it can be used only inside hospitals on the responsibility of clinical doctors because the use is not permitted by the Ministry of Health, Labour and Welfare.

Lipophilic atRA forms spherical micelles in the presence of sodium salt. We developed a new encapsulation technology called "NANOEGG®", which involves using the atRA micelles as molds and coating their surfaces with films of $CaCO_3$ (figure 1). Because the micelles are used as molds, the enclosure rate is almost equivalent to the concentration of atRA in the solution, or 100 %, when single dispersed atRA molecules are excluded. However, the technology cannot be used for materials that do not self-assemble, such as oleic acid and other fatty acids of low degrees of dissociation and very low ion absorption and retinol and other drugs that do not have dissociation functional groups. Theoretically, inorganic films cannot be either constructed on the surface of atRA particles.

The atRA capsules have a particle diameter of about 15 nm and many strings of water-soluble polyethylene glycol (PEG) polymer (of the PEG-type nonionic surfactants origin) sticking out

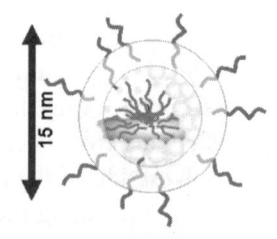

Figure 1 Schematic representation of NANOEGG®
The surface of a micelle of sodium salt of retinoic acid is coated with $CaCO_3$ and covered by many strings of PEG.

from the surface (figure 2 (b)). The capsules disperse in both oil and water transparently (figure 2 (a)). The $CaCO_3$ crystals do not dissolve in water, but are decomposed in the stratum corneum by enzymatic activities in the epidermis and dermis, slowly releasing their contents, because the surface of the nanoparticles, which has a large surface curvature of $1/R$ (R: radius of the particle), should have a vartrite or amorphous structure. The structure of NANOEGG® has been thoroughly investigated by small angle X-ray scattering measurement and transmission electron microscopic observation of freeze-fracture specimens, and the core-shell structure and the particle diameter have been determined (figure 2 (c,d)).

Because the developed NANOEGG® technology involves processing atRA, which processes lipophilic character but insoluble to water, into a molecular aggregate of a nanometer size that has an amphiphilic surface, the technology is expected to improve the penetrability into the skin and enable new injections to be designed. By the present, various methods for improving epidermal absorption have been developed (chemical[7,8] and physical methods[9,10]) for the three routes through which drugs can be administered (through keratinocytes, pores, and intercellular lipids). We have performed nanotechnological studies aiming to improve the penetrability through the intercellular lipids. Kontturi et al.[11] proposed that there are two possible routes through which drugs can penetrate into the stratum corneum: via aqueous pathways and through the lipid matrix. Therefore, we decided to design our capsules to (1) be capsules of nanometer size, (2) have amphiphilic surface like the stratum corneum and, as a result, can pass through both the aqueous pathways and lipid matrix, and (3) be formed not by enclosing target drugs as in conventional capsules but by the effective drug itself forming capsules, which is a high or 100% encapsulation system.

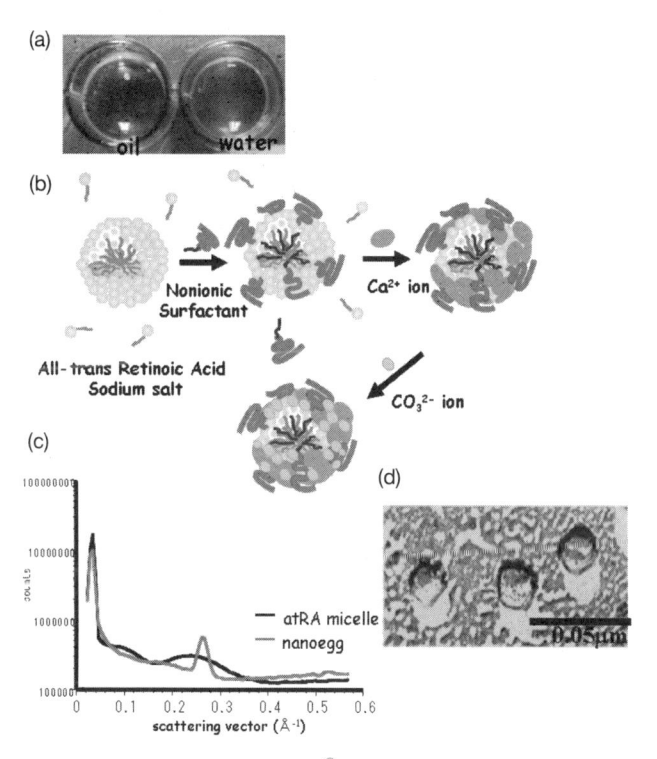

Figure 2 (a) Dispersibility NANOEGG® in water and oil, (b) Preparation method, (c) Small angle X-ray scattering pattern, (d) Transmission electron micrograph of a freeze-fracture specimen

8.3 Skin regeneration effect of NANOEGG® (reducing stains and wrinkles)

The skin consists of the epidermis and the dermis. In the epidermis, basal cells propagate and differentiate into prickle cells, granular cells, horny cells and finally into the stratum corneum, which, at the end, becomes loose and falls off from the skin (forming a cycle called the "turnover"). When the skin is irradiated by UV-A and/or UV-B, melanogenetic cells (melanocytes), which exist in the basal membrane at a ratio of one in every ten basal cells, produce melanin granules to protect the skin from damages by UV. The produced melanin is excreted from melanocytes and is transferred to prickle, granular and horny cells located in the upper layers. When the turnover of the skin is active, melanin is transferred upward efficiently, is discharged to the outside world together with old keratin, and does not deposit in the skin. However, when the turnover slows down due to aging, the balance between production and discharge of melanin is lost, and melanin accumulates in the epidermis. This is called stains (senile pigmented spots or solar pigmented spots). Wrinkles and deep lines are formed when aplasia of the stratum corneum occurs due to aging and/or exposure to the sun and the structure of collagen in the dermis is destroyed. The atRA ointment used in clinics in Japan and the US causes cutitis (inflammation accompanying erythema) when used. In many cases, it accompanies irritation and it is difficult to use over a long time because the stratum corneum peels off frequently. The pharmacological effects of NANOEGG® were examined also on these points.

We applied atRA or NANOEGG® vaseline ointment (many times a day, for five days, 30 mg of 0.1% atRA) on the shaved back of rats. The results of visual observation are shown in figure 3 (a). Harmful phenomena, such as inflammation, are shown with scores of 1 to 6. Smaller score values denote less adverse effect. The columns show that NANOEGG® caused less irritation than atRA at all concentrations. It showed that the skin irritating characteristic of atRA molecules was considerably improved.

When atRA is externally applied, the atRA molecules bind to atRA receptors (RAR) on keratinocytes and induce production of an EGF-like growth factor (HB-EGF: heparin-binding epithelial growth factor), which promotes differentiation and growth[12]. We measured the expression of mRNA HB-EGF (figure 3 (b)). We applied atRA or NANOEGG® vaseline ointment (30 mg) on 1 cm × 1 cm sections of the back of a mouse (5-weeks, male) frequently for four days. The atRA ointment showed an expression of about 2 times larger than that in the control, while NANOEGG® expressed 2.5 times larger amount of mRNA HB EGF than that by atRA. Increases in the production of the differentiation and growth factor should result in activation of keratinocytes, which is observed as acanthosis or thickening of the skin. We tested the effects by applying atRA and NANOEGG® ointment on the back of a colored guinea pig, and observed the degree of acanthosis and melanin excretion. The histological results are shown in figure 3 (c). Of all groups, the NANOEGG® application group showed the most advanced acanthosis and the largest HB-EGF expression; and almost no melanin was detected on the back as in the control. Photographs of the mouse skin specimens in which hyaluronic acid was stained are shown in figure 3 (d). The conditions of application were the same as those used for measuring HB-EGF expression in figure 3(b). A large quantity of stained hyaluronic acid was observed in the inter-keratinocyte spaces in the NANOEGG® application group. On the other hand, almost no hyaluronic acid was observed in the atRA application group. Based on the results of the fundamental experiments mentioned above, we performed a clinical test on volunteers (figure 3(e)). It involved applying 0.05 % atRA or NANOEGG® vaseline

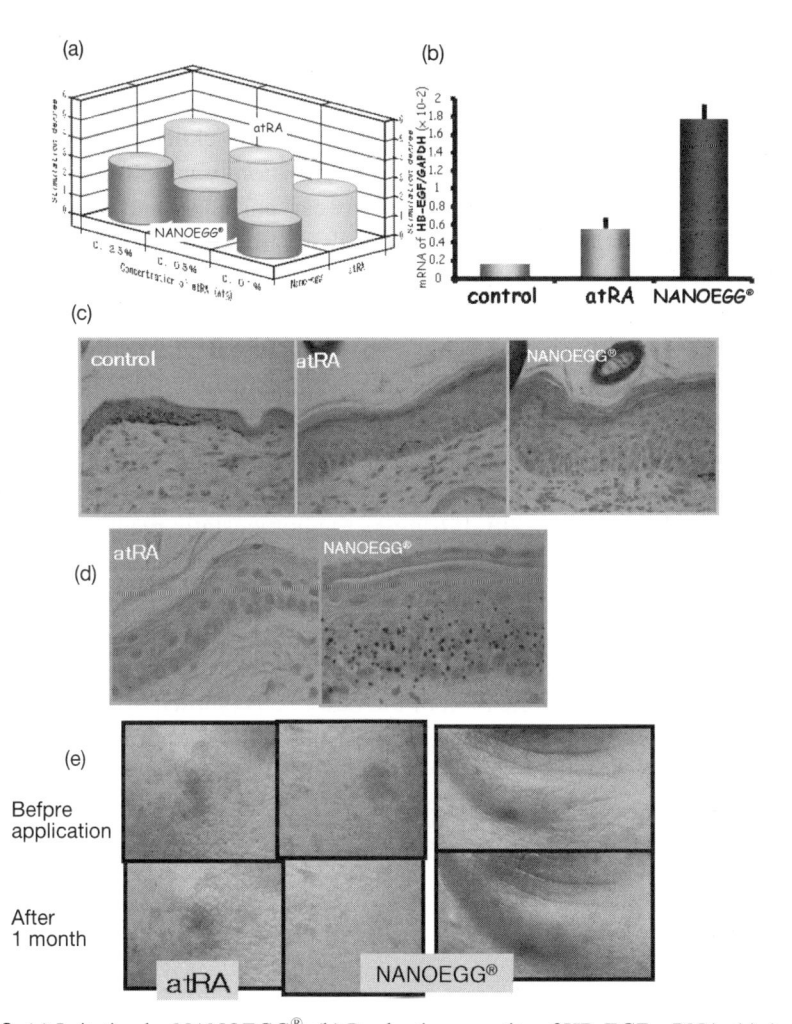

Figure 3 (a) Irritation by NANOEGG®, (b) Production quantity of HB-EGF mRNA, (c) Acanthosis, (d) Hyaluronic acid production, (e) Clinical test (stain and wrinkle improvement effect)

ointment every night for one month. The atRA application did not cause a big change, while NANOEGG® application caused stains to almost disappear, most of the lines at the corner of the eyes to be smoothened, and the skin to become firm, showing that NANOEGG® was very effective.

Because NANOEGG® disperses in oil and water transparently, it can be used as an injection agent. We are studying the feasibility of the technology for preparing injections for treating various intractable diseases. The drugs to be used must be amphiphilic and have at least one ionic functional dissociation group because NANOEGG® uses the drugs themselves as the mold. Formation of capsules using NANOEGG® for cosmetics and their pharmacological effects are also being investigated.

8. 4 Skin regeneration by bio-mimetic technology NANOCUBE®

A living body has homeostasis or ability to self regenerate. When the skin is injured, the skin regenerates autonomously. We developed NANOCUBE®, which is a bio-mimetic technology for inducing the homeostasis of the skin without causing macroscopic injuries and involves reproducing the intercellular lipids of stratum corneum[13]. NANOCUBE® is a soft matter (a general term for colloids, liquid crystals, emulsions *etc.*) that reproduces the mesoscopic internal structure of intercellular lipid of about 10–100 nm thickness. The dynamic property of a soft matter is generally $\tau = D^2/L$, where τ is the diffusion time between particles, D is the particle diffusion constant and L is the distance between particles; and it is $\tau \sim 10^{-11 \sim 12}$ s in a simple liquid. However, lyotropic liquid crystals show an extremely slow dynamics of 10^{-3} ls, are unbalanced, and undergo large structural changes by a small external force, showing nonlinearity. Application of NANOCUBE®, which has this property, on the skin induces decomposition or phase transition of the high-order structure of intercellular lipids (lamellar liquid crystal structure). Lamellar liquid crystal structures usually change into a temporary chaos structure by spinodal decomposition at 10^{-3}–10^{-6} S or into bicontinuous cubic liquid crystals (figure 4). Because both structures are biconnected structures, they can be deemed to be structures with irregularly and regularly arranged holes on all directions. This is an organic change of a nanometer level of the skin. It triggers the homeostatic reactions for repairing the injury and activates epidermic cells. The melanin deposited in the epidermis shifts upward as the epidermic cells multiply and differentiate and is discharged, resulting in erasing of stains (figure 5).

The epidermis thickens when NANOCUBE® is applied on the skin likely because NANOCUBE® promotes the multiplication of the basal cells (shown by positive immunostain of Ki-67, which is expressed during mitosis) and the differentiation of prickle cells and granular cells (shown by strong fluorescence of cytokeratin 1 and no increase in the expression of lorcrin from the wild-type)[13].

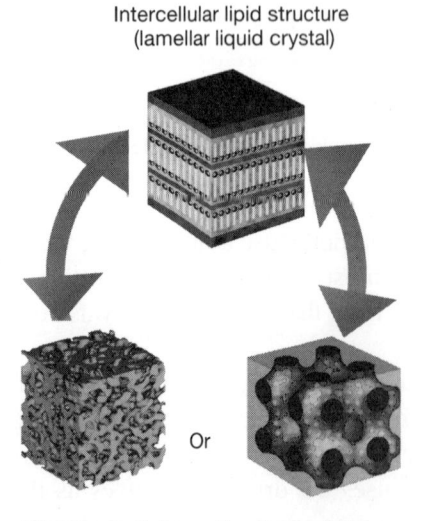

Figure 4 Schematic view of phase transitions of intercellular lipids and their structure after spinodal decomposition

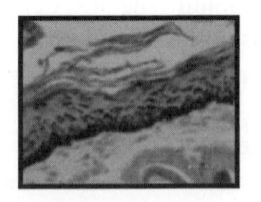

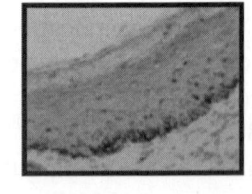

Figure 5 Improvement of pigmentation by NANOCUBE® application
30 mg of NANOCUBE® was applied on the shaved back of a colored guinea pig (5-weeks, male). The tissue was sampled four days after the application and was stained with Fontana-Masson. Prior to the application, UV-A and B were irradiated to the guinea pig to induce pigmentation. Black spots are melanin.

8.5 Mechanism of skin regeneration by NANOCUBE®

Cells are believed to produce hyaluronic acid to heal wounds[14]. If NANOCUBE® induces the state of wound healing, not only the production of hyaluronic acid but also the quantity of inflammatory cytokines (tumor necrosis factor (TNF)-α and interleukin 1 (IL) β etc.) should increase. The measured quantities of the cytokines are shown in Figure 6 together with the production of HB-EGF-like growth factor (HG-EGF: heparin-binding epidermal growth factor)[15] mRNA and transforming growth factor (TGF)-$\beta 1$[16] mRNA, which are cytokines involved in multiplication and differentiation of keratinocytes. As shown in the figure, NANOCUBE® application increased the production of HB-EGF and TGF-$\beta 1$ to four folds and also the production of TNF-α[17] and IL-1β, which are inflammatory cytokines. These results showed that the regeneration of the skin by NANOCUBE® application depended on the production of the cytokines that are related to multiplication and differentiation of cells and also suggested that it was possibly to be similar to the homeostatic reactions for healing wounds since the production of the inflammatory cytokines increased.

To understand what is taking place in the skin by NANOCUBE® application, we measured the transepidermal water loss (TEWL) after applying NANOCUBE® to a mouse (figure 7). The barrier function collapsed slowly after application until 6 hours and was restored in one day. Because 0.02 mg/cm^2/min is a normal TEWL, the collapse was not serious. The deterioration of the barrier function was likely attributable to changes of intercellular lipids in stratum corneum, which played a central role in the barrier function.

The phase transition of the intercellular lipids was probably caused either by the ingredients used in NANOCUBE® diffusing passively into intercellular spaces immediately after application or the constituents of intercellular lipids diffusing into NANOCUBE®. The ordered structure of the intercellular lipids started to collapse soon after application and changed into a chaotic structure by spinodal decomposition. With the passage of time, the lamellar structure of the intercellular lipids transferred into a bicontinuous cubic phase. However, this phase transition was transient, and the structure was restored to the original state in 24 hours. The changes in TEWL value (figure 7) and the SAXS data support this hypothesis[13].

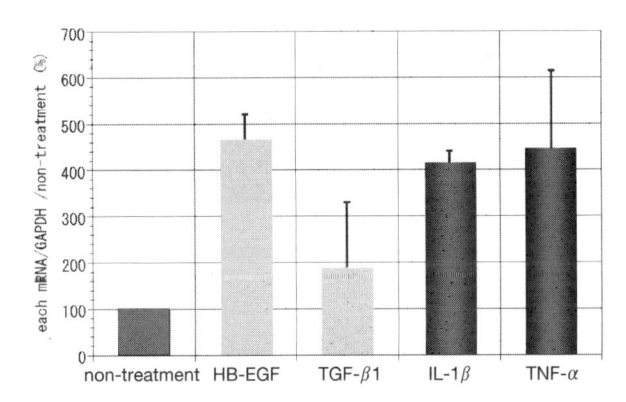

Figure 6 Changes in production of mRNA, HB-EGF, TGF-$\beta 1$, IL-1β and THF-α by NANOCUBE® application

The values were calculated by putting the values of no application to be 100%. The values were measured by applying 30 mg of NANOCUBE® on the back of ddY mouse and sampling the tissues two days after the application.

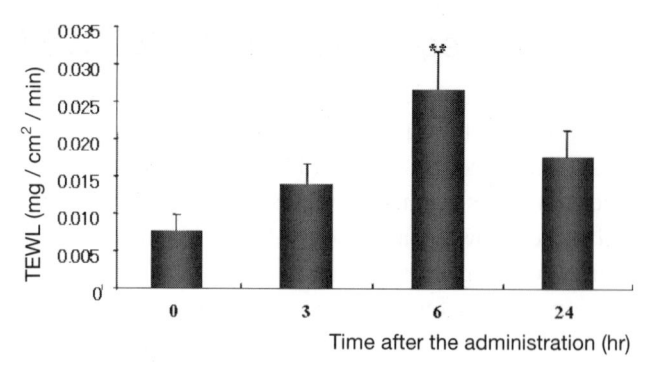

Figure 7 Changes in transepidermal water loss (TEWL) by NANOCUBE® application
NANOCUBE® was applied on the back of a rat, and TEWL was measured. Normal TEWL value is not exceeding 0.020 mg/cm^2/ min. The values were by putting the values of no application to be 100%. The values were measured by applying 30 mg of NANOCUBE® on the back of ddY mouse and sampling the tissues two days after the application.

8. 6 Effects of NANOCUBE® on human

Acanthosis promotes the discharge of melanin that is deposited on the basal membrane and reduces stains. The production of hyaluronic acid in the epidermis and in the dermis improves wrinkles. The application of NANOCUBE® to animals stimulated acanthosis and melanin discharge as shown in figure 5. Therefore, we actually applied NANOCUBE® on a human face. Placebo was applied on the left half of the face of a 42-years old woman, and lotion that contained 30 % NANOCUBE® was applied on the right half of the face every morning and evening for two months. The results are shown in figure 8. The placebo application caused almost no change. On the other hand, the NANOCUBE® application improved stains, which were likely to be solar pigment spots, a dark circle under her eye, and the line on the side of

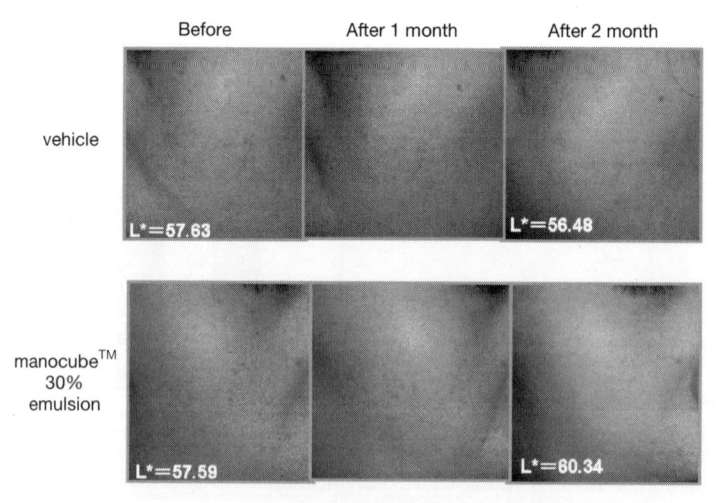

Figure 8 Effects of NANOCUBE® on human
Milky lotion containing and not containing 30% NANOCUBE® was applied on the left and right cheeks, respectively, of a 42-years old woman every day. The discharge ability of stains and L* values (lightness) in 2 months are shown.

the mouth notably. The L*value, which shows lightness, also changed sharply. The test showed that NANOCUBE® is also effective in humans.

8.7 Conclusions and the future prospects

The newly developed nanocapsule technology called NANOEGG® was shown to be a very effective transdermal delivery system. The stratum corneum, which acts as the natural barrier of the skin, has a lyotropic liquid crystal structure, and particles must diffuse efficiently to pass this structure. Our NANOEGG® technology not only improves the penetration into the skin but also gives us new knowledge on skin regeneration.

Skin regeneration has been promoted to improve stains and wrinkles by applying external preparations and cosmetics containing active drugs on the skin. The skin regeneration by NANOCUBE®, which was described in this chapter, involves a completely different concept. It uses no drugs and induces the regeneration of the skin by enhancing the natural autotherapeutic power of a living body. Using the phase transition ability of intercellular lipids, which are in charge of the barrier function of the skin, homeostatic signals were sent to the epidermal cells to activate the cells, which resulted in successful acceleration of the turnover of the epidermis. We are now conducting studies for improving the transdermal penetrability of fat-soluble, water-soluble, high-molecular-weight and low-molecular-weight drugs, proteins and genes using the phase transition of intercellular lipids. We are performing detailed investigations hoping that our technologies innovate skin studies and propel the development of useful tools in the aging society.

References

1. P. Elias K. R. Feingold, *Semin. Dermatol.*, **11**, 176 (1992)
2. K. Kontturi, L. Murtomaki, *J. Controlled Release*, **41**, 177-185 (1996)
3. Y. Yamaguchi, T. Nagasawa, N. Nakamura, M. Takenaga, M. Mizoghuchi, S. Kawai, Y. Mizushima, R. Igarashi, *J. Controlled Release*, **104**, 29, (2005)
4. L. H. Kingman, C. H. Duo, A. M. Klingman, *Connect. Tissue Res.*, **12**, 1139 (1984)
5. R. Tammi, J. A. Ripellino, R. U. Margolis, H. J. Maibach, M. Tammi, *J. Invest. Dermatol.*, **92**, 326-332 (1989)
6. S. L. Lamberg, S. H. Yuspa, V. C. Hascall, *J. Invest. Dermatol.*, **86**, 659-667 (1986)
7. M. Mezei, V. Glasekharam, *Life Sci.*, **26**, 1473 (1980)
8. W. Wohlrab, J. Lash, *Dermatologica*, **174**,18 (1987)
9. A. K. Banga, S. Bose, T. K. Ghosh, *Int. J. Pham.*, **179**, 1-19 (1999)
10. Y. W. Chien, *Insulin Ann. New York Acad. Sci.*, **507**, 32-51 (1987)
11. K. Kontturi, L. Murtomaki, *J. Controlled Release*, **42**, 177-185 (1996)
12. Y. Shiokawa, Collagen diseases rheumatology, Asakura Publishing Co., Ltd., Tokyo (1983)
13. Y. Yamaguchi, T. Nagasawa, A. Kitagawa, N. Nakamura, K. Matsumoto, H. Uchiwa, K. Hirata, R. Igarashi, *Die Pharmasie*, **61**, 112-116 (2006)
14. K. R. Taylor, J. M. Trowbridge, J. A. Rudisill, C. C. Termeer, J. C. Simon, R. L. Gallo, *J. Bio Chem.*, **279**, 17079 (2004)
15. S. W. Stoll, J. T. Elder, *Exp. Dermatol.*, **7**, 391 (1998)
16. M. T. Jennings, J. A. Pietenpol, *J. Neurooncol*, **36**, 123 (1998)
17. Y. Hatano, H. Terashi, S. Arakawa, K. Katagiri, *J. Invest Dermatol*, **124**, 786 (2005)

Chapter 9
Emulsification and Nanotechnology of Silicone and Application to Cosmetics

Hidetoshi Kondo

9. 1 Introduction

Silicone is water-repellent and gives a light touch, and thus is widely used in cosmetics. Silicone fluid of a viscosity of several to several hundred mm^2/s is used as emollient agents, those in the gum state of a viscosity of several million to ten million mm^2/s are used as agents for coating and preventing split hair and conditioning agents to make the hair shiny, and those that take cyclic forms (D4 to D6) and oils of a viscosity smaller than several mm^2/s are used as base materials and solvents of coloring cosmetics and suncare products using their volatility. Silicones have distinctive characteristics that cannot be seen in the other solvents due to their unique molecular structure.

Most silicones are poly-dimethylsiloxane and its derivatives. The structure of poly-dimethylsiloxane is shown in Figure 1. The principal chain consists of silicon and oxygen that are alternatively joined forming siloxane bonds and shows inorganic properties and polarity. The methyl radical, which is organic and non-polar, bonds to the two remaining bonds of the silicone. Having these units of mutually contradicting properties gives silicone its distinctive characteristics. The rotational energy of the principal chain is almost 0, the bond angle fluctuates within the range 130–160°, and the bond distance is large. Thus, the molecule is very flexible and has a large volume. The methyl radicals surrounding the principal chain have small intermolecular force and small cohesion power. The molecular structure results in low surface tension, viscosity that is little dependent on temperature, high moisture and gas permeability, and resistance to crystallization and solidification. The resultant properties are superior water-repellency, lubrication, coating and expansibility, and thus silicones are widely

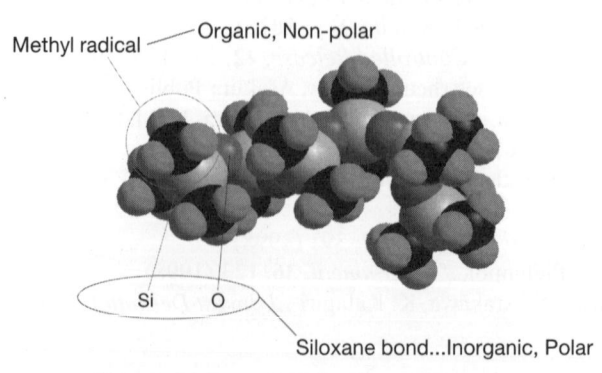

Figure 1 Structure of dimethylpolysiloxane

used as the oil base of cosmetics. The peculiar characteristics of silicones also obstruct stable inclusion in aqueous cosmetics using conventional hydrocarbon emulsifying agents. The below describes technologies used for dispersion and emulsification of silicones.

9.2 Emulsification of the silicone

Table 1 summarizes the various forms that can be taken by silicone in the water phase. Surfactants are necessary because dimethylpolysiloxane is water insoluble. However, due to its properties, silicone is difficult to prepare into a micelle solution using ordinary surfactants. To prepare emulsions, silicone can be emulsified either together with other oily ingredients in the emulsification process of cosmetics or in advance before mixing into the other ingredients. The former has disadvantages of limited emulsifying agents since they must be highly safe and not stimulant as they are to be contained in the cosmetics and need of taking a delicate balance between silicone and emulsifying agent in order to maintain the characteristics of cosmetic products. On the other hand, the latter can choose emulsifying agents just from the viewpoint of emulsifying silicone enabling free formulation and can process silicones that are difficult to mix directly and their derivatives in independent emulsification systems. A classification of silicone emulsions are shown in table 2. The emulsions are described sequentially in the following section.

9.2.1 Emulsion polymerization method (silicone fluid in water)

Emulsion polymerization involves emulsification and polymerization of monomers in micelle activators. Particles grow as polymerization progresses. In widely used processes, polymerization starts from small molecules such as cyclic silicone (for example, D4: octa methylcyclotetrasiloxane). The final particle diameter is affected by the ratio of emulsifying

Table 1 Emulsification and dispersion methods and characteristics

Solubility and dispersibility	Particle size (nm)	Appearance	Thermodynamic stability
Micelle solution	5–15	Transparent	Stable
Solubilizing solution	10–20		
Microemulsion	20–50	Semitransparent to tranceparent	
	50–100	Semitransparence	
Emulsion	100–1000	Pale blue to Milky white	Unstable
	1000 <	Milky white	

Table 2 Silicone emulsions and characteristics

		Particle size (nm)	Polymerism reaction (Fluid viscosity)	Emulsifying agent
Si/W	Emulsion polymerization	30–200	Dehydration condensation (< million mPa·s)	Anionic, Cationic
	Suspension polymerization	200–2000	Dehydration condensation (< million mPa·s)	Anionic, Cationic
		1000 <	Addition polymerization (< ∞)	Nonionic, *etc.*
	Mechanical emulsification	200 <	No reaction (< million mPa·s)	Nonionic, *etc.*
W/Si	Mechanical emulsification	100–1000	No reaction (< approx. 100 mm^2/s)	Silicone

* Si/W: Silicone oil in water type, W/Si: Water in silicone oil type

agent to oil, shear conditions during emulsification, and the polymerization temperature, and is several tens of nanometers the minimum. The reaction results in a transparent emulsion[1,2].

Because this method involves emulsification of materials of relatively low viscosity, it is easy to obtain small particle diameters, does not need high shear emulsifiers under certain conditions, and results in highly viscous polymer emulsion that cannot be mechanically emulsified. It is also possible to introduce functional organic groups and produce copolymers with organic polymers. The resultant emulsions have ionicity because condensation polymerization of silanol is ionic reaction. The emulsions become anionic or cationic depending on the system to combine and the kind of the functional group introduced. High polymerized anionic emulsions and cationic emulsions of amino modified silicone are used mainly for hair care products[3,4].

9. 2. 2 Suspension polymerization methods (silicone fluid in water)

Two polymerization reactions are mainly used in suspension polymerization: the condensation reaction and the addition reaction. The condensation reaction involves dehydration condensation reaction of silicone polymers that have hydroxyl groups on both ends[5], and the addition reaction involves polymerization by hydroxylation addition of a polymer that has a hydrogen (SiH) and a polymer that has an unsaturated radical (Si-Vi)[6].

In either case, the size of emulsification particles does not change during polymerization since the reactions occur within the particles. Therefore, the particle size is determined by the viscosity and polarity of oil, the ratio between the emulsifying agent and oil, and shear conditions during emulsification. Particle diameter is not limited in the addition type. On the other hand, in the condensation reaction, which takes place near the interface, the rate of reaction depends on the particle surface area, and thus the size of particles that are used in practice is about 100–1000 nm.

9. 2. 3 Mechanical emulsification methods (silicone fluid in water)

There are two mechanical emulsification methods: (1) atomizing oil particles in water using the shear power of an emulsifier, and (2) mixing oil, emulsifying agent, and a small amount of oil to prepare a gel of W/Si and then converting the phase into Si/W by adding water and stirring[7,8]. The methods are highly depended on the shear power of the emulsifiers, and there are limits on the particle diameter and viscosity of emulsified oil. For example, highly polymerized materials such as gum need to be pretreated to lower the viscosity by mixing with low viscous oil, but still the resultant particles are as large as several submicrons the smallest. In case of polar oils, such as amino-modified silicone and carboxyl-modified silicone, transparent emulsions of particle diameters of several ten nanometers can be obtained.

A suitable emulsifying agent is POE alkyl ether of an HLB of around 10 HLB. Ionic emulsifying agents and polyether-modified silicone may also be used to improve stability[9].

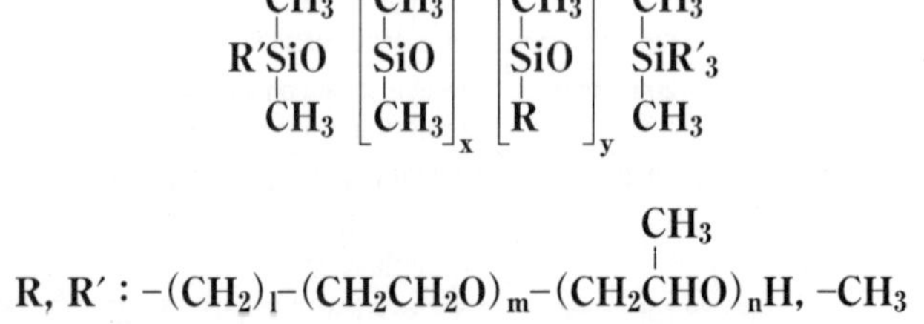

$$R'SiO\begin{matrix}CH_3\\|\\|\\CH_3\end{matrix}\left[\begin{matrix}CH_3\\|\\SiO\\|\\CH_3\end{matrix}\right]_x\left[\begin{matrix}CH_3\\|\\SiO\\|\\R\end{matrix}\right]_y\begin{matrix}CH_3\\|\\SiR'_3\\|\\CH_3\end{matrix}$$

$$R, R' : -(CH_2)_l-(CH_2CH_2O)_m-(CH_2\overset{\underset{\mid}{CH_3}}{C}HO)_nH, -CH_3$$

Figure 2 Structure of polyether-modified silicone

Silicone emulsions prepared using this method is easy to prepare into cosmetics because they can be added into formulations at the room temperature. To maintain the original particle size, emulsions should be added at a last process of manufacture at a low temperature and stirred at a low speed. There are diverse emulsion products available, which differ in the kind of oil and emulsifying agent and particle size, and those appropriate for usage and purpose can be chosen.

9. 3 Polyether-modified silicones

Of silicone fluids, those that can be directly mixed into an aqueous system are polyether-modified silicones. They are polydimethyl polysiloxane with parts of the methyl group replaced by polyoxyalkylene, and include copolymers with hydrophilic polyoxyethylene and hydrophobic polyoxypropylene. The solubility in water changes by the weight ratio of polyoxyethylene in the molecular weight, and decreased (from soluble to dispersion and insoluble and separation) as the ratio decreases as in hydrocarbon system surfactants.

Observation of dispersion in polyether-modified silicone has shown that they self- assemble into spherical, cylindrical and lamellar forms. The structures vary depending on the structure of polyether-modified silicone, and it has been reported that the structures agree with the molecular packing model calculated from the cross section of the hydrophilic group and of the volume and length of the hydrophobic group[10,11].

Polyether-modified silicone was developed as foaming agents for producing urethane foams. Today, they are used for wide applications, such as paint additives, dispersants, emulsifying agents, spreading agents, anti-fogging agents and surface modifiers. It is superior to hydrocarbon system surfactants in adsorbing performance of the oil-water interface due to its high flexible principal chain and in reducing surface tension due to its side chain methyl radicals that have low surface tension. Products of various properties can be designed because three factors of the polymerism degree of the main chain, the number of the side chain polyether groups added and the polymerisatim degree of the polyether basis can be changed arbitrarily[12].

As described above, polyether-modified silicones are used as a supportive emulsifying agent to improve the stability. The following sections describe the use as the main emulsifying agent for preparing thin nanodispersion and vesicles.

9. 4 Nanodispersion of silicone fluids

Silicone fluid is mixed into thin aqueous formulations such as lotion or torner usually by dissolving and dispersing water-soluble polyether-modified silicone. It has been difficult to stably mix water-insoluble dimethyl silicone fluid. Today, low viscous dimethyl silicone fluid can be easily nano-dispersed by delicately adjusting the pro-silicone pro-water balance[13]. Minute emulsion of a particle diameter of about 10 nm can be prepared by dissolving silicone fluids and polyether-modified silicones in ethanol and adding the solution in a water phase. It is likely because the dissolution retentivity decreases as the ethanol, which is the solvent, is diluted with water, resulting in separation of the oil phase and stabilization by the surfactant. Unlike conventional microemulsions, they need no special devices such as high pressure homogenizers, but special attention should be paid on the conditions of manufacture, such as temperature and stirring conditions.

The process enables low viscous silicone fluid to be mixed in transparent to semitransparent lotion. Together with the use of silicone system surfactants instead of the conventional hydrocarbon system surfactant of high HLB, today lotion that is not sticky but smooth can be

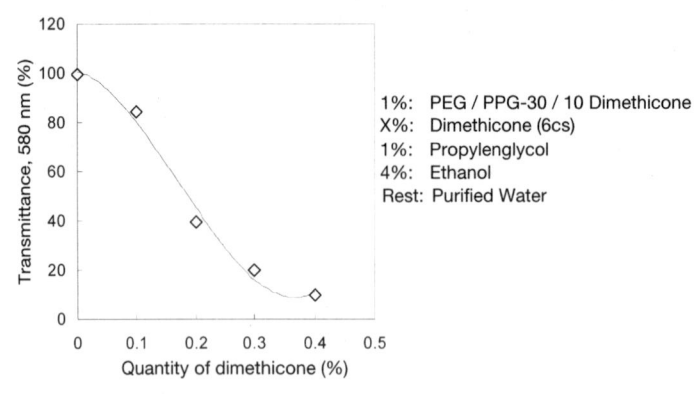

Figure 3 Relationship between the quantity of dimethylsilicone fluid added and light transmittance

Figure 4 Application to cosmetics: Lotion with dispersed silicone fluid

prepared.

9.5 Silicone vesicles

As described above, polyether-modified silicones self-assemble in various forms in water. Polyether-modified silicones, which can form lamellar liquid crystals, can produce vesicles surrounded by septa of bimolecular films[14].

The form of assemblies depends on the structure of polyether-modified silicones. The forms are shown in figure 3 for the weight ratio of the silicone fraction (Y axis) and the modification rate of the polyether side chain (the ratio of Si with modified side chain) (X axis). As the ratio of silicone increases and as the ratio or modified side chain decreases, the form changes from micelle to self-assembled vesicle and finally to compulsorily formed vesicle.

Self-assembled vesicles can be obtained by mixing polyether-modified silicone and water at a suitable concentration and stirring lightly. Although single-layered, multi-layer, large and small vesicles are formed intermixed, the vesicles become uniform in size (of about 30–100 nm) and in form (single-layered) by applying supersonic waves and high shear force[15]. However, it easily receives effects from the outside environment and changes its form or is destroyed by temperature, concentration and presence of other surfactants, and thus the use as stable drug

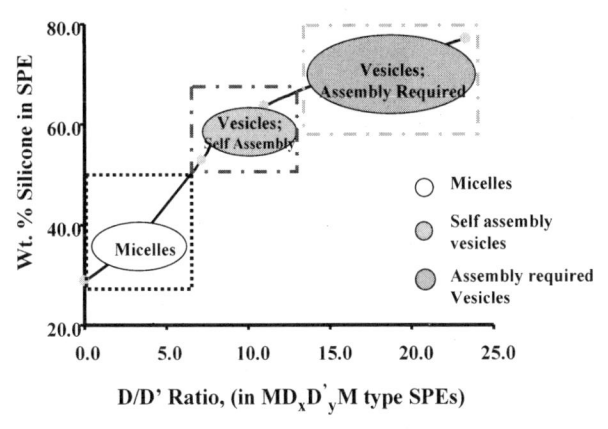

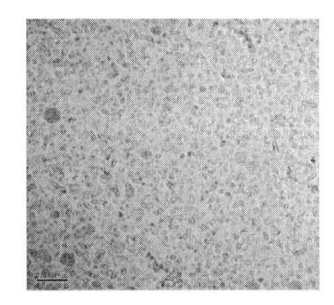

Figure 5 Relationship between the structure and form of self-assembly of polyether-modified silicone

Figure 6 A Cryo-TEM image of compulsorily formed silicone vesicles

is difficult.

Hydrophobic polyether-modified silicones, which contain a large percentage of silicone, are insoluble to water and do not self-assemble into vesicles. Particles of a uniform size of about 10 nm can be obtained by applying high shear force. Alcohol is then removed if necessary[16]. Such a compulsorily formed vesicle resists effects of other surfactants, especially nonionic surfactants, and is possibly mixed stably in cosmetics[17].

The thickness of surfactant bimolecular films is 10–20 nm, and the diameter of a smallest single-layered vesicle is about 100 nm on average. Because water insoluble materials can be enclosed in the oil phase within the bimolecular films and water-soluble materials can be enclosed in the water phase within the vesicles and between the bimolecular films, the vesicles are likely to be applicable as nanocapsules for delivering and protecting active agents in various medicines, vitamins, sunscreening agents, whitening agents and moisturizing agents.

9.6 W/Si nanoemulsions

Minute dispersion of the silicone fluid in a water phase in forms nanoparticles has been described. Finally, minute dispersion of water phase in silicone fluid is described. Generally, emulsification of silicone in cosmetics denotes W/Si emulsification, in which polyether-modified silicone is an indispensable emulsifying agent. The resultant emulsion has a light touch and water resistance because the outside phase is silicone fluids. Another characteristic is a high internal water phase ratio of around 70 %. The particle size of the water phase is usually at least 1 μm, and the touch when applied on the skin, viscosity and stability depend on the size[18].

Transparent gel is obtained by dispersing a certain kind of polyether-modified silicone, which has a specific structure, in silicone fluid and adding and uniformly dispersing a small amount of water, which increases the viscosity of the system. A three-component system diagram is shown in figure 7. As shown in figure 8, the viscosity suddenly increases at around 1 % and reaches a constant value at around 3 %, changing the system into milky white cream of W/Si emulsion[19]. As shown in a schematic diagram in figure 9, the entire system is believed to become viscous because polyether groups of the polyether-modified silicone dissolved in silicone fluid form an artificially bridging structure by hydrogen bond via water molecules, resulting in a structure resembling nanoparticles being dispersed in the silicone fluid[20].

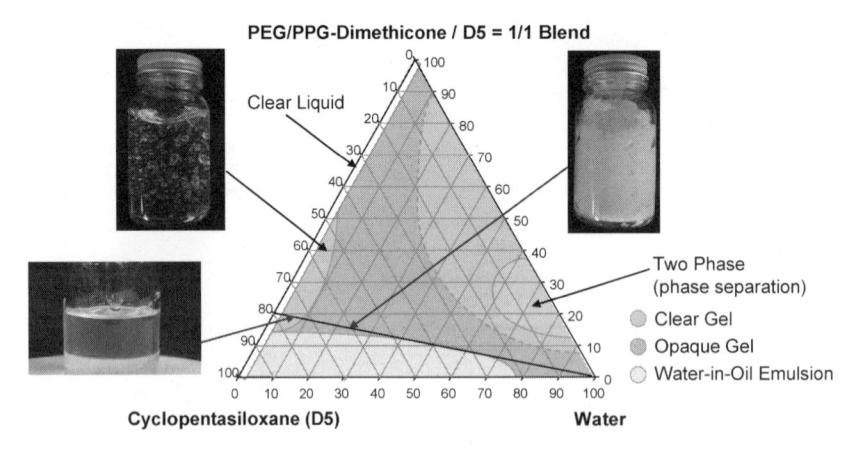

Figure 7 Three-component system diagram of polyether-modified silicone, D5 and water

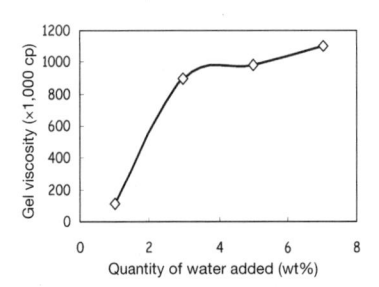

Figure 8 Relationship between the quantity of water added and gel viscosity (polyether-modified silicone: D5 = 80:20)

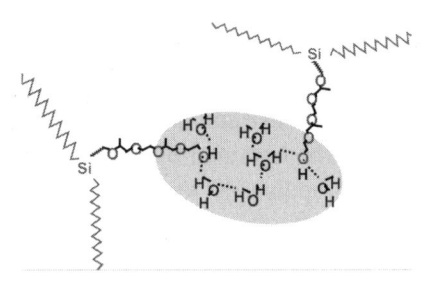

Figure 9 Schematic diagram of gel consisting of polyether-modified silicone, D5 and water

9.7 Conclusions

This chapter described methods for preparing silicone emulsion, which is to be added into cosmetics, and technologies for directly emulsifying and dispersing silicone in cosmetics, as methods of micro- and nano-dispersion of silicone fluid. Especially the latter is useful to use polyether-modified silicone as emulsifying and dispersion agents.

Silicones have unique properties that cannot be seen in organic materials and thus are used in diverse fields. Emulsification technologies should be developed to make best use of the properties of silicones. Thus, development of silicone surfactants, such as polyether-modified silicone, and application technologies is indispensable. Silicone surfactants have many structural change factors, and a number of variations can be formed. Understanding their surfactant properties will certainly lead to discoveries and creations of new functions and applications. Development of new technologies such as silicone vesicles is a good example. The author expects much on future development in this field.

References

1. Publication of unexamined application 63-13052, Publication of unexamined application S 62-

141029
2. Publication of unexamined application H 4-227932
3. Publication of unexamined application S 63-130518
4. Publication of unexamined application S 63-307810
5. USP 3 294 725 (SP), Publication of unexamined application Hei 11-71459
6. Publication of unexamined application 2000-281523, Publication of unexamined application 2000-34205
7. Publication of unexamined application 8-198969
8. Publication of unexamined application S 63-551230
9. Publication of unexamined application H 4-285665, Publication of unexamined application H 148010
10. Iwanaga et al., *Surface*, **37** (4), 38 (1999)
11. R. Hill et al., *Langmuir*, **9** (11), 2789 (1993)
12. Harashima et al., *SCCJ*, **27** (3), 484 (1993)
13. Publication of unexamined application
14. S. B. Lin et al., "Development of A Novel Silicone-based Vesicle Delivery System", 23rd IFSCC (2004)
15. Publication of unexamined application Hei 7-323222 (R. Hill Vesicle)
16. WO 2005103157, WO 2005103118, and WO 2005102248 (SBLin Vesicle)
17. S. B. Lin et al., "Novel Silicone Vesicles for Delivery of Cosmetic Active", 24th IFCSS Congress Abstract 19B-60 (2005)
18. Kondo, Fragrance Journal extra edition No.19, 112 (2005)
19. Hamachi, The 3rd Symposium of the Society of Silicon Chemistry, Japan, Abstracts (1998)
20. Matsumoto, Rheology for Colloidal Chemistry, Maruzen Co. Ltd., p.104 (2003)

Chapter 10
Bio-Drug Delivery System

Laurent Martin

10. 1 Introduction

The development of the first liposomes in the 1930s led to their first use for cosmetic applications in 1984 by Louis Vuitton Moet Hennessy and L'Oréal. Later, the development of the first microspheres and microcapsules in the 1960s set the stage for their first industrial development in 1986 by BASF, and their first use for cosmetic applications in 1987. With the passage of time it has become evident that encapsulation is an excellent method to modify the bioavailability of many beneficial active compounds and to overcome the inherent instabilities of some active compounds. Encapsulation technologies have provided consumers with visible innovations in formulations containing already known and commonly used ingredients. This leap in technology has provided additional marketing flexibility to increasingly educated consumers.

Today, encapsulation of active ingredients is becoming imperative in the design and formulation of cosmetic products. The primary challenge in adding an active compound to a formulation is in the short duration of the compounds effectiveness following its application to the skin. Typically, if the active is not protected by encapsulation, it may react spontaneously with other ingredients in the formulation, as well as with oxygen in the air. Such interaction can greatly inhibit the active compound's bioavailability to the skin. It is often difficult to use higher levels of the active compound in the formulation because this may cause skin irritation or toxicity as a result of the strong cellular metabolization. Repeated product application using low active levels may effectively solve such problems, but application of a cream every two to three hours does not fit with the limitations of daily life and, therefore, this approach is not a sensible solution.

The enzymatically activated encapsulation technology that is described in this chapter is a true biodegradable-drug delivery system (Bio-DDS). In this system, there is no modification of the physical or chemical properties of the formulation. Further, the system is environment insensitive and no change in the environment can induce release of the active compound. This specific encapsulation approach provides formulations that are stable in the challenging environment that the cosmetic chemist must address. This approach provides a high level of comfort that the active compound will only be released when the formulation containing the encapsulate is applied to the skin.

Enzymatically activated encapsulates can be considered as very small reservoirs comprised of biodegradable polymers. Such systems have considerable advantages. For example, they are biocompatible and provide a non-toxic transport system for the actives chosen. They act as an active compound protector, and enable the development of a progressive release system for active ingredients. The approach also provides a tool for modulation and control of active ingredient penetration. It is a perfectly adapted delivery system for innovative and effective cosmetic formulations.

10. 2 Limits of current technologies

The term encapsulation covers a wide variety of technologies. Distinctions within this broad class of technology can be made depending upon the industrial area of application involved. In the cosmetic field, many techniques are available and this makes the choice of an optimal approach for the formulators quite difficult. In order to narrow the choices of technique to the most appropriate ones, there are two questions that are important to consider when selecting the most appropriate encapsulation technology.

10. 3 Overview of trigger release mechanisms

The most important question to answer before selection of the appropriate encapsulation method is, "How will the release of the encapsulated active compound be triggered?" Most of the encapsulation technologies employed in the pharmaceutical, food, or glue industries are not appropriate for use in the cosmetic field. Very acidic pH such as in the stomach, which is a pH of 2, very high temperature such as the 80°C reached during cooking for release of flavor, or strong mechanical strength, are not relevant for cosmetic purposes. For this reason, a variety of specialized technologies have been employed to trigger the release, for cosmetic applications of useful active compounds.

10. 3. 1 Release by enzymatic digestion

Biodegradable polymers can be employed as constituents of the microsphere or microcapsule membranes. These polymers can be used to trigger the release of encapsulated active compounds. The goal, in this case, is to obtain release of the entrapped material after application of the particles onto the skin surface. Active release is directly linked to enzymatic digestion of the polymers. In this approach, the triggering mechanism is very specific because the encapsulated active compound is only released when the microsphere or microcapsule is in contact with some enzymatic activity on the skin's surface. With the exception of cosmetic formulations purposely containing enzymatic hydrolytic activities, microspheres or microcapsules are able to release their encapsulated active compounds when they are in contact with skin only. This concept is the basis of a patented delivery system, also called "Bio-DDS." This technology has been developed and promoted by BASF since 1988.

10. 3. 2 Enzymatic release: an exact approach

Protease enzymes are naturally present on the skin surface, examples include stratum corneum trypsic enzyme (SCTE) and stratum corneum chymotrypsic enzyme (SCCE). Both of these enzymes have been widely described in the literature. Other enzymes present on the skin's surface are proteases that are secreted by microorganisms.

The selection of polymers that can be easily degraded by one of these specific enzymes, or microorganisms is a "smart" way to target the delivery of an active compound to the required site of action.

10.4 Micro- and macrosized particles for enzymatically activated technologies

Many applications have been developed for microcapsules and microspheres in cosmetic applications. Some of BASF's tradenames for these micro delivery systems are known as Thalasphere®, Phytosphere®, and Cylasphere®. Using these technologies, it is possible to topically target and delay the delivery of active compounds such as vitamins, bactericides, collagen boosters, fragrances, anti-oxidants, and UV filters. Examples of targeted delivery include cosmetic treatment of oily skin, dandruff treatment, and bactericidal activity targeted on the microorganisms involved in anti-perspirant and deodorants. Other examples include anti-aging formulations and delayed release of fragrances. Some of the components of the membranes that have been used to create BASF's micro-delivery systems are marine collagen, wheat protein, soy protein, lupine protein, pea protein, locust bean gum, acacia, cyclodextrins, oligosaccharides, guar, and cellulose derivatives.

10.4.1 Marine collagen

Historically, the first biodegradable polyaminopolymer that was used in BASF's fabrication process of a microsphere or capsule delivery system was collagen. This polymer is now extracted from the white side of flat fish and demonstrates extraordinary film-forming properties. This capability makes this polyamino-polymer an excellent candidate for membrane formation during encapsulation process. Marine collagen has a very high quality in terms of its mechanical resistance, and also provides mildness during skin application. These unique properties are very difficult to mimic with any other polymer, and the material is widely used in the cosmetic field.

The first step in the encapsulation process of a hydrophilic or lipophilic active compound using marine collagen is to solubilize it in a water-based gel of marine atelocollagen. The resulting solution is then poured into a low-viscosity fatty ester. In some cases, this emulsification process may be enhanced by the addition of a surfactant, under controlled stirring. Once the initial emulsion droplets are obtained, the interfacial polymerization is then initiated at the oil/water interface by adding a dicarboxylic acid, such as azelaic acid or sebacic acid, in their reactive forms of di-anhydrides or di-halides. The carboxylic acids then react with the lateral amino groups of the collagen and form covalent amide groups. This reaction forms a membrane around each droplet. Thus, once the interfacial polymerization is complete, these droplets are transformed into individualized microspheres or microcapsules that contain the active ingredient. The microparticles are then separated from the fatty ester media by centrifugation and washed intensively using fatty esters. Therefore, they are washed with water until all the impurities are removed and a concentrated suspension is obtained. The resulting microparticles are then suspended in a selected media that is compatible with the intended cosmetic formulation.

Using the technology described above, the resulting products are known as Thalasphere®. Thousands of different active compounds have been encapsulated with this method, and the resulting products are widely used in various cosmetic formulations. Examples of these include oil-in-water or water-in-oil emulsions, water-based or oil-based gels, and anhydrous preparations as well. Examples of the latter include lipsticks, mascara, foundations, and various creams and gels.

Thalasphere are sensitive to degradation by specific proteases, such as trypsin and chymotrypsin. These proteases are naturally present on the skin's surface. Thalasphere technology is often selected when a formulator wants to provide the controlled release delivery of an active material in a skin care product.

10.4.2 Plant proteins

Applying the same technical principals used to create the Thalasphere, BASF created the Phytosphere® technology using plant proteins for the formation of the microsphere barrier. Various plant proteins have been used for this technology and they include high molecular weight wheat protein, pea protein, lupine protein, soy protein, and corn protein.

In these cases, as previously described, the amino groups of the proteins are reacted with reactive forms of the dicarboxylic acids, in order to form amide linkages. The microspheres and microcapsules thus obtained using different membranes have a different sensitivity to proteases that are present on skin's surface. For example, when the membrane of the Phytosphere is formed with wheat protein, it is more sensitive to the enzymes released from the bacteria typically found under the arms. This property could be useful when developing a product for anti-perspirants or deodorants. Different membranes could then be selected depending on the expected properties of the final formulated product.

10.4.3 Polysaccharide-based encapsulation

The Phytosphere technology has been extended to include the use of natural biodegradable polymers containing hydroxyl groups. Examples include plant-based polysaccharides or oligosaccharides. Various molecules used for this approach include locust bean gum, guar gum, acacia gum, cellulose derivatives, alginate derivatives, and cyclodextrins themselves. These latter materials typically used as molecular encapsulates can also be used as polymers in the Phytosphere interfacial polymerization process. This approach provides a double encapsulation system that is particularly efficient for the slow release of low molecular weight and partially hydrophobic active compounds. In this case, as described previously, the hydroxyl groups of the proteins are reacted with reactive forms of dicarboxylic acids and form ester bonds. By means of this approach, it is possible to design membranes that have a specific sensitivity to different proteases found on the skin's surface. For example, when the membrane of the Phytosphere is formed with locus bean gum, it is more sensitive to the amylase enzyme. Amylase is present in our saliva; therefore this type of Phytosphere could be useful in a product developed for the lip area.

10.4.4 Nanoencapsulation

When the technologies presented above have been adapted for the production of particles that have sizes below one micrometer, the resulting products are called nanoparticles. The main difference between the above mentioned process conditions and the ones required to produce nanosized particles is the process used to produce the nanoemulsion. High shear rate or homogenization equipment is commonly used in this process in order to produce particles below 1 μm. In view of their small size, the nanoparticles are quite difficult to separate and wash from their suspension media. The final nano-Thalasphere and nano-Phytosphere AQ- (s) or emulsions or solutions or suspensions? are commonly used in cosmetic applications when penetration into the hair follicle or long-lasting release properties are desired since their ultra small size makes them valuable in this regard.

10.5 Properties and performance of micro- and nanospheres and capsules

The release mechanism of the active compound is time dependant and is, therefore, potentially useful for a wide variety of applications.

AQ. Suggested rewording.

10.5.1 Enzymatic digestion *in vitro*

It is possible to quantify the sensitivity of a micro- or nanosphere membrane with respect to different enzymes and characterize the time-dependent degradation in vitro. Using microscopic evaluation, different enzymes are combined, one at a time, with different microspheres. This technique enables an evaluation of the sensitivity of each microsphere to different enzymatic degradation.

The time dependant enzymatic digestibility of biodegradable microspheres, using in vitro enzymatic digestion, may be followed using light microscopy.

Enzymatic sphere digestion. When microspheres are incubated in the presence of different enzymes such as proteases, these microspheres are digested and their active compound is released. This phenomenon is observed in figure 1 a-c using light microscopy under mid-range enlargement (× 20) at T = 3 and T = 24 hours after incubation.

Ex vivo experiments may also be performed to follow the enzymatic digestion process using the Franz cell diffusion technique. The Franz cell diffusion device has two compartments, the donor compartment and the receptor compartment, separated using a human skin biopsy. The penetration versus time of active compounds, free or encapsulated, can be followed using standard analytical techniques on the samples taken on receptor media (figure. 2).

Using this technique, the microspheres are applied on the surface of a human skin biopsy. This experiment can be conducted with or without enzymes that are able to degrade the membranes of the microspheres. After twenty-four hours, the amount of active compound that is able to diffuse through the skin is then quantified by means of the

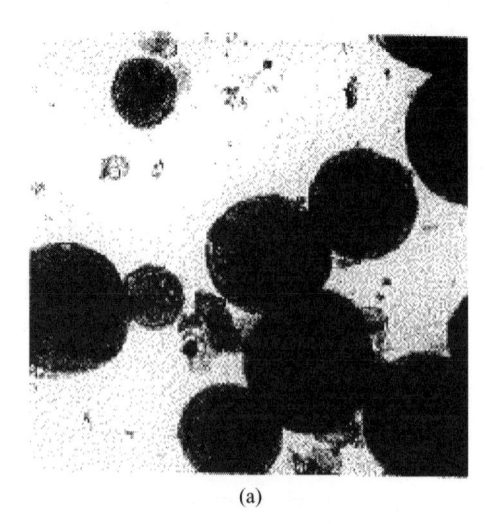

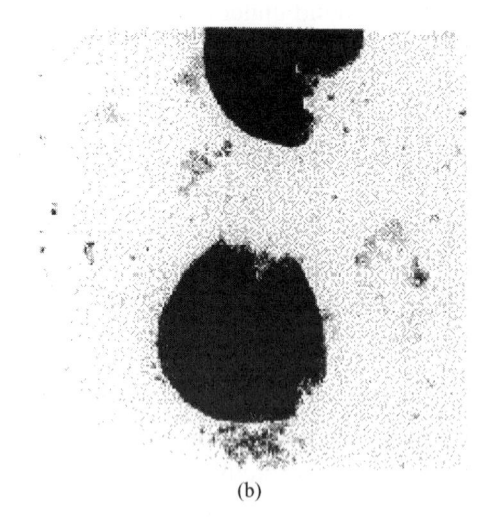

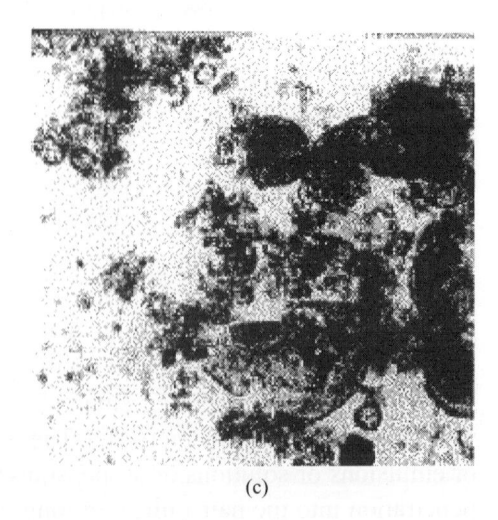

(a)

(b)

(c)

Figure 1 Enzymatic sphere digestion at (a) T = 0, (b) T = 3, and (c) T = 24

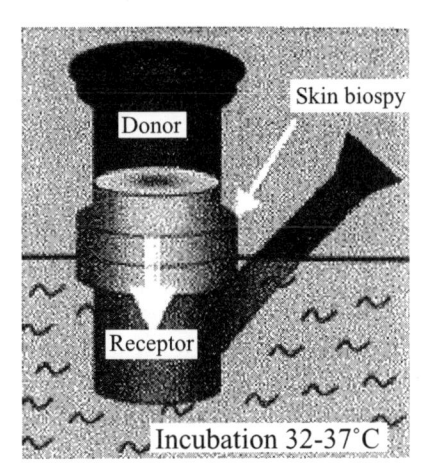

Figure 2 Franz cell diffusion device

most appropriate analytical technique. Most often, high performance liquid chromatography (HPLC) is used to measure the quantity of an active compound that has either penetrated through the skin or has been stored by the skin.

10.5.2 Penetration vs storage

One of the main capabilities of BASF's encapsulation technology is to modify and control the degree of skin penetration by the encapsulated active compound. In this regard, two main properties may be evaluated: improvement of penetration and sustained delivery effectiveness.

Penetration improvement. There is often a need to improve skin penetration of an active compound when it is lipophilic or insoluble. Usually, such materials are either unable to cross the barrier function of the epidermis or they are unable to reach the target site deeper in the dermis. One example of such a target site is the melanocyte cells and the situation where a degree of pigmentation modulation is desired. Another example is the targeting of the fibroblast cells involved in connective tissue formation through cell multiplication, or collagen synthesis. Using a Franz cell diffusion device quantification of the degree of penetration by encapsulated active compounds can be evaluated. With this method, the active compound is usually tested "free vs encapsulated." Typically, the active compound is applied in the donor compartment on the surface of a human skin biopsy. Thereafter, the concentration of actives is measured in the receptor compartment by an appropriate analytical technique. Penetration is followed for twenty-four hours, and, the concentration of the "free" active is compared to the encapsulated active. Improvement in penetration is often due to the nature and quality of formulations used to transport active compounds across the skin's barrier. Significant improvements in penetration can be obtained by using vectors of penetration that are based on deformable lipidic or amphiphilic structures such as liposomes, secondgeneration liposomes, nanoemulsions and inverted liposomes. Figure 3 demonstrates the enhanced Figure penetration of an active compound that is specific for tyrosinase inhibition. Looking at the tyrosinase inhibition in the in vivo model, it has been possible to demonstrate that 0.3% of an active compound able to whiten the skin is four times more effective for this activity when used encapsulated than when used free.

In this figure (figuer 3), a comparison has been made between the diffusion of an active by

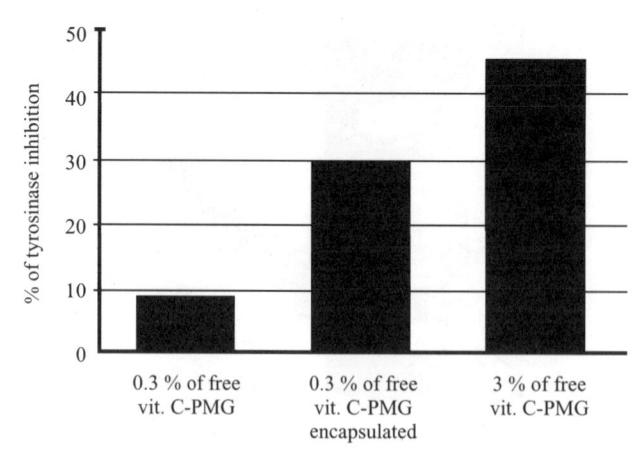

Figure 3 Percent tyrosinase inhibition comparing free vs encapsulated vitamin C-PMG

itself, compared to the diffusion through skin, of the same active compound that has been incorporated into a biodegradable protein-based liposome. As seen from the figure, a 40% improvement was detected using the Franz cell diffusion device.

Biodegradable protein-based liposomes, also known as second-generation liposomes, may be observed using transmission electronic microscopy (figuer 4).

Sustained delivery. In personal skin-care formulations when the active compound penetrates the skin too rapidly and in this process induces skin intolerance and irritation, there is a need to slow and control the rate of penetration. Such systems are called sustained delivery systems. They are primarily used in the pharmaceutical field to delay the rapid and deep diffusion of an active compound, sometimes called "burst effect" when such behavior produces severe and undesirable side effects.

Quantifying sustained delivery profiles. Measurement of sustained delivery behavior for an encapsulated active compound can be performed in vitro using a Franz cell diffusion technique, or in vivo using tape stripping. These techniques are able to quantify the amount of active compound stored in the upper part of the epidermis. As we have stated previously, such studies are normally conducted to compare the concentration of free active compound to the concentration of encapsulated active compounds ability to penetrate, to be stored and to be slowly released.

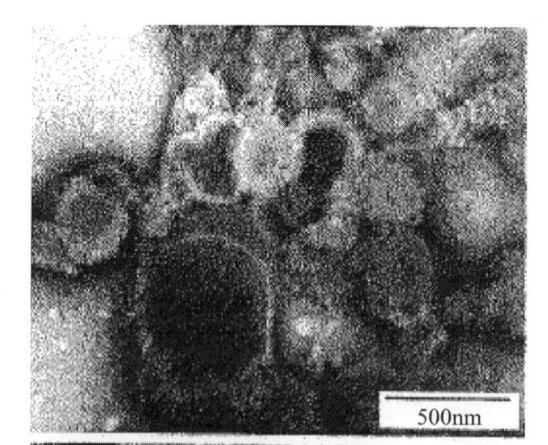

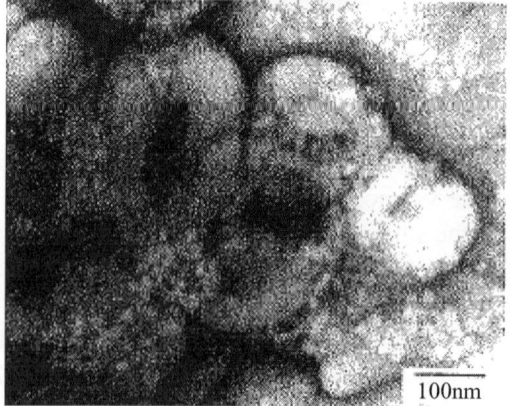

Figure 4 Transmission electronic microscopy of second-generation liposomes

The first step is to measure penetration. This is measured by the amount of active compound that can pass through the skin biopsy into the receptor compartment in the first 24 hours. In the second step, the skin biopsies are then rinsed, and the amount of active compound able to pass across the biopsies is collected in the receptor compartment for the next 24 hours. The amount of active released after rinsing is called "release." The release is a measurement of the amount of active compound stored in the skin after the twenty-four hour contact and the rinse period. During the third step, the Franz cells are dismounted again and the skin biopsies are extracted with a solvent that is able to remove any trace amount of active compounds stored in the skin. The amount of active compound that can be extracted with the solvent is called "stored." figures 5 and 6 demonstrate diffusion, release, and storage of the active compound in the skin biopsy, when comparing free versus encapsulated concentration of active compound.

Figure 5 shows that, using a Franz cell diffusion device, it is possible to follow the penetration of vitamin A into the skin versus the time. In this figure, it is possible to see that encapsulation reduces the penetration of this active compound providing a reservoir effect and a drug release property of this molecule into the skin.

Figure 6 is a comparison of the release and storage of free and encapsulated retinol. After twenty-four hours of diffusion using a Franz cell diffusion device (figure 2), the human skin biopsy is dismounted and the quantity of retinol able to release by the skin during the next twenty-four hours is measured (release). After this time, an ethanolic extraction is performed on the biopsy and the quantity of retinol is evaluated (storage). This experiment demonstrates the increase of retinol content into the skin for those two parameters when retinol is used encapsulated compared to the free form.

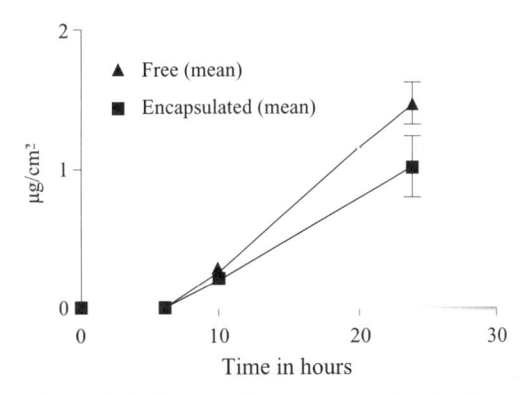

Figure 5 Diffusion of free vs encapsulated retinol

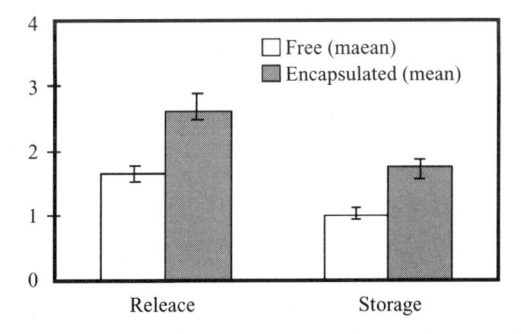

Figure 6 Comparison of release and strong of free and encapsulate retinol

Figures 5 and 6 demonstrate the excellent ability of the right delivery system to slow down the diffusion of a very irritating well-known active ingredient used for its efficacy in the personal care field. Retinol, when applied directly to the skin is too harsh and will cause irritation. Using the Franz cell diffusion device, it is possible to demonstrate the ability of the delivery system to slow down the diffusion of the active for a progressive and mild delivery of the retinol through the skin. The storage and release parameters are both strongly improved allowing the retinol to become available to the skin over a long period of time for greatest efficacy.

The release and storage improvements observed in figure 6 are due to the nature and quality of formulations based on microcapsules used to transport the active compounds. For a given formulation, significant improvements can be obtained using delivery systems that are based on non-deformable micro-/nanospheres or micro-/nanocapsules. These non-deformable structures stay on the surface of skin, thereby providing long-term storage on the top of the skin's structure.

The size of the particles is very important in controlling the increased storage capacity of the skin. Using microcapsules, with a median size of 50 μm, the storage capacity of the encapsulated active compound in the skin is approximately several hours. Longer storage times can be obtained for as much as a twenty-four hour period using nanocapsules with a size of about 500 nm.

Nanoencapsulation made with non-deformable structures is different from liposomes, which are deformable structures with regard to their penetration effectiveness. Nanocapsules do not have the ability to enhance the penetration of active compounds but they are able to enhance the long-term delivery of the active compound, as well as the storage capacity of the skin for an active compound. This distinction is based on the ultra small size of the nanocapsule and hence their ability to migrate deeper into the stratum corneum than the larger microspheres or microcapsules.

10.5.3 Pharmacokinetic

It is also possible to follow the biodelivery system effect of the different microencapsulate technologies when comparing the effectiveness of a concentration of encapsulated active compounds versus a concentration of free active compounds by means of in vivo experiments. The highest performing, and most relevant, tests are kinetic studies and these are usually performed in the pharmaceutical field. Using pharmacokinetics, the distribution in the body of a radio-labeled active compound, and the clearance of a selected radio-labeled active compound, both encapsulated and free, are followed over time in an animal. Figure 7 shows an experiment that has been done using this methodology.

Three formulations (oil-in-water emulsion) containing free or micro- or nanoencapsulated paba was applied once on animal skin. The pharmacokinetic of this radioactive compound versus time was followed in the urine of animals (figure 7). Used free, the major amount of the active is secreted within the first two days, which indicates a massive penetration and elimination of this compound sometimes called a burst effect. The encapsulation of this active compound into micro- and nanocapsules allows a strong reduction of this release and a slow delivery.

The bioavailability of the active into the skin was also measured (figure 7). Seventeen days after the single application, human skin biopsies were analyzed for their active compound content. This figure demonstrates that a stronger content of active compound could be detected in the skin when nanocapsules have been used compared to the quantity that is able to be detected using microencapsulated active compounds or free ones.

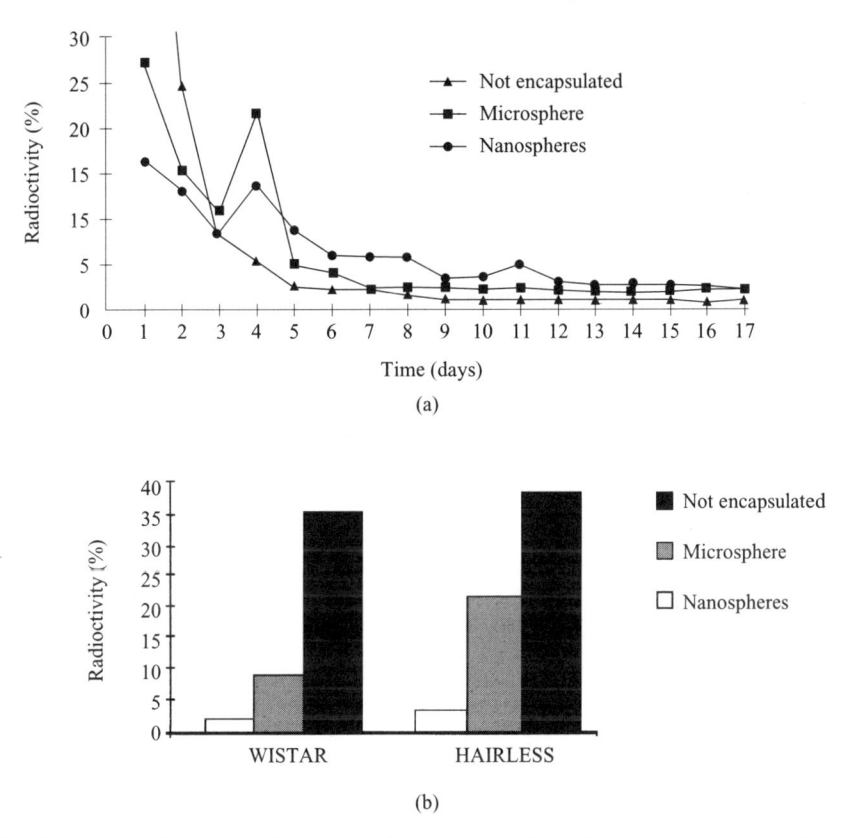

Figure 7 Pharmocokinetic results of an encapsulated compound (radioactive paraminobenzoic acid) used free and in micro-and nanosize particles. Radioactivity recuperated into the (a) urine and (b) skin.

The results demonstrate that not only the release effect of the radio-labeled compound is stronger with micro- or nanocapsules, but that the incorporation of this compound inside the tissue is greater. This study demonstrates there is an increase in the bioavailability of the active compound in the area of the skin where the active compound has been applied.

10.5.4 Membrane selection
Membrane design for the microencapsulation of active compounds into microencapsules or rigid nanoencapsules can be powerfully tailored. This approach is based on knowledge of microorganisms or enzymes present at the site where the encapsulated active compound is to be applied and released when the encapsulate encounters the microorganism enzymes that will digest the encapsulate.

Figure 8 demonstrates the sensitivity of five different membranes to degradation by the enzyme trypsin. This enzyme is naturally present on the skin's surface. Different microspheres were prepared using different polymers such as acacia, locust bean gum, cyclodextrins, marine collagen, or wheat protein. After incubation with trypsin, active compound release was evaluated and the polymers could be ranked from the susceptibility of enzymatic hydrolysis.

It is clear from this study, that a microsphere membrane based on wheat protein is far more sensitive to degradation by trypsin than one based on locust bean gum, cylclodextrin, collagen, or acacia gum based spheres. Thus, by tailoring the delivery system to the enzymes typically present at the targeted delivery site, a whole new technology has emerged that provides

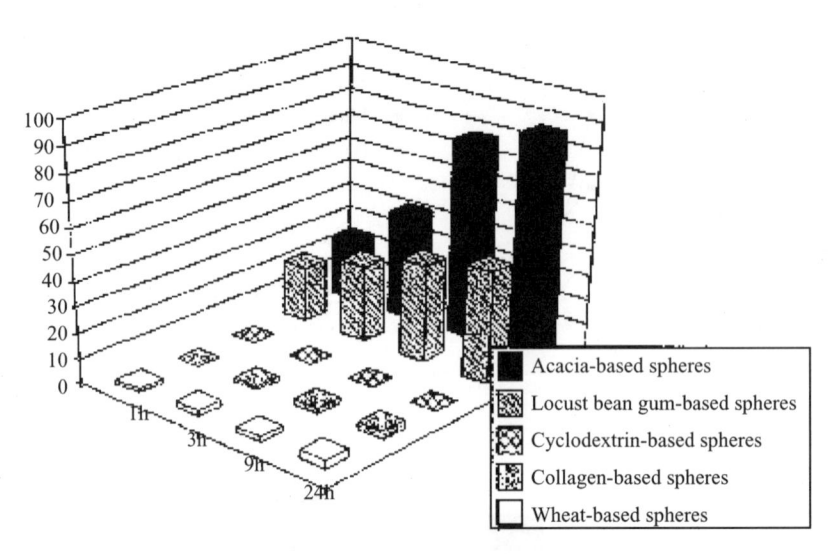

Figure 8 Enzymatic hydrolysis of different types of microspheres by trypsin (3.32 MU/g of wet spheres).

formulators with a novel approach and a range of new product distinctions.

10. 6 Perspectives and conclusions

Today, cosmetic chemists cannot use encapsulated active compounds without first thinking about how to trigger their release, and then thinking about how to target the release. Triggering and vectorizing represent the second and third generation of drug delivery systems (DDS) in the pharmaceutical field. Cosmetic chemists and biologists regularly follow the development and evolution of these new drug delivery systems. There is now a significant interest, and enhanced understanding in the use of such delivery systems because stronger and stronger active compounds are now being used. Some of these may present some undesirable side effects. Cosmetic delivery systems based on enzyme-triggered technology are now available as tools that should be present in formulator's minds when vectorization, penetration, slow delivery, and better tolerance of active compounds are desired.

Acknowledgements

This article was created thanks to the active support of BASF Beauty Care Solutions' current and former colleagues including Mr. Eric PERRIER, Ms Janice HART and Dr. Isabelle BONNET.

Chapter 11
Lamellar Vesicles of Monoglyceride

Toshiro Sone

11. 1 Introduction

The skin of mammalians maintains its barrier function and hydration by stratum corneum lipids (SCL) that take a lamellar structure[1–5]. The lipids are secreted from keratinosomes, when granule cells become horny cells in the epidermis, and form stratified lamellar sheets among the horny cells. The author[6] has noticed the importance of the lamellar structure lipids in controlling the water content in the skin based on the observations that removing SCL from the human skin using an organic solvent resulted in decreases both water content and holding capacity of the stratum corneum and that the water content was restored only when the removed SCL was returned to the skin while keeping the lamellar structure (figure 1).

The main component of SCL is ceramide. However, ceramide alone cannot form vesicles that have a lamellar structure. Lamellar vesicles are formed only when the ceramide is combined with acidic lipids, such as free fatty acids[7,8]. The author investigated the effects of each component of SCL on the skin and vesicle formation, and found that amphiphilic monoglyceride is highly capable of forming vesicles[8] and the resultant vesicles showed a

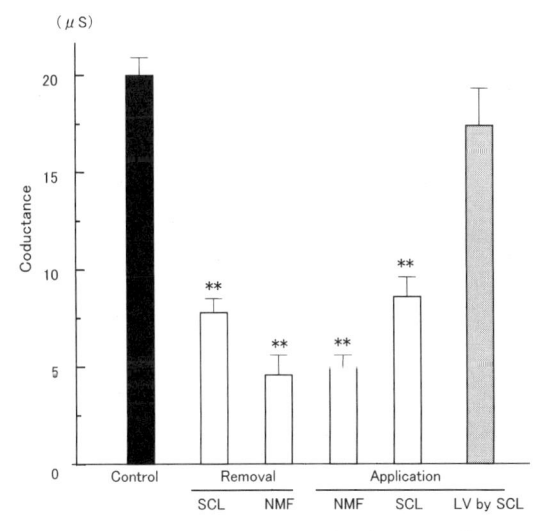

Figure 1 Role of intercellular lipids in retaining hydration in stratum corneum
Control: Before removing SCL and NMF using solvent, SCL: Stratum corneum lipids, NMF: Natural moisturizing factor (water-soluble monomeric substances such as amino acids and organic acids), LV: Vesicles prepared by applying ultrasonic waves to SCL in the presence of Ca^{2+} Significantly different from the control: **$p<0.001$ (n=5)

11.2 Screening and particle diameters of monoglycerides

All monoglycerides, which are saturated fatty acids containing 8–10 carbons, form lamellar vesicles when supersonic waves are applied while suspended in distilled water at 60°C. When the lamellar vesicles were applied to the skin of the inner forearm, all these vesicles, except those of monotridecanion (C13) and monostearin (C18), increased the water content of the stratum corneum, which was determined by measuring the skin conductance, from those prior to the application and of the control (table 1). Application of lamellar vesicles to which Ca^{2+} was added resulted in a higher increase from that caused by the application of lamellar vesicles alone. This was likely because the addition of Ca^{2+} caused lamellar vesicles to fuse and a multilayered lamellar structure to be formed. Addition of Tween 60, which is a hydrophilic surfactant, prevented the lamellar structure from forming and resulted in almost no change in water content of the stratum corneum.

Observation of these specimens under a polarization microscope in crossed Nicols showed

Table 1 Water content in stratum corneum when lamellar vesicles of monoglyceride were applied

Monoglycerids	Conductance (μS)		
	Before	10min.	30min.
Control (CaCl$_2$ 10mM)	20.0±1.0	19.3±1.5	19.3±1.3
Monocaprylin (C8:0)	19.3±1.9	41.0±3.5**	36.0±4.7*
Monocaprylin-Ca^{2+}	19.0±1.0	68.0±5.4**	42.0±2.6**
Monononanoin (C9:0)	19.7±1.2	24.3±1.5*	19.3±1.5
Monononanoin-Ca^{2+}	20.7±1.2	39.7±1.2**	27.3±1.5**
Monocaprin (C10:0)	21.0±2.0	37.0±3.0**	27.0±1.5*
Monocaprin-Ca^{2+}	22.3±0.6	61.3±6.0**	37.0±1.0**
Monoundecanoin (C11:0)	20.7±1.2	38.9±6.0*	22.2±3.4
Monoundecanoin-Ca^{2+}	21.3±0.6	46.8±5.3**	29.3±1.5*
Monolaurin (C12:0)	20.7±0.5	30.3±2.6**	12.7±1.5
Monolaurin-Ca^{2+}	21.0±1.0	38.7±4.5**	19.0±2.0
Monotridecanoin (C13:0)	19.7±1.5	15.3±1.5	10.3±1.2
Monotridecanoin-Ca^{2+}	19.7±0.6	15.7±0.6	8.7±1.5
Monomyristin (C14:0)	19.7±1.2	26.3±3.8	12.3±1.2
Monomyristin-Ca^{2+}	20.0±1.0	36.3±3.1**	29.3±1.2**
Monopalmitin (C16:0)	20.0±1.0	24.3±1.5*	17.7±0.6
Monopalmitin-Ca^{2+}	20.3±1.5	58.7±1.5**	24.3±1.0*
Monostearin (C18:0)	19.6±0.6	14.7±0.6	15.3±0.6
Monostearin-Ca^{2+}	20.0±1.0	34.3±2.1**	18.7±1.5
C16:0-Ca^{2+} [a]	20.7±0.6	41.0±1.0**	24.0±1.7
Tween60-Ca^{2+}	20.3±0.6	22.0±2.1	19.3±0.6
C16:0-Ca^{2+}+Tween60 [a]	19.7±1.2	22.4±2.9	19.0±1.0

The monoglycerides were applied on the skin at 200 μg/cm^2.
a): 100 μg/cm^2 of both monopalmitin (C16:0) and Tween 60 were applied.
Significantly different from the control: *$p < 0.05$, **$p < 0.001$ (n=5)

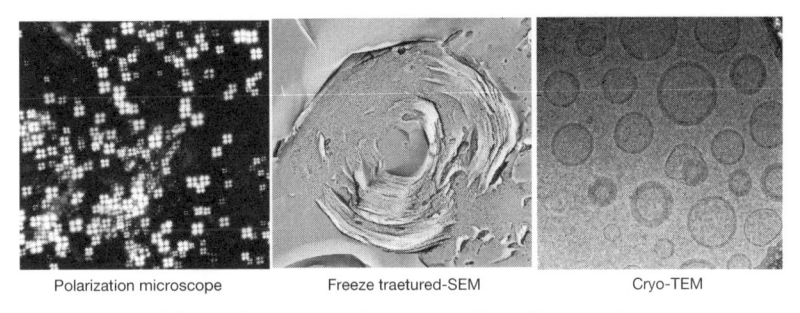

Figure 2 Microscopic images of lamellar vesicles

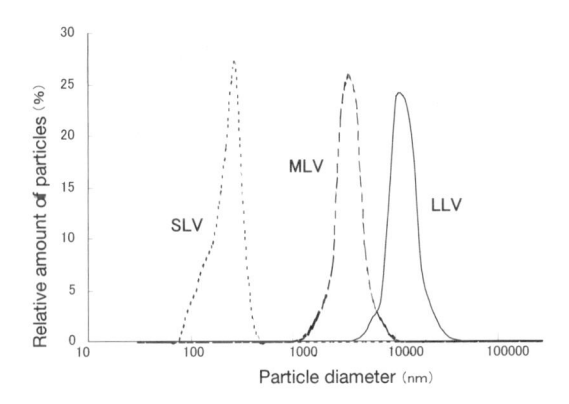

Figure 3 Particle size distribution of lamellar vesicles

organizational images peculiar to lamellar vesicles[11,12]. Electron microscopic observation of the freeze fracturing method[13] and Cryo-TEM also confirmed the formation of multilayered lamellar structures (figure 2).

The diameters of lamellar vesicles vary depending on the number of carbons in the monoglyceride fatty acid residue, the kind of emulsifier, and the amount of energy used in preparation. Monopalmitin lamellar vesicles (LLV), which were prepared by applying supersonic waves under the presence of Ca^{2+}, had an average particle diameter of $10\mu m$, and those (MLV) prepared in the absence of Ca^{2+} had an average particle diameter of $3\mu m$. They also differed in grain size distribution, and largest vesicles were 20–30μm in the former and were at least 50 nm in the latter. Monomyristin vesicles (SLV) prepared in the absence of Ca^{2+} had an average particle diameter of as big as 200 nm (figure 3)[10].

11. 3 Differences in effectiveness to the skin by particle diameter

Ultraviolet rays (UV) of 40 mJ/cm^2/day were irradiated to hr-1 hairless mice for four days. Then lamellar vesicles of various sizes were applied to the skin of the back of the mice to investigate the differences in the efficacy to the skin by particle diameter.

11. 3. 1 Barrier function

Trans epidermal water loss (TEWL) was measured and used as an index of the skin barrier function[14]. TEWL increased sharply by ultraviolet irradiation, showing decreases in barrier function. Application of lamellar vesicles restored the barrier significantly. The degree of restoration was the highest with LLV, followed by MLV and SLV in this order. When the

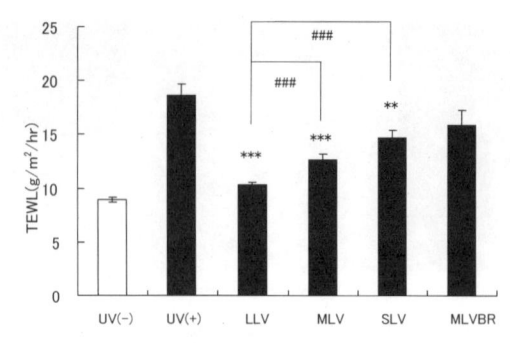

Figure 4 Effects of lamellar vesicles on the epidermal barrier function
UV(-): No UVirradiation, UV(+): UV irradiation, LLV: Ultraviolet irradiation and LLV application, MLV: UV irradiation and MLV application, SLV: UV irradiation and SLV application, MLVBR: UV irradiation and MLVBR application
Significantly different from UV (+): ***$P<0.001$, **$0.001<P<0.01$ (n=6)
Significantly different from LLV: ###$P<0.001$ (n=6)

lamellar structure of MLV vesicles was disintegrated by applying repetitive temperature load of low and high temperatures (MLVBR), almost no improvement effect was observed (figure 4). As in TEWL, the water content of the stratum corneum was the highest in LLV, followed by MLV and SLV, in this order.

11. 3. 2 Skin surface morphology

Irradiation of ultraviolet rays caused wrinkles to form and deteriorated the texture of the skin. Application of LV was visually observed to sharply restore the damages (figure 5). Image analysis of silicon rubber replicated samples of the skin[15] showed increases in wrinkle area by ultraviolet irradiation. In contrast, various kinds of LV showed superior improvement effects by application. The effects were more notable in MLV and SLV than LLV. On the other hand, the improvement effect of MLVBR was low (figure 6). The improvement of the skin texture was similar to the wrinkle improvement effects.

11. 3. 3 Transglutaminase I (TGase-1)

The skin tissue was sampled from the back of the mice by freeze grafting and was immunostained with anti-94G[16] antibody, which recognizes the tissue parts at which an epidermis crosslinking enzyme transglutaminase I (TGase-1) is activated. Stainability was found to dcrease by UV irradiation. Particular, the localization of the activated enzyme near the cellular membrane, which was observed in the non-irradiated skin, was disturbed by UV irradiation. Application of LV sharply restored the distribution. LLV was effective especially in the granular layer of epidermis, and MLV and SLV were effective in the stratum spinosum and the stratum basale epidermis. The thickening of the epidermis caused by UV irradiation was restored by SLV, MLV and LLV, in this order. On the other hand, MLVBR showed almost no effect (figure 7).

As shown by the experiment, the effects of LV in restoring skin damage varied depending on particle size and the part of the skin. The differences in effect were likely because LV of large particle sizes of several micrometers tended to stay and restore the activity of TGase-1 in the stratum corneum and the granular layer, resulting in restoration of the barrier function, while those of small particle sizes of several nanometers penetrated to the stratum basale, restoring the TGase-1 in the layer, resulting in restoration the differentiation of epidermal

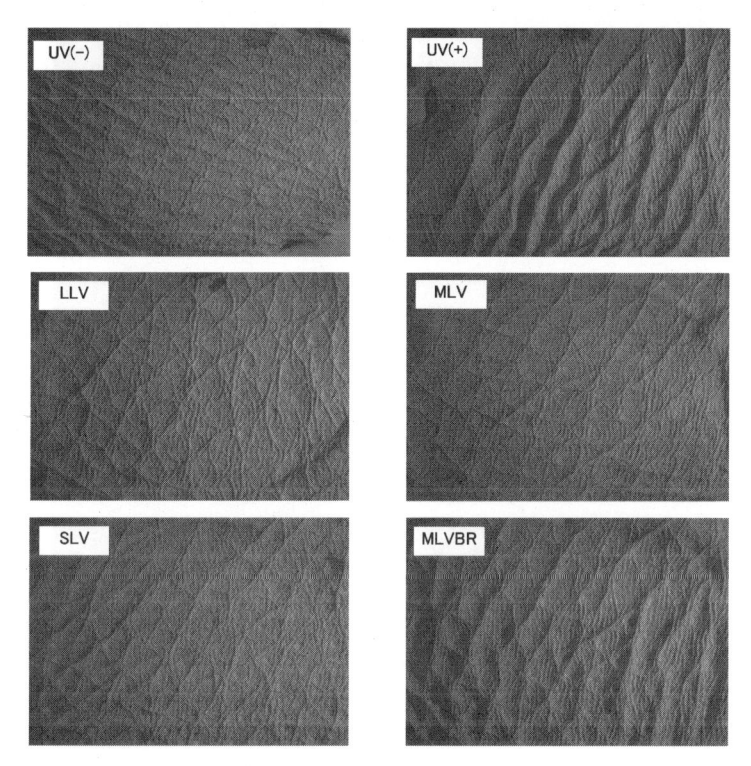

Figure 5 Effects of lamellar vesicles on the skin surface (replica images)
UV(-): No UVirradiation, UV(+): UV irradiation, LLV: Ultraviolet irradiation and LLV application, MLV: UV irradiation and MLV application, SLV: UV irradiation and SLV application, MLVBR: UV irradiation and MLVBR application

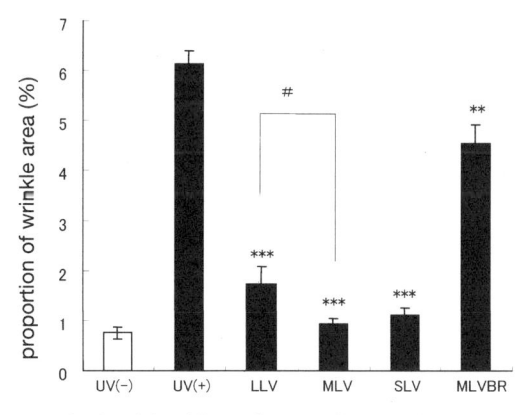

Figure 6 Image analysis of the skin surface (replica) applied with lamellar vesicles
UV(-): No UVirradiation, UV(+): UV irradiation, LLV: Ultraviolet irradiation and LLV application, MLV: UV irradiation and MLV application, SLV: UV irradiation and SLV application, MLVBR: UV irradiation and MLVBR application
Significantly different from UV (+): *** $P<0.001$, ** $0.001<P<0.01$ (n=6)
Significantly different from LLV: ### $P<0.001$ (n=6)

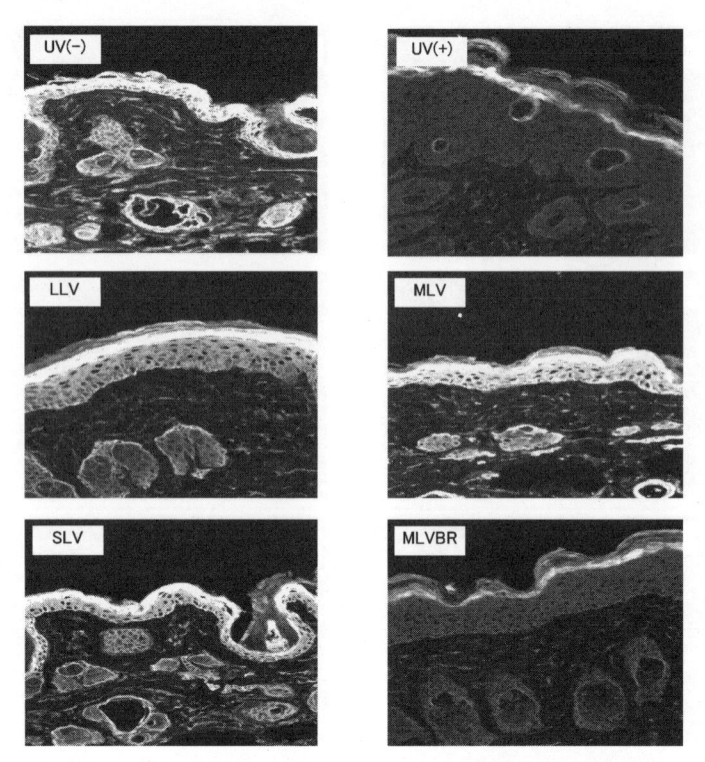

Figure 7 Effects of lamellar vesicle application on acanthosis and TGase1 activity

cells, skin texture and wrinkles. SLV is small and different in fatty acid composition from the others. SLV showed a high improvement effect of the skin because the fatty acid composition was appropriate for activating TGase-1[17] and also possibly because the TGase-1 induced involvement of anti-inflammation and other factors.

11. 4 Stabilization

As described above, LV sharply loses its skin restoration effects when the lamellar structure in the vesicles is destroyed[6,8,10], and thus stability is crucial. Stability of vesicles and the lamellar structure was investigated.

11. 4. 1 Phase transition temperature

LV undergoes phase transition from liquid crystal to gel or to solid crystal when temperature falls[18]. The effects of cholesterol on phase transition were investigated as cholesterol is widely used to control the phase transition of phospholipids.

Changes in calorie of lamellar vesicles during temperature rises were measured using a differential scanning calorimeter (DSC). At cholesterol 0 %, the phase transition temperature (T_c) was 53.59°C, and the calorific value of the endothermic reaction was −3.91J/g. The peak of the DSC curve decreased in size as the amount of cholesterol increased, and disappeared at a cholesterol content of 2.0 % (figure 8). When the temperature was lowered, an exothermic peak was observed. The exothermic peak disappeared also at a cholesterol content of 2.0 %. The phenomena, which is widely known to occur in bimolecular films of phospholipids[19], were also observed in monoglycerides.

Wu et al.[20] measured the lateral diffusion coefficient of bimolecular films of phospholipids

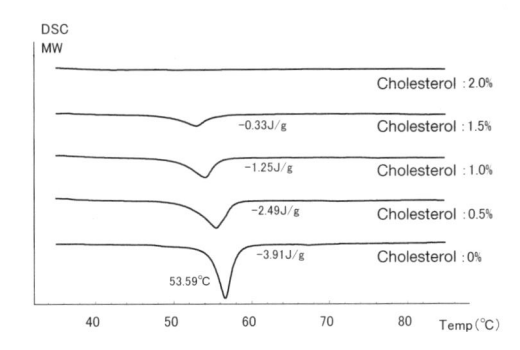

Figure 8 Relationship between the amount of cholesterol added to lamellar vesicles and DSC curve (temperature rising process)

and reported that the molecular movement was large at temperatures above T_c, suddenly decreased at and below T_c, and addition of a certain amount of cholesterol moderated the changes near T_c to a line of gentle inclination. The molecular movement of lipids composing LV was predicted to also show similar trends. In practice, a polarization microscopic observation showed that the addition of a certain amount of cholesterol restrained crystal formation and stabilized the lamellar structures of the vesicles over a long period of time.

11.4.2 Protective colloid action

When stored at the room temperature, the lamellar structure of LV disintegrates gradually[8]. The disintegration of the lamellar structure was estimated to be attributable to coagulation caused by interactions among LV particles. Some surfactants and electric charged lipids, which are widely used aiming to stabilize vesicles, may destroy the vesicles on the contrary [21,22]. The effects of water soluble polymers[23,24], which are mainly used to prevent coagulation of powder particles, on the stability of vesicles were investigated. The investigation showed that polyvinylpyrrolidone (PVP) was effective in stabilizing LV[25]. The ζ electric potential of LV was about -30 mV. The absolute power of the ζ electric potential decreased by an addition of PVP. The changes in the ζ electric potential were large when PVP of a large molecular weight was added. This shows that addition of PVP caused the mobility of the particles to reduce and decreased the interaction of the particles (figure 9). The author then investigated how PVP was adhered on the surface of LV.

PVP and most other polymers adhere to particles by taking a form that can be divided into three segments: (1) the train part, which is directly in contact with the particle, (2) the loop, which is a part of the polymer chain between trains, and (3) the tail, which is the end of the polymer chain extending from the train. The more the loops and tails, the more the polymer chain projects out in the fluid three-dimensionally, and the more the contact among particles is inhibited.

ESR spectrum is used to understand the morphology of adsorption. In ESR spectrum, loops and trails appear as sharp peaks because their radical signals caused by rotational movement are strong while the trains appear as broad peaks because they are almost immobile[26,27]. The ESR spectrum of a spin-labeled PVP showed sharp peaks (figure 10(a)) in both adsorbed and isolated types. Thus, PVP was likely to have adhered to LV in a form in which the majority was loops and tails (figure 10(b)). The PVP polymer chains were likely to have projected out in the solution and prevented vesicles from being in contact with another, and the protective colloidal action reduced the interactions among particles and coagulation.

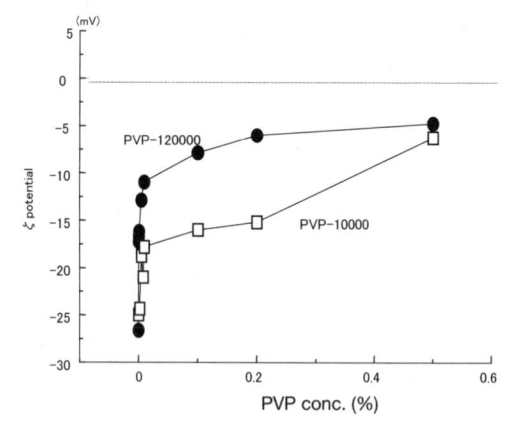

Figure 9 Changes in ζelectric potential by addition of PVP to lamellar vesicles

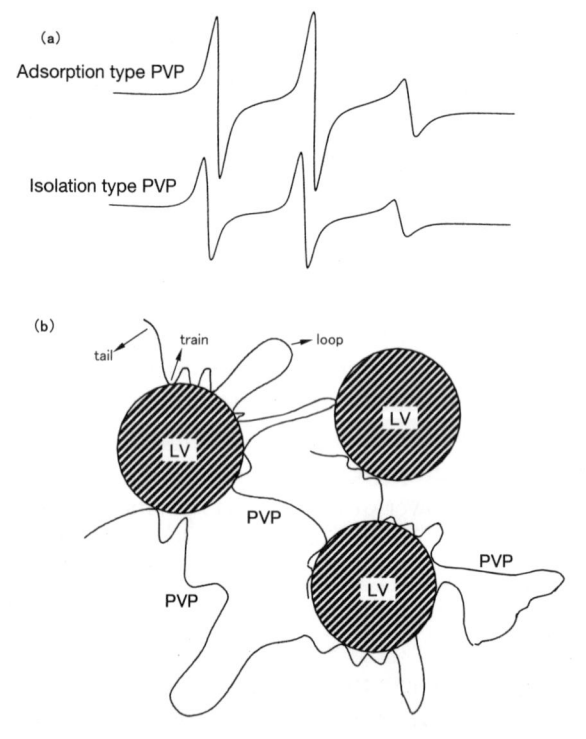

Figure 10 ESR spectrum of spin-labeled PVP (a) and adsorption pf PVP to LV (b)

11.5 Conclusions

The effects of LV on the skin (anti-aging effects) were found to differ depending on the particle size. LV should be applicable to multi-purpose cosmetics by using lamellar vesicles of sizes appropriate for the aimed performances. LV of wide grain size distribution can be mixed to develop cosmetics that are effective to the entire epidermis. Because the lamellar structure of LV must be maintained over a long period of time for the LV to show its efficacy, evaluating the stability of the vesicles is indispensable for applications of LV in cosmetics. LV has also been found to be effective against deposition of yellow pigments[9], such as bilirubin, as well as the efficacies described in this chapter. LV can be used to develop a novel whitening agent different from conventional agents that restrain melanin production. LV is also a promising carrier for delivering drugs of various kinds.

References

1. P. W. Wertz et al., *Science*, **217**, 1261 (1982)
2. L. Landman, *Invest. Dermatol.*, **87**, 202 (1986)
3. D. C. Swartzendruber et al., *J. Invest. Dermatol.*, **88**, 709 (1987)
4. K. Madison et al., *J. Invest. Dermatol.*, **88**, 714 (1987)
5. G. Imokawa et al., *J. Invest. Dermatol.*, **96**, 845 (1991)
6. T. Sone et al., *Int. J. Cosm. Sci.*, **21**, 23 (1999)
7. H. Iwai et al., Substance of the 45th Divisional Meeting on Colloid and Interface Chemistry, the Chemical Society of Japan, p344 (1993)
8. T. Sone et al., *J. Jpn. Cosm. Sci. Soc*, **18**, 186 (1994)
9. T. Sone et al., *Fragrance Journal* **12**, 76 (2006)
10. R. Iizuka et al., *Fragrance Journal* **7**, 60 (2003)
11. F. B. Rosevear, *J. Am. Oil Chemist' Soc.*, **31** 60 (2003)
12. F. B. Rosevear, *J. Soc. Cosmet. Chem.*, **19**, 581 (1968)
13. H. Utsumi, Liposomes, Nankodo Co., Lid., p53 (1988)
14. T. Ager et al., *Clin. Exp. Dermatol.*, **15**, 29 (1990)
15. K. Mastumoto et al., *J. Jpn. Cosm. Sci. Soc.*, **28**, 118 (2004)
16. R. Iiziuka et al., *J. Invest. Dermatol.*, **121**, 457 (2003)
17. P. M. Steinert et al., *J. Biol. Chem.*, **271**, 26242 (1996)
18. K. Larsson et al., *Z. Phys. Chem.*, **56**, 173 (1967)
19. B. Ladbrook et al., *Biochim. Biophys. Acta*, **415**, 29 (1975)
20. E. S. Wu et al., *Biochemistry*, **16**, 3936 (1977)
21. D. Lichtenberg, *Biochem Biophys. Acta*, **821**, 470 (1895)
22. A. Helenius et al., *Biochem. Biophys. Ata*, **415**, 29 (1975)
23. H. L. Jakubauskas, *J. Coatings Technology*, **58**, 71 (1986)
24. T. Ishitaba et al., emulsifications·dispersions technical applied handbook, Science Forum Co. Ltd., p442 (1987)
25. T. Sone et al., *Colloids and Surfaces*, **108**, 273 (1996)
26. O. H. Griffith et al., *Acc. Chem. Res.*, **2**, 17 (1969)
27. K. Esumi et al., Substance of the 45th Divisional Meeting on Colloid and Interface Chemistry, the Chemical Society of Japan, p118 (1993)

Chapter 12
Microcapsule Drug: Active-ingredient Delivery System

Takayuki Tsuboi

12.1 Active-ingredients Delivery System (ADS)

In the field of medical pharmacy, systems called drug delivery systems (DDS) have recently been developed to control the movement of drugs in the body in terms of both quantity and time. The systems help specific actions to take place effectively, reduce the side effects and other adverse effects of the drugs, and are expected to improve the quality of life of the patient, reduce the medical costs, and develop drugs that are easy and not unpleasant to use. DDS is attracting attention also in the field of skin care as the system enables active ingredients to be efficiently sent to the parts of the skin that need the treatment and thus to show their full efficacy. The ADS theory was developed on delivering active ingredients to the place of need in a controlled manner, and micro capsules were developed based on the theory.

The characteristics of the skin vary greatly depending on person, and the state of the skin is affected by various conditions such as season, parts, and hour (morning and evening), and is not always constant. Thus the ways of the actions of the active ingredients and their effects are different, too. For example, the skin can be moist and soft in the morning but can be either extremely oily or dry in the evening depending on the midday environment. The skin becomes moist and soft by taking bath, but gradually loses the moisture, and returns to the bare skin state characteristic to the person in the morning. Just giving moisture is enough if the skin is healthy. However, the skin damaged by aridity and/or ultraviolet rays (UV) is in deteriorated conditions of "having insufficient moisture", "rough", and "prone to irritation".

Application of skin care cosmetics containing active ingredients on the skin after washing is insufficient because sweat and sebum secreted from the skin wash away the cosmetics and block the ingredients from reaching the skin parts that need them. The conditions of the skin vary greatly depending on part, such as the T and U zones of the face, and thus the degree of damage caused by drying and UV vary by part. The skin of a person can be a mosaic of the healthy and damaged parts. However, unlike skin care cosmetics for treating stains and wrinkles, which can be applied to specific target parts of the skin, dry and sensitive skin parts are difficult to care separately from the rest of the skin because they are not visibly clear.

ADS systems deliver various ingredients for mitigating skin troubles to the places that need them. Unlike spot care products for treating stains, a system is need to efficiently deliver active ingredients to parts of damage and provide sufficient moisture to the healthy skin parts to enable them defend themselves. For example, whitening agents do not need to be applied on the parts without stains, and it is desirable to deliver as much agents as possible to the stained parts.

The skin consists of the stratum corneum, epidermis, and dermis layer, and keeps the barrier function by keratinocytes and intercellular lipids. The barrier protect the skin from outside irritants, but the barrier also prevents active ingredients for keeping the skin healthy from reaching the skin. The skin with the healthy barrier may not feel irritation by an ingredient, but once the barrier is lost, the skin can be irritated by the same ingredient. The parts that have lost the barrier need ingredients to regain homoestasis, but must be cared very carefully to not be further irritated.

The ADS theory is a new form of skin care and aims to deriver active ingredients to the skin part that needs them within the necessary period of time, treat the damage efficiently, and minimize the adverse effects on the skin.

In the ADS theory, active ingredients must show their efficacies on the skin parts of specific conditions and must be temporarily encapsulated when added to the formulations. It is difficult to chemically modify active ingredients so as to have the ADS function, and the development should be costly. Active ingredients need to be encapsulated also to prevent deterioration within the formulations. Encapsulation also reduces the influences of the ingredient to the formulations, and enables even ingredients that are physically difficult to mix into the formulations.

Factors that show time historical changes on the skin and factors that are different between healthy and damaged parts of the skin include enzymes, ions, temperature, pH, minerals, oil, water content, and the number and species of indigenous flora on the skin. Factors that may cause skin troubles include drying and UV are given, and it is desirable that active ingredients start functioning by these factors. Active ingredients can be efficiently supplied to the skin parts of need by hand massage and shiatsu therapy.

Another condition for supplying active ingredients to the skin when necessary is encapsulating the ingredients stable within the formulation in one way or another. Possible tools are micelles of several millimeters from several nanometers, liposomes, and capsules. Selectivity of the target is possibly achieved by using appropriate kinds and sizes. Several tens of micrometers to several millimeters are appropriate for targeting the skin, and it is necessary to prepare them into a size of several micrometers to several nanometers to target pores and the stratum corneum. Recently, materials that have various physical properties have been developed that can be used for skin care, and there is a possibility of constructing the ADS theory that has various functions from every viewpoint.

Changes that are closely related to the state of the skin include the quantities of oil and water and their balance. Water always transpires from skin, but the quantity of transpiration is controlled by the barrier function of the skin. The quantity of transpiration increases when the barrier function deteriorates due to roughness, resulting in water loss from the skin. Water content in the skin plays a key role in maintaining the skin homoestatic, and the dry skin ages fast by various irritations. The transpiration of water from the skin is controlled by the structural barrier of the skin, and oil forms a film on the skin and functions as a hydrophobic barrier. This barrier prevents various stimulants from contacting the skin directly and controls water transpiration. The balance between water and oil is also important. In the "dry skin" state, both water and the oil are insufficient; in the "mixture skin" state, water is insufficient while oil is superabundant; and in the "oily skin" state, oil is superabundant. When the skin is extremely dry, the skin is at the "sensitive" state and is highly sensitive to various kinds of irritation. These states may coexist, and the ADS theory needs to be applied in manners appropriate to these changes, states, and the skin environment.

12.2 ADS - Microcapsules

"ADS-Oligonol (ICHIMARU PHARCOS Co., Ltd.)" is microcapsules with ADS functions and is used in cosmetics. This material focuses on treating the roughness and dryness of the skin based on the ADS theory. Encapsulation of oligonol, which is a physiologically active substance, was achieved by giving the ADS function to the microcapsules.

Oligonol reinforces whitening and antioxygenation by being activated to levels higher than those of ordinary polyphenols and binding with cysteine, which are achieved by reducing the molecular size of antioxidant proanthocyanidins, produced from Lychee fruits (figure 1). Polyphenols, such as proanthocyanidins, are widely used for skin care as they have very strong antioxidation characteristics and improve the skin environment[1-3]. However, polyphenols undergo self oxidization while restraining the oxidation of the sebum, due to their high oxidation characteristics, and are easy to self oxidize also within formulations, such as cosmetics. Polyphenols are structurally water soluble up to a certain molecular weight but may form precipitation within formulations, and care should be taken in using polyphenols in cosmetics. ADS-Oligonol overcomes these disadvantages by the use of microcapsules (figure 2) and can deriver oligonol to the parts of the skin in need.

The ADS theory of ADS-Oligonol is based on the disturbed ion balance on the damaged skin. Unlike on the healthy skin, where cations and anions are kept in balance, anions are dominant on the damaged skin (figure 3). The anions trigger the microcapsules to break and release the oligonol content, which mitigates the trouble of the skin (figure 4).

Microcapsules are widely used in various fields, including cosmetics, food, insecticides, and articles for daily use, such as ink, copy paper and electronic paper[4]. Materials constituting microcapsules are also diverse, ranging from natural and synthetic polymers to inorganic substances such as silica. In cosmetics, natural and other biocompatible polymers are widely used as they are believed to be soft to the skin. Each of these polymers has peculiar properties, enabling the ADS theory to be widely applied. The materials constituting microcapsules

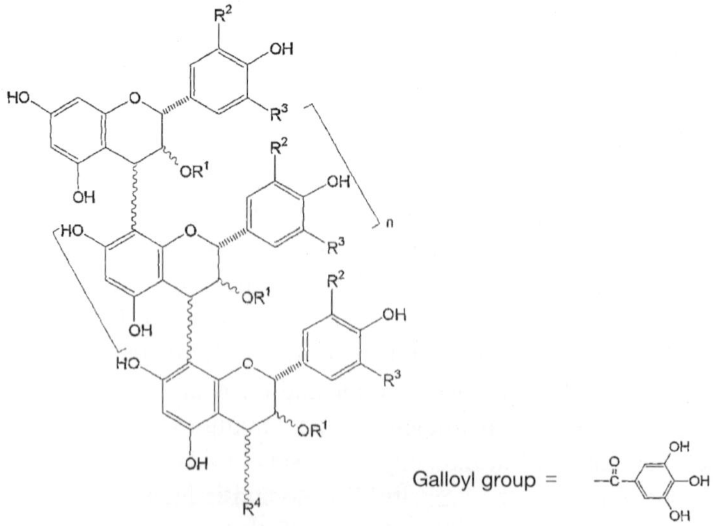

Figure 1 Structure of Oligonol-CS
R_1 = H or Galloyl group
R_2, R_3 = H or OH
R_4 = Cysteine

Chapter 12 Microcapsule Drug: Active-ingredient Delivery System 215

not only function to encapsulate ingredients but can also show their own functions, such as moisture retention.

The ADS microcapsules of ADS-Oligonol consist mainly of natural polymers, and a measure is taken to improve the affinity with lipids. The natural polymers used are alginic acid, which is contained in quantity in seaweeds, hyaluronic acid, which keeps water in the skin, γ- polyglutaminic acid, which is a sticky substance produced *Bacillus natto*, and chitin and chitosan, which are found in the shells of crabs and lobsters and also used for surgical thread. They are all ionic polymers and are electrically charged either positively or negatively.

Figure 2 Electron micrograph of ADS-Oligonol

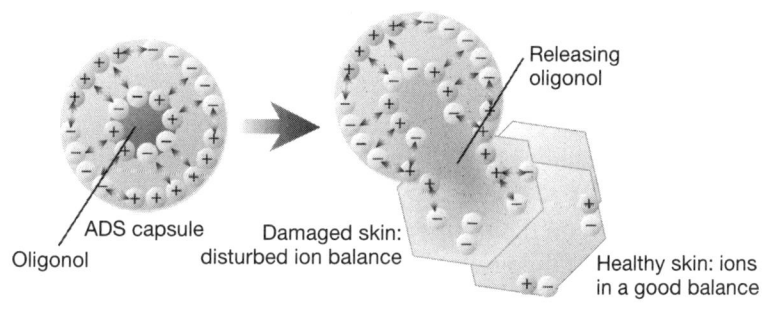

Figure 3 ADS mechanism of ADS-Oligonol

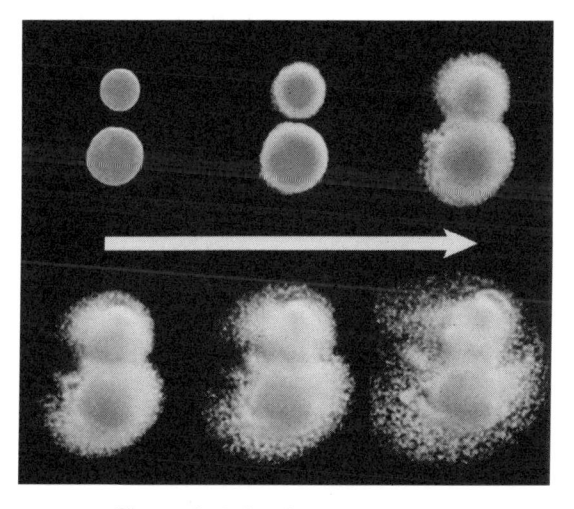

Figure 4 Collapsing ADS capsules

The microcapsules are produced using the coacervation method, which uses the interactions between the charges. The physical properties of the film of the capsule vary depending on the molecular weight and units constituting the polymers. For example, alginic acid consists of β-D-mannaric acid (M units) and its isomer α-L-guluronic acid (G units), and the physical properties of the capsule vary depending on their ratio. Alginic acid that contains more M units than G units forms a flexible film, and a hard film is formed when the amount of G units is large. The molecular weight of chitosan, which is positively charged and interacts with anionic polymers to form films, also affects the strength of the films. The larger the molecular weight of chitosan, the stronger the interaction with alginic acid, and the stronger the resultant film. On the contrary, chitosan of a small molecular weight forms a weak film. Chitosan is produced by deacetylating chitin. The amount of remaining acetyl radicals also affects the strength of the film. With a larger the amount of acetyl radicals remaining in the amino groups, the chitosan is less positively charged, and the interaction with alginic acid becomes weaker. The films of microcapsules should be strong to a certain extent so that they are stable in formulations, such as cosmetics, but should easily break on the skin. Therefore, it is necessary to keep the capsule strength appropriate for the purpose by using appropriate alginic acid and chitosan of the appropriate molecular weight and appropriate degree of deacetylation.

The ADS capsule of ADS-Oligonol is mainly composed of a compound film of alginic acid and chitosan. However, films composed of only alginic acid and chitosan shrink and become hard when they dry, disabling active ingredients to be released on the skin. To prevent drying, the microcapsule film also contains hyaluronic acid and γ-polyglutaminic acid, which retain humidity and prevent the film from shrinking by ionicly interacting with chitosan and forming a film.

Microcapsules consisting of only alginic acid lose their ability to release the encapsulating ingredient when they dry, but the compound films, such as hyaluronic acid and γ-polyglutaminic acid, do not shrink by drying enabling capsules to break and radiate the encapsulating ingredients on the skin. The capsules break responding to dryness as well as by ionic stimulants.

As described above, sebum on the skin serves as the barrier and protects the skin. Because the ADS capsules of the ADS-Oligonol consist of water-soluble macromolecules, the surface of the capsules is hydrophilic. A hydrophobic surfactant is included in the capsule film to facilitate the capsules to contact the hydrophobic sebum barrier of the skin.

ADS-Oligonol facilitates polyphenol to be mixed in cosmetics, which is otherwise difficult. It can also efficiently care troubled parts on the skin by the ADS function and proves moisture to the skin.

"MPG (ICHIMARU PHARCOS Co., Ltd.)" is ADS capsules based on a different ADS theory (figure 5). MPG is prepared by surrounding fine particles of polyethylene and cellulose by water-soluble or insoluble polymers or wax. MPG is characterized by its flexibility in adjusting the

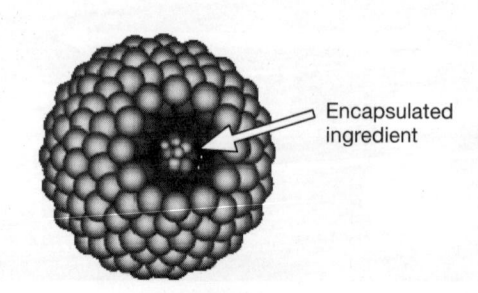

Figure 5 MPG

capsule intensity depending on formulations to be encapsulated and in encapsulating arbitrary ingredients regardless of their hydrophilicity and solubility to oil. The capsules should be designed so as to collapse on the skin by the frictional force applied by a finger while spreading on the skin. When the capsules are too strong, the ingredients cannot reach the target skin parts, and the capsules may irritate and deteriorate the skin. On the other hand, capsules too weak may collapse in the middle of the production process of cosmetics or during storage. In other words, ADS capsules must be stable during production, collapse smoothly while used, and release the encapsulated active ingredients at the target parts of the skin.

Since many kinds of materials can be used as the nuclei and films constituting MPG capsules, those that are appropriate for the ingredients to encapsulate and their properties, such as W/O or O/W emulsion systems, monophily of W and O, and the polarity, can be selected. For example, synthetic polymers such as non-water-soluble acrylic acid, oils and fats and cellulose derivatives can be used for the film for encapsulating plant extract, vitamin C and other water-soluble ingredients in the W phase in an isolated manner. On the other hand, water-soluble natural polymers, cellulose derivatives and PVA synthetic polymers should be used to encapsulate oil-soluble ingredients such as the Vitamin E and Coenzyme Q10 in the O phase. Capsules that collapse responding to sebum can be prepared by using components affinitive to sebum for the film, which swells and dissolves by sebum releasing the encapsulating agents.

12.3 Possibilities of ADS

The ADS theory and its application for skin care were described. The theory can also be used for hair care and makeup. Hair is easily damaged by bleaching and hair dye, which involves conversion of the cysteine component of the hair into cysteic acid and increases in anionic characteristic stronger than those on the skin. Stimulation response would be more efficiently shown because changes of the target factor are bigger, resulting in better manifestation of the ADS function.

ADS is also possibly used for applying agents that have side effects. For example, antimicrobial agents are effective for improving the skin environment but also irritate the skin. Encapsulating antimicrobial agents in ADS capsules, micelles or liposomes, which are decomposed by enzymes of the target bacteria, can be a system that shows bacteria-specific activities. For example, the bacteria that cause acne are anaerobic and grow in pores by feeding on sebum. The bacteria metabolize sebum, for which enzymes are essential. Micelles and liposomes composed of lipids that are decomposed by the enzymes will collapse at the contact with the bacteria losing the molecular association and release the agent that kills the bacteria. ADS, which carries the active agents to the necessary place at the necessary time, will be effective while minimizing the stimulation on the skin.

ADS is not a theory of a certain special form, but is a new form of cosmetics that is efficient and soft to the skin, which is developed by understanding daily changes and troubles of the skin. New ADS can be designed using various technologies in the field of cosmetics, such as emulsification technology, micelle liposome formation, and encapsulating technologies and taking the properties and behaviors of active ingredients and environment on the skin.

References

1. T. Tsuboi *et al.*, Proceeding of the 125th Annual Meeting of the Pharmaceutical Society of Japan, 4, p.113 (2006)
2. K. Yamada *et al.*, *Aesthetic Dermatology*, **16** (3), p.229 (2006)
3. A. Iddamalgoda *et al.*, ASCS2007 Program & Abstract, p.118 (2007)
4. M. Koishi *et al.*, Make+Use Microcapsules p.6, Kogyo Chosakai Publishing (2005)

Chapter 13
Hyaluronic Acid
of Super Low Molecular Weight

Kohei Watabe

13. 1 Introduction

Hyaluronic acid has been widely used for producing cosmetics and medicines for at least 20 years. It is attracting attention these years as an additive into supplements and foods. Cosmetics that contain hyaluronic acid are characterized by elastic texture and high moisturizing effect, which are attributable to the viscosity of hyaluronic acid. The high moisturizing effect is believed to be a result of hyaluronic acid incorporating water, covering the skin surface, providing moisture to the skin, and preventing water from evaporating from the skin. However, conventional hyaluronic acid molecules are as large as 100 000 to 2 million and cannot penetrate into the skin. Thus, the moisturizing effect is lost once the skin surface is washed.

The authors developed hyaluronic acid of low molecular weights, which can penetrate into the stratum corneum once applied and retain the moisturizing effect even the skin is washed.

13. 2 What is hyaluronic acid?

Hyaluronic acid was first isolated by Meyer *et al.* in 1934 from the corpus vitreum of the eyes of a cow, and was named after hyaloid (vitreum in Greek) and uronic acid, which is its constituent unit[1]. Hyaluronic acid is a chain of two-sugar units, each of which is D-glucuronic acid and D-N-acetylglucosamine connected by β-1,3 glycosidic bond. The units are connected by β-1,4 bonds. The molecular weight is 10 million the maximum, and it is a largest polymer in the living body[2]. For example, a hyaluronic acid molecule of a molecular weight of 10 million is composed of 20 000 two-sugar units (figure 1).

Hyaluronic acid is present in vertebrates and some microbes. In the animals, hyaluronic acid is contained in every connective tissues and organs, and is especially abundant in the skin, synovial fluid, eye balls and umbilical cord[3].

Hyaluronic acid is produced either by extracting from cockscombs (cockscomb extraction

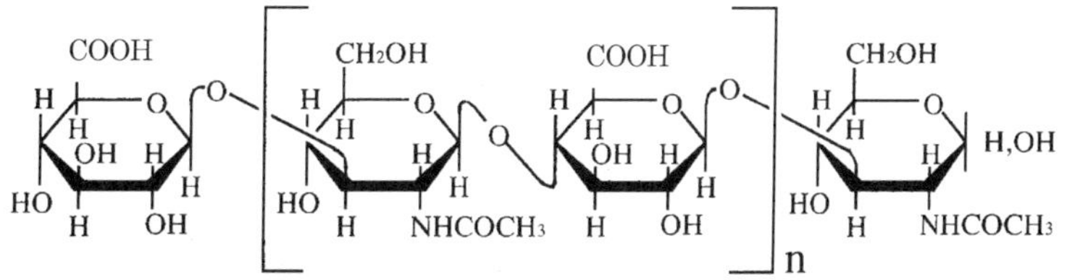

Figure 1 Chemical structure of the hyaluronic acid

method) or by culturing lactic acid bacteria that produce hyaluronic acids (fermentation method), and the fermentation method is the mainstream today.

13. 3 Properties of hyaluronic acid

Most prominent properties of hyaluronic acid are high water holding capacity and viscoelasticity, which are entirely attributable to its molecular structure. Hyaluronic acid exists in a state in which the hydrogen ion of the carboxyl group of D-glucuronic acid, which is one of the sugar constituents, is detached. In other words, the large molecule has a number of negative electric charges, which attract water molecules and retain within itself[4]. This physical property varies greatly depending on the molecular weight of the hyaluronic acid.

13. 4 Use of hyaluronic acid in cosmetics

Hyaluronic acid is widely used in lotion, milky lotion, essence, foundation and face mask to increase the moisture retaining capacity. The molecular weights of marketed hyaluronic acid products are diverse various (ranging from approximately 100 000 to approximately 2 million), enabling those of appropriate molecular weights to be chosen. For example, use of hyaluronic acid of a large molecular weight can add elastic texture to cosmetics, and that of a small molecular weight results in a smooth texture.

13. 5 Development of hyaluronic acid of super low molecular weight

In this chapter, hyaluronic acid of super low molecular weight is defined as those of a mean molecular weight less than 10 000.

Development of hyaluronic acid of a super low molecular weight has been tested by using hyaluronidase, which is an enzyme that decomposes hyaluronic acid, but not on an industrial production basis. It has been known that adding acid to an aqueous solution of hyaluronic acid of a large molecular weight lowers the molecular weight of the hyaluronic acids, but the process of producing hyaluronic acid powder of low molecular weight from an aqueous solution is very complicated and is low in yield. An ideal industrial method for producing hyaluronic acid of low molecular weight should be different from the conventional method, which involves decomposition using hyaluronidase, and avoid dissolving hyaluronic acid in water.

The authors found that dispersing high molecular weight hyaluronic acids in ethanol hydrate, adding hydrochloric acid and heating produced hyaluronic acid of a mean molecular weight less than 10 000[5]. The method can be applied for mass production of hyaluronic acid of very low molecular weights (the "super low molecular weight hyaluronic acid").

13. 6 Properties of the super low molecular weight hyaluronic acid

The molecular weight of the so produced super low molecular weight hyaluronic acid (trade name: "HYALO-OLIGO™", Q. P. Corporation) did not exceed 10 000 on average. The mean molecular size in linear chain was about 15 to 30 mm, and it consisted of 15 to 30 of the two-sugar unit shown in figure 1.

13. 6. 1 Low viscosity

The super low molecular weight hyaluronic acid has a very low molecular weight and is highly soluble in water. Thus, its aqueous solutions have viscosities similar to that of water. The solutions adapt to the skin in a manner different from water and give a new texture (figure 2).

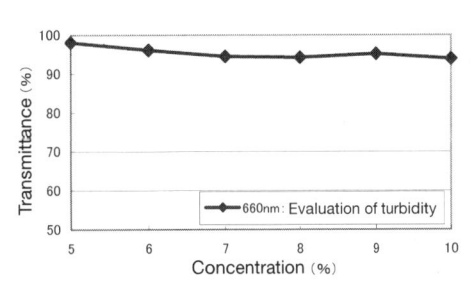

13.6.2 Superior solubility

When the super low molecular weight hyaluronic acid was dissolved in water so as to account for 10%, the solution was shown to be sufficiently transparent by transmissivity (660nm: evaluation of cloudiness). It dissolved in water immediately after it was added to water (figure 3).

13.6.3 High stability of viscosity (to heat and pH)

A 1% aqueous solution of the super low molecular weight hyaluronic acid was heated to 80°C in a hot bath. Even after 5 hours, the super low molecular weight hyaluronic acid kept at least 90% of the initial viscosity, showing high stability to heat. The viscosity did not lower even at pH 3 to 10, showing very high stability to changes in pH (figures 4 and 5).

13.6.4 Compatibility

The super low molecular weight hyaluronic acid showed very high compatibility with representative polyhydric alcohols, viscosity agents, organic acids, minerals and moisturizing agents. It produces less precipitates of quaternary ammonium chloride than large molecule hyaluronic acid.

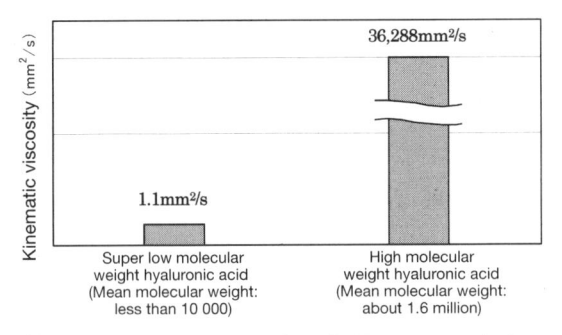

Figure 2 Kinematic viscosity of 1% aqueous solution of hyaluronic acid

Figure 3 Solubility of super low molecular weight hyaluronic acid

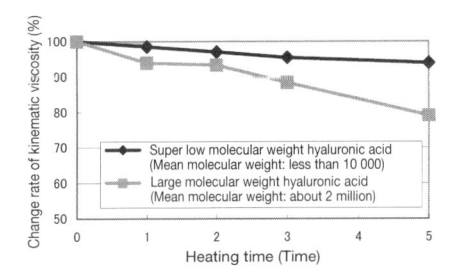

Figure 4 Stability of aqueous solution of super low molecular weight hyaluronic acid against heat

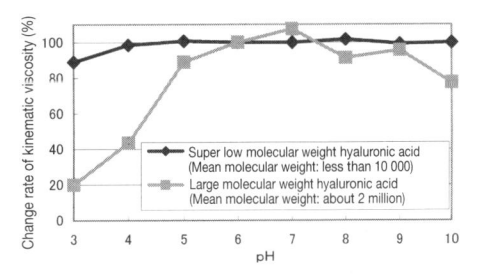

Figure 5 Stability of aqueous solution of super low molecular weight hyaluronic acid against pH changes (The kinematic viscosity at pH6 was put as 100%.)

13.7 Checking the performances of super low molecular weight hyaluronic acid

13.7.1 Penetration into the skin

Because super low molecular weight hyaluronic acid has is very small, it is very possible to show a high moisturizing effect by penetrating into the skin. The authors checked the skin penetration of the super low molecular weight hyaluronic acid.

Pieces of lint sheet (1 cm ×1 cm), which were soaked with distilled water for control, 100 mg of hyaluronic acid of about 1.8 million molecular weight, or super low molecular weight hyaluronic acid of about 6000, were attached to the skin of hairless mice and left for 1, 3, and 5 hours (nine groups in total). After the predetermined period, the pieces were removed and the skin sections were thoroughly washed with water. The skin sections were excised and homogenized. The homogenate was subjected to size exclusion chromatography to determine the amount of hyaluronic acid that remained in the skin. The test involved using fluorescent-labeled hyaluronic acid and measuring the fluorescence.

The skin specimens treated with the super low molecular weight hyaluronic acid showed a significantly stronger intensity of fluorescence than those treated with large molecule hyaluronic acid and water. A comparison of the peak patterns showed that the rise in fluorescence intensity was attributable to the super low molecular weight hyaluronic acid that penetrated into the skin (figure 6).

The quantity of hyaluronic acid that penetrated into the skin was the largest at three hours after application and similar at five hours. The amount penetrated into the skin was 0.1 to 0.2 mg per gram of the skin (figure 7).

The test showed that reducing the mean molecular size to about 10 000 facilitated hyaluronic acid to penetrate into the skin.

13.7.2 Moisturizing effect

The moisturizing effect of super low molecular weight hyaluronic acid was examined by applying the hyaluronic acid on the human skin.

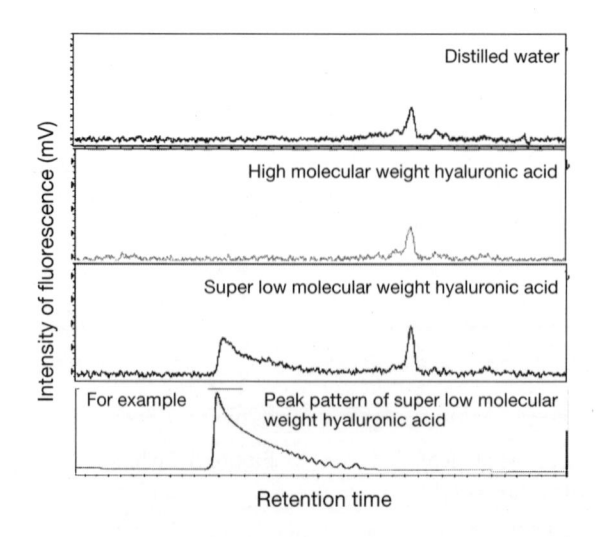

Figure 6 Fluorescence intensity of the hyaluronic acid origin in the murine skin tissue

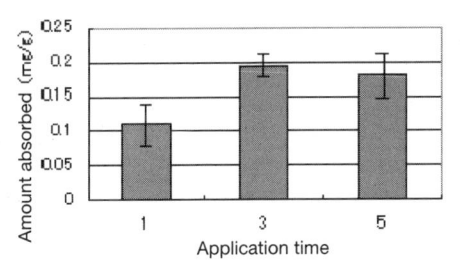

Figure 7 Quantity of super low molecular weight hyaluronic acid penetrated into the murine skin
(Hyaluronic acid content in the human skin: 0.1 to 0.3mg/g[6])

3 cm × 3 cm gauze piece was soaked in 1 mL of 1% aqueous solution of a super low molecular weight hyaluronic acid (molecular weight: about 6000) or distilled water for control and was applied on the human skin of the forearm for three days, eight hours a day. After removing the gauze, the electric conductivity at the skin surface was measured using a skin surface hygrometer (SKICON-200, IBS) to monitor time historical changes in the moisture content of the skin.

In one to three days of application, the moisture content increase by about 20 to 30μS (figure 8). The moisture content, which reflected the conditions of the skin (table 1), showed that the super low molecular weight hyaluronic acid effectively improved the dry skin.

13. 7. 3 Differences in moisturizing effect by the molecular weight of hyaluronic acid

Differences in moisturizing effect by the molecular weight of hyaluronic acid were examined by applying hyaluronic acid on the human skin.

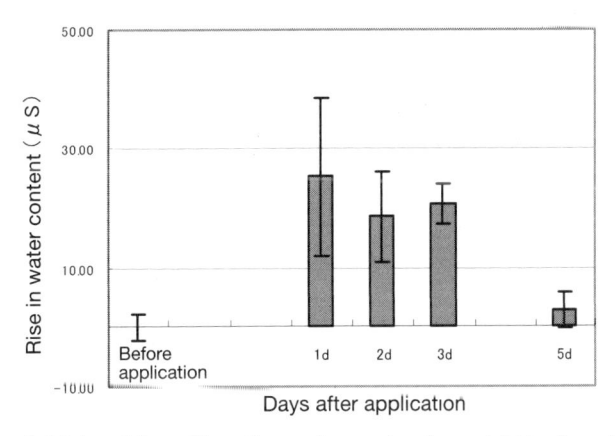

Figure 8 Moisturizing effect of super low molecular weight hyaluronic acid

Table 1 Water content in the skin[7]

State of the skin	Quantity of stratum corneum water (μS)
Very dry skin	14–20
Dry skin	40–50
Hydrated skin	200

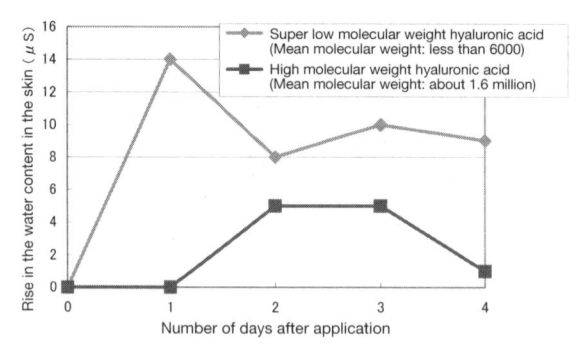

Figure 9 Difference in moisturizing effect by molecular weight of hyaluronic acid

3 cm × 3 cm gauze piece was soaked in 1 mL of 1% aqueous solution of a super low molecular weight hyaluronic acid (molecular weight: about 6000), 1% aqueous solution of a high molecular weight hyaluronic acid (molecular weight: about 160 million), or distilled water (control) and was applied on the human skin of the forearm for 24 hours. After removing the gauze, the skin was washed with water, and the electric conductivity was measured using a skin surface hygrometer to monitor time historical changes in the moisture content of the skin.

The super low molecular weight hyaluronic acid was shown to have a double improvement of the moisture content from that of the high molecular weight hyaluronic acid (figure 9).

It was also confirmed that even three days after the application, the moisture content was larger in the super low molecular weight hyaluronic acid than in the high morecular weight hyaluronic acid, showing a longer effect.

The test suggested that the super low molecular weight hyaluronic acid not only covers the skin surface but also penetrates and is retained long in the skin and gives moisture from the inside of the skin.

13. 7. 4 Penetrability into hair

Super low molecular weight hyaluronic acid was expected to penetrate not only into the skin but also into hair and give moisture and flexibility to the hair. The authors examined the penetrability of super low molecular weight hyaluronic acid into hair.

Damaged hair was soaked in a 1% aqueous solution of a super low molecular weight hyaluronic acid, which was fluorescent labeled, at 40°C for 5 to 15 minutes. Assuming daily hair washing, the hair was rinsed and dried after soaking. Cross sections of the hair were observed under a fluorescent microscope.

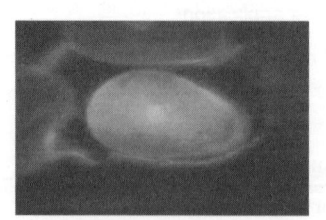

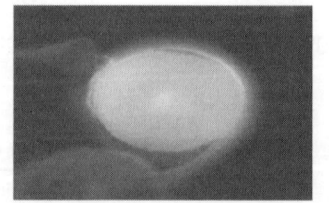

Figure 10 Penetrability of super low molecular weight hyaluronic acid into hair
The penetrability is shown with the degree of color development.

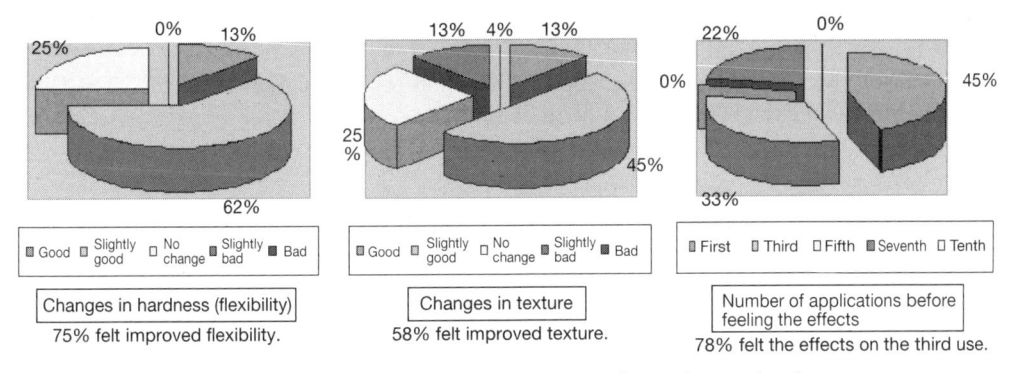

Figure 11 In-house monitoring test on use of super low molecular weight hyaluronic acid for treating the hair

Changes in hardness (flexibility)
75% felt improved flexibility.

Changes in texture
58% felt improved texture.

Number of applications before feeling the effects
78% felt the effects on the third use.

The super low molecular weight hyaluronic acid was shown to have penetrated into the hair (figure 10).

An in-house monitoring test was conducted on application of a 1% aqueous solution of a super low molecular weight hyaluronic acid. About 80% of the monitors realized improvement in the feeling and flexibility of their hair (figure 11).

As shown by the tests, super low molecular weight hyaluronic acid is very attractive for hair caring in and out the bath.

13. 8 Future prospects

Super low molecular weight hyaluronic acid has high moisturizing effect and is little viscous in aqueous solution. Thus, it can be used in formulations that are difficult with high molecular weight hyaluronic acid. The use in sunscreen products and mist spray is likely to grow, as well as in highly moisturizing cosmetics combined with high molecular weight hyaluronic acid.

13. 9 Conclusions

The production of hyaluronic acid started over 20 years ago using the cockscomb extraction method. Microbe fermentation enabled its mass production and reduced the price of hyaluronic acid, which was an expensive moisturizing agent, enabling the use in a number of products. Very advanced techniques are needed to strictly regulate the molecular weight of hyaluronic acid, and methods for freely regulating the size are being developed.

Finally, products that use the functions and concept of hyaluronic acid will increase not only in the field of cosmetics but also in medical supplies and food. The market of hyaluronic acid will be increasingly active and attract attention.

References

1. K. Meyer, J.W. Palmer, *J. Boil Chem.*, **107**, 629-634 (1934)
2. T.C. Laurent, JR. Fraser , *FASEB J.*, **6**, 2397-2402 (1992)
3. K. Abu, E. Hasegawa, Mucopolysaccharide Experimental Method (1), NANKODO Co., Ltd., p.6-7 (1972)
4. Y. Kanou, *FRAGRANCE JOURNAL*, **15**, p.69-74 (1996)

5. Q. P. Corporation, Patent Publication Number 2006-265287
6. M.O. Longas, C.S. Russell, X.Y. He, *Carbohydr Res.*, **159** (1), 127-36 (1987)
7. K. Takausu., Measurement of the Skin, Collection of Evaluation Manuals, TECHNICAL INFORMATION INSTITUTE Co., Ltd., p.78 (2003)

[Powders and Fine Particles]

Chapter 14
Pearl Pigments Coated with Barium Sulphate

Sadaki Takata

14.1 Introduction

Makeup cosmetics can be classified into base makeup cosmetics for preparing the skin beautiful and color makeup cosmetics for controlling the impression of the face by coloring and producing textures. Base makeup cosmetics cover freckles and skin color ununiformly, conceal unevenness of the skin, such as wrinkles and sweat pores, and make the skin look uniform. Color makeup cosmetics adjust the look of the parts of a face, such as the eyes by drawing outlines, and change the impression by coloring. Powder materiel is one of the most important materials for producing makeup cosmetics. Technology development is remarkable in late years for producing cosmetic powders of improved and additional functions, backed by application of various nanotechnologies.

This chapter describes the development and background of new functional hybrid powders for makeup cosmetics, which were developed by controlling the surface morphology of powders.

14.2 Evolution of hybrid powders by surface modification

Recent evolution of makeup cosmetics, particularly foundations and other base makeup cosmetics, is remarkable. Its technological theme is controlling light with powders. There are two general ways to control light. One is to control color, and another is to control the reflection of light.

Color pigments, such as iron oxide, have been widely used to control colors with powders. Today, titanium dioxide coated mica, or so-called pearl pigment, plays a big role. Titanium dioxide coated mica, which is prepared by coating a substrate of mica with a layer of titanium dioxide by controlling the thickness on the order of 100 nm, gives an interference color that depends on the thickness. Color control that uses the interference light of the pearl pigment has participated in the recent evolution of phototechnologies of base makeup cosmetics.

Conventional makeup cosmetics used the colors and textures of their powder constituents to make the skin look beautiful. The color of the skin was adjusted using red, yellow and other color pigments. Such a color adjusting method, which uses color pigments, involves mixing different colors, which causes subtractive color mixture and resultant drops in value and chroma causing the face to look dark, and is not appropriate for translucent and bright finish. Therefore, titanium dioxide coated mica attracted attention, which gives interference light depending on the thickness of the titanium dioxide coat. Color adjustment by interference light is additive color mixing, which can adjust colors without reducing brightness and solve the problems of color adjusting using color pigments. However, this pearl pigment, which emits interference light, has a plate shape and a high specular reflection. Thus, when used in cosmetics, it results in a very shiny makeup. To reduce shininess, hybrid powders of titanium dioxide coated mica covered by small-size spherical resin particles, such as PMMA (polymethyl acrylate), were developed[1]. The needs of modifying the surface of plate-shaped powders, such as mica and pearl pigments, to control their reflective properties have increased, and surface modification technologies for various powders have been developed. Figure 1 (a) is a SEM image of a hybrid powder of titanium dioxide coated mica covered by PMMA particles[1], which is the first result of such technological developments. The coat of PMMA particles changes the light reflection property of the powder to some extent, making a portion of the light to diffuse on the surface and not to specular reflect; but the hybrid powder is not durable against physical forces. The development of this powder led to further development of surface modification technologies. A SEM image of hybrid powder in which coral-shape crystals of aluminum hydroxide grow on the surface of titanium dioxide coated mica[2] is shown in figure 1(b), and a hybrid powder consisting of titanium dioxide coated mica covered by fibrous crystals of zinc oxide[3] is shown in figure 1(c). These hybrid powders do not only diffuse light at the surface but are also stable when they are processed into cosmetics due to their improved durability by the coat of the inorganic compounds. The hybrid powder shown in figure 1(c) can also solidify sebum and prevent smearing of makeups because the fiber-shaped zinc oxide reacts with the fatty acid of sebum and forms mineral residue. These powder compositing technologies that involve surface modification not only adjust shininess but can also add new functions to powders.

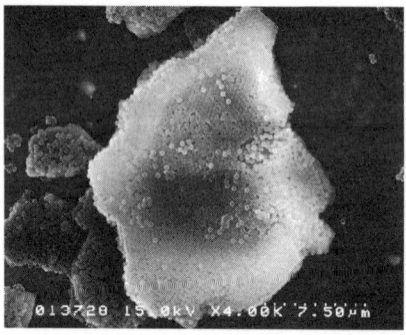

(a) Hybrid powder of mica substrate coated with spherical PMMA particles

(b) Hybrid powder of titanium dioxide coated mica substrate coated with crystals of aluminum hydroxide

(c) Hybrid powder of titanium dioxide coated mica substrate coated with fibers of zinc oxide

Figure 1 Various hybrid powders for adjusting the luster of plate-shaped substrate

Technologies have also been developed for forming various forms of crystals of inorganic compounds on the surface of powder, and have enabled new optical functions to be developed by advanced control of light reflection properties.

14.3 Development of functional powder by controlling the shape of powder surface using barium sulphate

These technologies for controlling the shapes of powder surface have been found to improve functions further by using barium sulphate as a coating compound. A process of manufacturing hybrid powder coated with barium sulphate using the surface modification technologies is shown in figure 2. First, fine seed particles of an inorganic compound are adhered on the surface of the substrate powder, which is then subjected to covering reactions under presence of various complexing agents. Crystals are then grown from the seed particles by controlling their shapes. The shapes of the crystals are determined by the seed particles and complexing agents used and also by the conditions of reaction (pH, temperature, reaction rate, *etc.*), enabling coats of various forms to be formed on powder.

14.3.1 Development of hybrid powder with a new optical property of obfuscating sagging on a face

The color of the skin, sweat pores, wrinkles and other minute unevenness can be adjusted by applying makeup cosmetics. However, bags and sagging skin caused by aging are difficult to conceal.

To find an approach to solve the problem, we investigated how sagging on a face are recognized visually. Computer graphics of faces of women of different ages are shown in figure 3. They are virtual images prepared by averaging eight faces typical of each age by morphing image analysis. The images show morphological changes accompanying aging. To facilitate understanding, the blightness distribution of the faces in their twenties and sixties was analyzed on a gray scale of 18 (figure 4). Shadows are cast from the wings of the nose to the lower part of cheeks around lips due to the sagging skin. The shadows make us recognize sagging on faces. Means to eliminate the shadows were investigated. The viewpoint adopted was reflex boards used by photographers when taking photographs of models. Reflection boards reflect light to diverse angles, and the diffusion light erases the shadows of sagging and wrinkles on faces, hiding sagging, and makes the skin to look bright and the face to look younger. Aiming to control the reflection properties of powder so as to have a similar effect, the surface of powders was designed. As a result, a hybrid powder of mica titanium dioxide coated with small pieces of barium sulphate was synthesized using technologies for controlling the surface shape

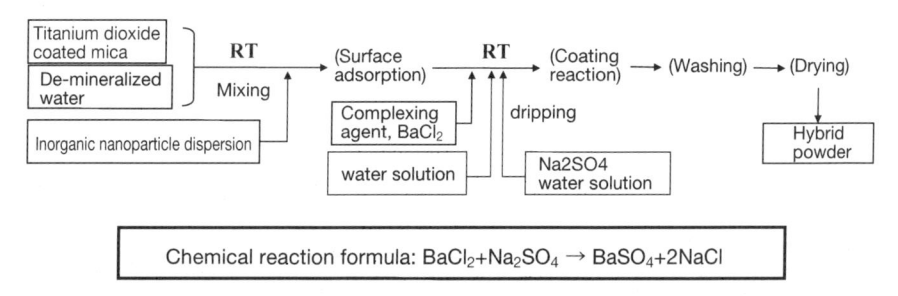

Chemical reaction formula: $BaCl_2 + Na_2SO_4 \rightarrow BaSO_4 + 2NaCl$

Figure 2 Process of covering powder with form-controlled coats and reactions involved

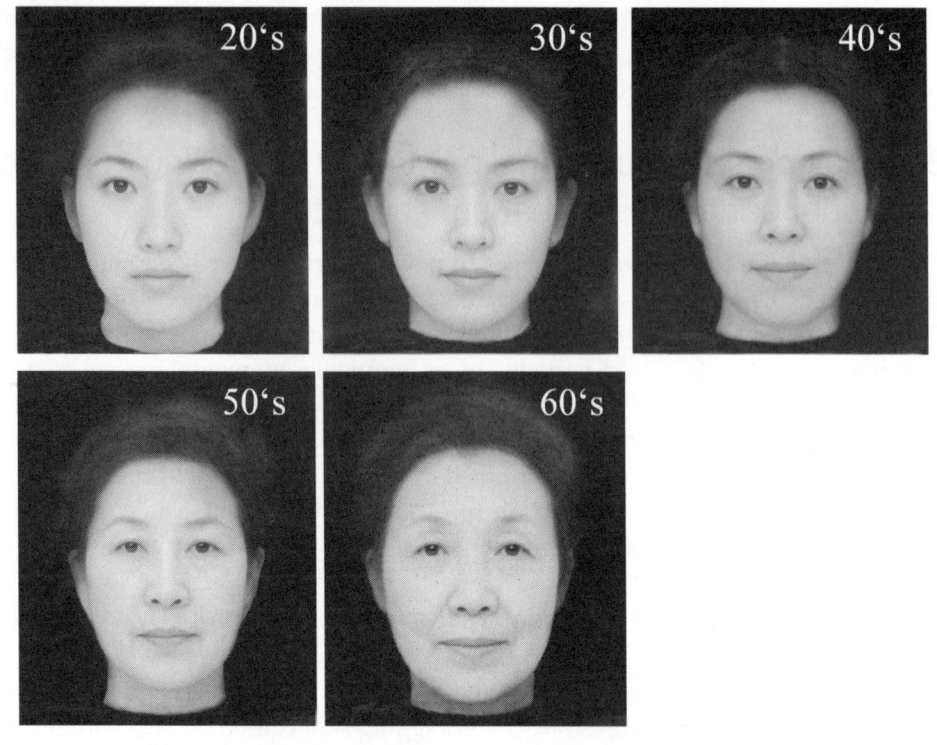

Figure 3 Typical faces of women in each age prepared by averaging the faces of eight women by morphing image processing

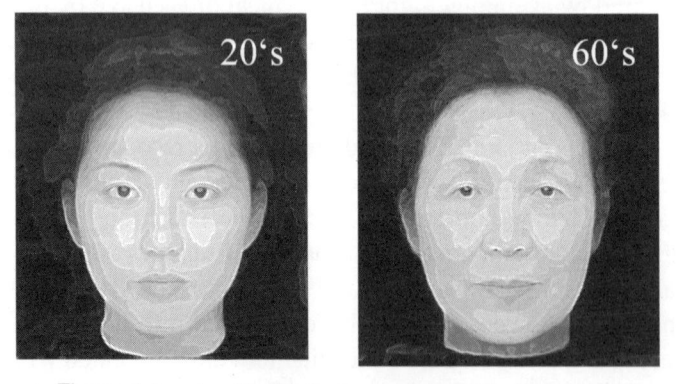

Figure 4 Luminosity distribution analysis of averaged faces of women in their 20's and 60's on a gray scale

and coating of powder (figure 5)[4,5].

In order to erase the shadows of sagging, powders must have a new light reflecting property. On parts where light is strongly irradiated from the front and the powder is irradiated vertically, such as on cheeks and forehead, the substrate titanium dioxide coated mica of the powder should emit red interference light by specular reflection to make the skin look firm. On parts prone to shading, where light comes aslant, the powder should emit white light diffused by micrometer-size reflex boards. Using a gonio-spectrophotometeric color measurement meter, the optical properties of the newly synthesized hybrid powder and titanium dioxide coated

Figure 5 Hybrid powder prepared by titanium dioxide coated mica with small pieces of barium sulphate

Table 1 Optical properties of hybrid powder and substrate titanium dioxide coated mica on reflectance difference by the incidence angle of the light

	$R_{22.5}$	R_{65}	$(R_{22.5} - R_{65})/R_{22.5}$
Hybrid powder	0.9316	0.7529	0.2371
Mica titanium	1.3364	1.334	0.0018

mica, which was the substrate of the new hybrid powder, were evaluated by irradiating red light of wavelength of 650 nm from the front (22.5° from the perpendicular line) and an oblique angle (65°), putting the specular reflectance to the irradiations as $R_{22.5}$ and R_{65}, respectively, and calculating the relative reflectance as $(R_{22.5} - R_{65})/R_{22.5}$. The results are shown in table 1 for the hybrid powder and titanium dioxide coated mica. The intensity of specular reflection did not depend on the incidence angle of the light in the plate-shaped powder of mica titanium dioxide coated mica; $R_{22.5}$ was equal to R_{65}, and the relative reflectance $(R_{22.5} - R_{65})/R_{22.5}$ was approximately 0. On the other hand, the hybrid powder showed a large reflectance $R_{22.5}$ value, showing high specular reflection to the light from the front, a smaller R65 value because the powder diffused light when the light was irradiated aslant. In other words, this hybrid powder had a peculiar optical property of changing reflection between specular reflection and diffusion reflection by the incident angle of the light. When the hybrid powder was mixed in a foundation and applied on faces, the skin of the subjects looked firm, and the shadows of sags were reduced. The makeups made their faces look younger than their actual ages.

14. 3. 2 Development of hybrid powder of ideal reflection characteristics by numerical computation

The reflection properties of powder have been controlled by coating plate-shape substrates with spherical particles of PMMA and changing the surface reflection characteristics from specular reflection to diffuse reflection. The powders have been designed by determining coating conditions experimentally, such as particle size coating rate, and their light diffusion properties were not theoretically ideal. Recent progress in computational technologies, such as dramatic improvements in operation speeds and high performance computers, has enabled optimum shapes of coats to be designed and hybrid powders to be synthesized by computer simulation of surface coating. A model of powder shown in figure 6 was prepared to understand the optimum light diffusion properties, and a theoretical optical structure was designed by

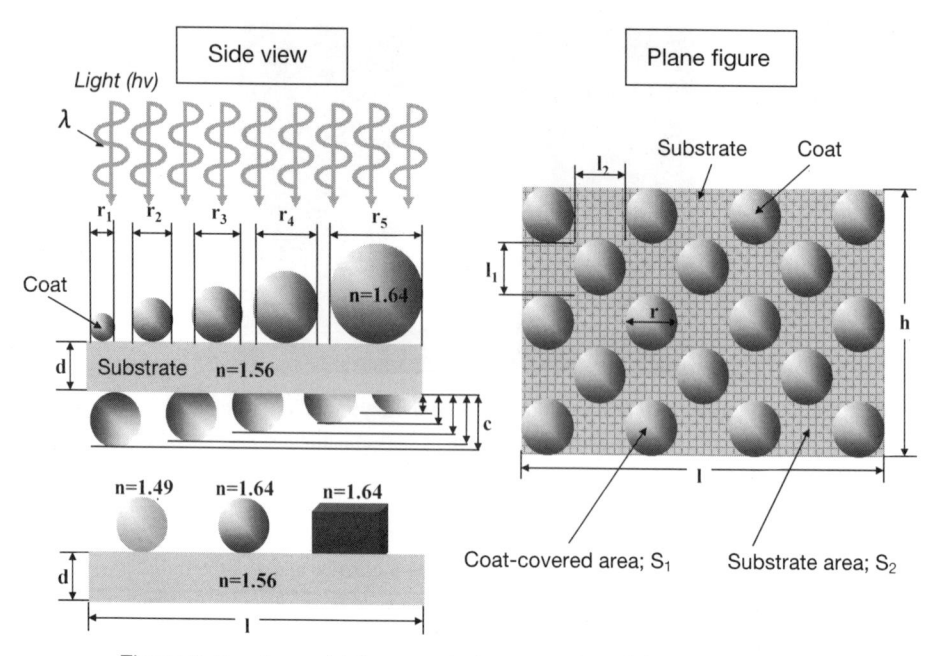

Figure 6 Powder model for numerical computation of the FDTD method

computational numerical analysis[6]. The FDTD (Finite Differential Time Domain) method was used for numerical computation. The algorithm of the FDTD method consists of three steps of (1) differences in time space, (2) time space in an electromagnetic field, and (3) configuration space in the electromagnetic field. The analytical domain was established so as to surround the wave source and the diffusion body and was divided into small rectangular parallelopiped cells. The differential equations of Maxwell, which were differenced by electric and magnetic fields on the surfaces and edges, were numerically solved in time space. When there was an object of any form in the analytical domain, the electromagnetic field of reflected waves could be analyzed. Because the FDTD method is an analytical method for a closed domain like the finite element method in principle, it is necessary to set the virtual borders so that waves do not reflect on the outer walls of the domain when solving problems in an open domain. An advantage of this method is that it can calculate not only the results but also the changes in electromagnetic fields in time spaces until the results are obtained. It can also visualize the states of magnetic fields changing from transient to steady states. The only disadvantage is that it is difficult to know the behaviors of photons at a point at infinity.

To simplify the conditions of FDTD numerical computation, the substrate of hybrid powder was decided to be flaky mica of high optical transparency (refractive index: 1.56, average thickness: 0.4μm) instead of ordinary mica, and barium sulphate (refractive index: 1.64), which is easy to control the shape, was used as the coating agent. The effects of 1) the shape of coating particles, 2) the size of the coating particles, and 3) covering ratio on resultant hybrid powder were simulation analyzed using the model shown in figure 6.

The calculated effects of the shapes of coating particles are shown in figure 7. The upper figures shows the diffusion of photons, and the lower figures describe the structure of hybrid powder (cross section of vertically positioned mica substrate coated on one or two sides). Photons were beamed from the left perpendicularly to the substrate, and the diffusion in the domain is shown with patterns. Mosaics denote photons overlapping with each other in a wave

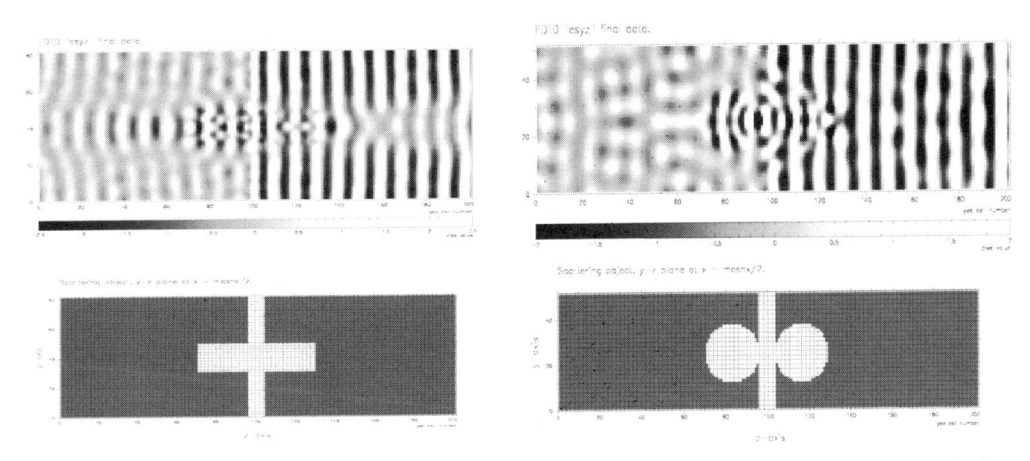

Figure 7 Effects of coating particle shapes on hybrid powder calculated by FDTD numerical analysis
Top: Visualized photon diffusion, Bottom: Section of coating particles on substrate

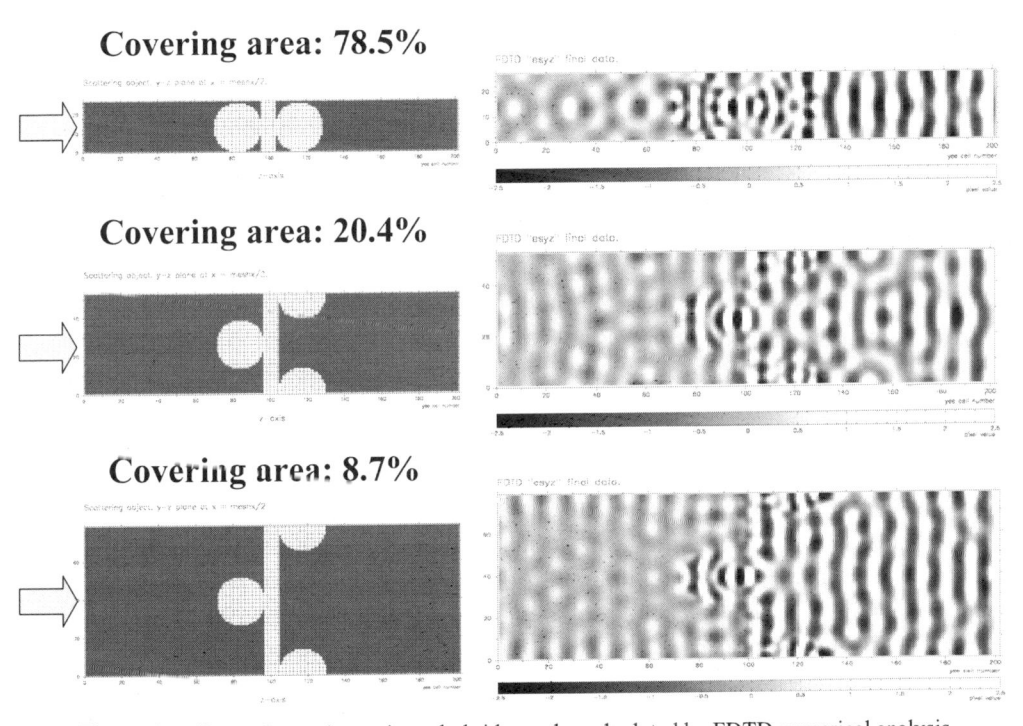

Covering area: 78.5%

Covering area: 20.4%

Covering area: 8.7%

Figure 8 Effects of covering ratio on hybrid powder calculated by FDTD numerical analysis

motion and thus light diffusive property. Stripes denote photons not overlapping with each other and thus specular reflection or straight penetration of light. The analysis showed that the particles were better be spherical than columnar to show mosaic patterns of photons and to gain light diffusive properties. Similarly, the sizes of coating particles (ϕ: 0.7, 1.3, and 2.0μm) were computationally analyzed. A diameter of 1.3μm was found to be optimum for gaining light diffusive characteristic. The adhesion state of the spherical particles on the substrate was found to not affect the reflection properties, which were similar when the particles were in a spherical shape or half spherical shape. The effects of covering ratio by spherical barium

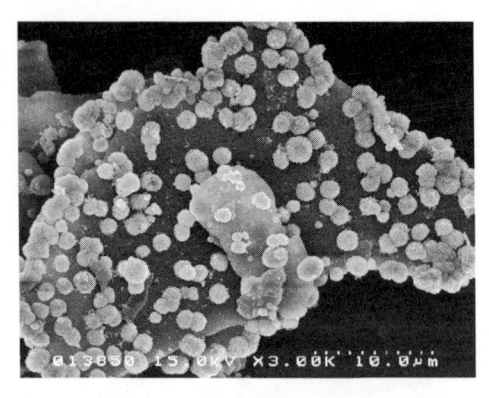

Figure 9 Hybrid powder designed by numerical computation and synthesized by the surface shape control technology

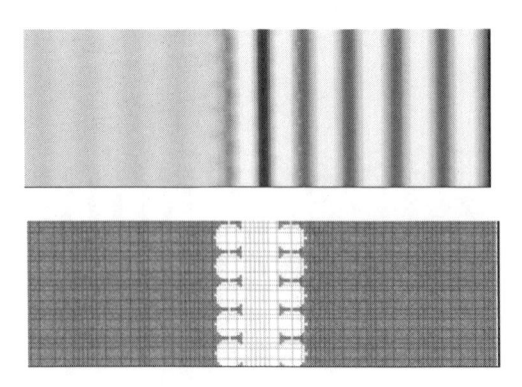

Figure 10 FDTD numerical analysis of a hybrid powder coated with PMMA

sulphate particles are shown in figure 8. At a covering ratio of 78.5 %, diffusion of reflected light was high, but almost all transmitted light was straight. At a covering ratio of 8.7%, the diffusion was low for both reflected and transmitted lights. At 20.4 %, diffusion was the highest for both reflected and transmitted lights. Based on the results of the computer simulation analysis of the optical model of hybrid power, an ideal structure of an optical hybrid powder model, which has high diffusive properties for both reflected and transmitted lights, are 1) spherical or semispherical coating particles, 2) coating particle diameter of about 1.3μm, and 3) covering ratio of about 20.4%.

Hybrid powder that has the ideal structure determined by the numerical analysis was synthesized by the surface shape control technology (figure 9)[7-9]. The same coating conditions can also be used for substrates of titanium dioxide coated mica.

In skin application tests of foundations that contained the synthesized ideal hybrid powder, the powder-containing foundation was shown to be superior to a foundation that did not contain the powder in concealment of unevenness of the skin, sweat pores, wrinkles, other shape-related defects, stains, freckles, and other color-related defects and in producing uniform and fine-texture makeups. Even compared to foundations that contained PMMA hybrid powder (figure 1(a)), the foundation containing the new powder was superior in concealing sweat pores, stains and freckles. The PMMA hybrid powder tested had fine spherical particles of PMMA of about $\phi 0.3\mu$m uniformly and fully coating the substrate and diffuses light by the spherical resin powder. The new powder was shown to have higher concealment effects of sweat pores, stains and freckles and produced uniform makeups.

The results of a computer simulation analysis of conventional PMMA hybrid powder are shown in figure 10. A stripe pattern appeared for the wave motion states of photons, showing high specular reflection and straight transmission of light. This suggests that the coat of the fine spherical particles was optically a single surface for photons due to the approach effect. On the skin, the powders, which are contained in the foundation, are arranged at random facing diverse directions, and thus scatter light as a whole and conceal sweat pores, stains and freckles to some extent. The effects are likely to be particularly notable in conventional thick hybrid powders. However, the conventional hybrid powders have specular reflection properties in the primary particle level, and this is a reason why they are inferior to the newly developed hybrid powder in concealment effect and texture.

14.4 Conclusions

The powder compositing technologies described in this chapter not only have improved conventional functions of powders but also have possibilities of developing new functions. In this sense, nanotechnologies are playing an increasingly important role in the field of cosmetics. Innovations of powder functions by nanotechnologies have caused dramatic evolution of makeup cosmetics, which now can reproduce the beautiful bare skin. However, we should not forget that the world of beauty that can be created with cosmetics has a sensual side that cannot be described by physical and optical properties. The world of makeup is artificial beauty, and we should also pursue beauty that is not imitations of the bare skin. Pursuit for "makeup skin exceeding the bare skin" combined with psychological studies on feeling beauties will lead to new technological innovations.

References

1. Patent No. 3425011
2. Publication of unexamined application No. 2001—146238
3. Patent No. 3671013
4. K. Yagai, K. Ogawa, T. Kanemaru, K. Joichi, N. Kunizawa, R. Takano, Optical rejuvenating makeup using an innovative shape-controlled hybrid powder, *IFSCC Magazine*, **9**. 2, 109–113 (2006)
5. Patent No. 3671045
6. Publication of unexamined application No.2006—16539
7. K. Ogawa, K. Yagi, S. Takata, K. Fuzima, Preprints of 3rd Meeting on Society of Nano Science and Technology, p.320 (2005)
8. Publication of unexamined application No.2004-00080
9. K. Ogawa, K. Yagi, H. Hata, Y. Miura, K. Nakata, Takata S., Fuzima K., *IFSCC Magazine*, **10**, 3, 199–206 (2007)

Chapter 15
Shape-controlled Composite Powder of Titanium Dioxide and Sericite

Kazuyoshi Tsubata, Hiroyuki Asano

15. 1 Introduction

Damages to the skin by ultraviolet rays, such as photoaging, stains, and wrinkles, are attracting attention these years[1,2]. The amount of ultraviolet rays has increased in the middle latitude areas of the southern hemisphere[3,4]. In Australia, measures are taken to minimize the exposure of the skin to ultraviolet rays, such as wearing hats and sunglasses. In Japan, fair skin has been favored as there is an idiom of "Fair skin hides all defects". The skin should be blocked from ultraviolet rays not only from the viewpoint of having fair skin but also to prevent skin damages in the future. To keep the skin beautiful even after aging, the cumulative exposure to ultraviolet rays should be reduced as much as possible. An effective measure is to daily apply sunscreens and base makeup cosmetics that block UV, such as foundations. According to a dynamic statistics survey of the Ministry of Economy, Trade and Industry, foundations are more shipped thus more frequently used than sunscreens. Thus, the cumulative exposure to UV can be reduced by increasing the UV blocking performance of foundations. There are two kinds of UV blocking agents: organic UV absorbents and inorganic UV scattering agents. Of these, organic absorbents are unstable in products (causing separation and strange smells) and are little compatible with other materials, and the use is limited. They cannot be freely used particularly in powder foundations. Thus, to increase the UV blocking performances of powder foundations, unlike in sunscreens and emulsion-type foundations, a large amount of inorganic UV screening agents, such as titanium dioxide and zinc oxide, should be added. However, large contents of titanium dioxide and zinc oxide result in bad touches when applied on the skin[5]

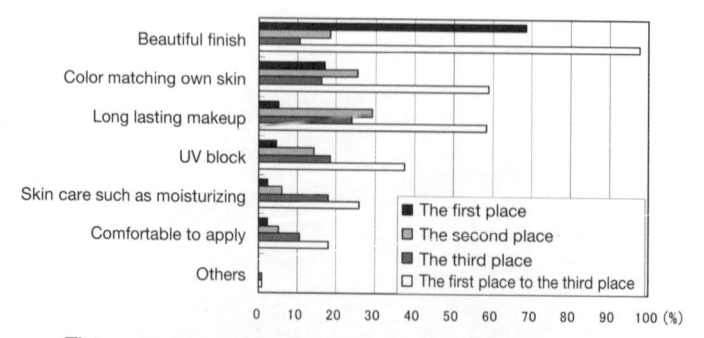

Figure 1 Effects considered when purchasing foundations (N=134, questionnaire survey by Menard)

giving "powder like", "creaking" and "heavy" touch due to the strong cohesive characteristics of the particles. They also result in unnaturally whitish makeups and unnaturally highlighted faces in photographs due to their high refractive indices. Foundations should also have coloring and psychological effect demanded to cosmetics, such as giving "beautiful appearance", "the color that matches the skin" and "comfort using" (figure 1). To respond to these demands, it is necessary to develop cosmetics that both block UV and have coloring and psychological effects of cosmetics by controlling the adverse effects of fine particles of titanium dioxide and zinc oxide on the feeling and finish.

Surface coating with silicone, fatty acids and silica and compositing in board-shaped powder have been investigated to improve the transparency and UV blocking performances of fine particles of titanium dioxide and zinc oxide[6-8]. However, few studies have been conducted on giving smooth feeling when applied on the skin. This chapter describes an investigation on composite powder that both blocks UV and is soft to use, which was prepared using sericite, which is produced in Furikusa, Aichi Prefecture, and gives a smooth touch and silky luster. The relationship between the optical properties of the composite powder and the shape of titanium dioxide is also described, which was revealed during the investigation.

15.2 Sericite of Furikusa, Aichi Prefecture[9, 10]

Mica powder is transparent due to its small refractive index and adheres well to the skin due to its flaky shape. It is widely used in cosmetics as an extender pigment to give luster and smoothness. Sericite (muscovite) has a particularly beautiful luster (like that of silk) and soft and silky touch, is easy to mold because the particle size is smaller than mica (muscovite), and is thus widely used in press mold cosmetics such as foundation cakes.

Mica crystals can be classified into some polytypes by the laminar structure taken by two tetrahedral sheets: such as 1M type (single monoclinal lattice), 2M1 type (double monoclinal lattices), 2M2 type (double monoclinal lattices) and 3T type (triple triclinical lattices). Sericite of the Furikusa, Aichi Prefecture, is almost pure 2M1 type, which is rare worldwide, and is precious. It has a hexagonal shape, high aspect ratios, and the mean particle diameter of about 10μm. It has high purity and crystallinity and a whiteness level of over 88.0 by the Hunter method (figure 2).

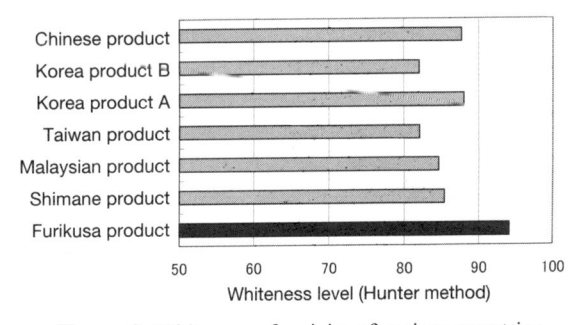

Figure 2 Whiteness of sericite of various countries

15.3 Preparation of composite powder and physical property evaluation

15.3.1 A preparation of composite powder

Sericite was prepared by processing sericite raw produced in Furikusa, Toei, Kitashitara, Aichi Prefecture, as described below:

The raw sericite was first sorted, and was loosened and dispersed in water. Impurities were removed by a sedimentation process. The particle size distribution was adjusted using a wet cyclone classifier. It was then repetitively washed with water, spin dried, dried, crushed and adjusted in white powder of a mean particle size of about $10\mu m$[10].

The resultant sericite was dispersed in methanol and was transformed into slurry by adding a titania-based compound (ammonium citratoperoxotitanate) or fine particles of titanium dioxide. It was then spray dried using a Parvis mini spray dryer (GB-22, YAMATO SCIENTIFIC CO., LTD) and baked. The resultant sericite was then coated with titanium dioxide. Sericite was coated with needle-shaped, membranous-shaped, granular-shaped and indefinite-shaped

a. Sericite of Furikusa, Aichi Prefecture

b. Sericite coated with needle-shaped titanium dioxide

c. Sericite coated with membranous-shaped titanium dioxide

d. Sericite coated with granular-shaped titanium dioxide

e. Fine particles titanium dioxide + sericite (dry mixed and crushed)

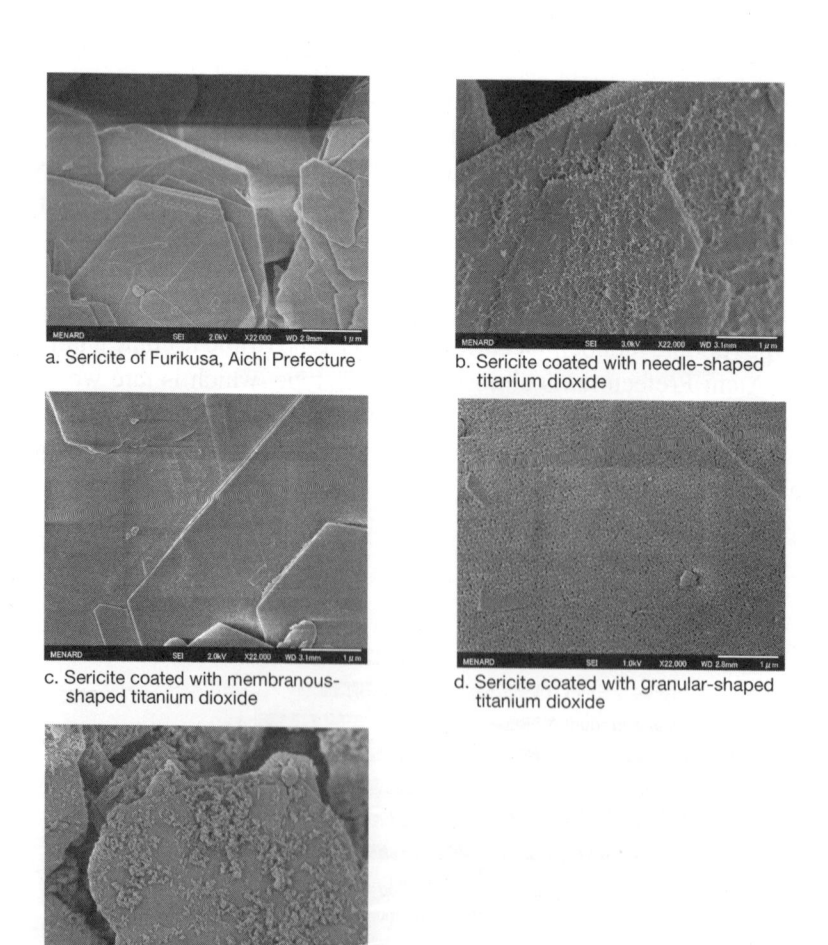

Figure 3 Scanning electron microscope photographs of powders

Chapter 15 Shape-controlled Composite Powder of Titanium Dioxide and Sericite

Table 1 Prescription of samples used for transmittance measurement

Raw materials	Quantity combined
Silicone KE-1300T	75
CAT1300	7.5
Silicone SH-200/10CS	12.5
Composite powder	5
Total	100

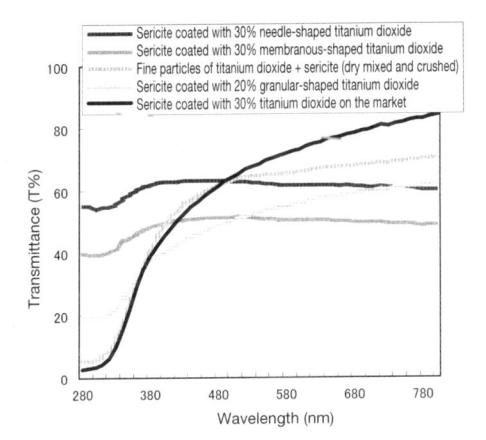

Figure 4 Transmittance of ultraviolet visible ray of sericite coated with titanium dioxide

titanium dioxide by adjusting the mixing time, reaction temperature, density, droplet sizes during the spray drying process, drying temperature and baking temperature. Scanning electron microscope photographs of these composite powders are shown in figure 3. The powders were evaluated for transmittance of ultraviolet-visible rays, light reflecting properties, and spreading and adhesive characteristics on the skin.

15.3.2 Transmittance of ultraviolet and visible light

Silicone resins and the composite powder were weighed in a container as shown in table 1, placed in a vacuum to remove bubbles, and mixed using a Hoover-type Muller (number of revolutions: 50 times × 2). The mixture was cast on glass plates using a film applicator and was solidified at 60°C. The transmittance of ultraviolet and visible light was measured using a spectrophotometer (UV-160A, Shimazu Corporation, wavelengths: 290 to 800 nm). The untreated sericite was used as the baseline, and a mixture of sericite and fine particles of titanium dioxide, which was dry mixed and crushed, was used as the control (figure 4). The transmittance of the ultraviolet region was the lowest (highest UV blocking performance) in the control, followed by sericite coated with 30% titanium dioxide on the market, sericite coated with 20% of granular-shaped titanium dioxide, sericite coated with 30% of titanium dioxide membrane, and sericite coated with 30% needle-shaped titanium dioxide, in this order. From the results and the electron microscope photographs, sericite that was entirely covered by granular-shaped titanium dioxide was likely to have high UV blocking performance.

15.3.3 Light reflecting properties

Measurement samples were prepared by affixing double-side adhesive tapes on the black part of OPACITY CHARTS supplied by LANETA COMPANY, INC. and applying the composite powders using a makeup brush. Reflection intensity was measured using a digital variable gloss meter (SUGA TEST INSTRUMENTS CO. LTD, UGV-5DP, incidence angle: 45°, angle of acceptance: 0 to 85°) (figure 5).

Reflected light intensity at an incidence angle of 45° was the strongest in the control, followed by sericite coated with 30% titanium dioxide on the market, sericite coated with 20% of granular-shaped titanium dioxide, non-treated sericite, sericite coated with 30% needle-shaped titanium dioxide, and sericite coated with 30% membranous-shaped titanium dioxide, in this order, showing that the intensity of regular reflection was related to the shapes of titanium dioxide. Sericite coated with needle-shaped or membranous-shaped titanium dioxide, which showed low regular reflection intensities, scattered more light than non-treated sericite,

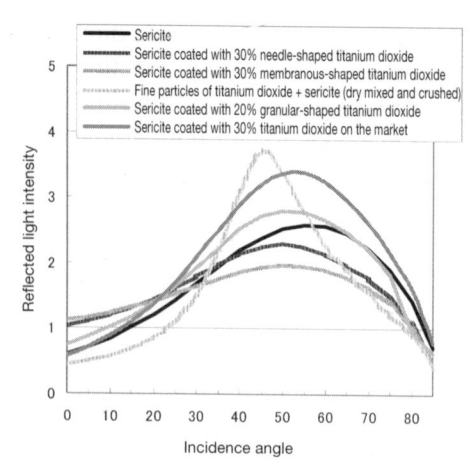

Figure 5 Varied angle reflection intensity of ultraviolet and visible rays in titanium dioxide-coated sericite

showing that the coating added soft-focus characteristics to the mica. On the other hand, the control and titanium dioxide-coated sericite on the market showed strong regular reflection intensities. The regular reflection intensities were higher in specimens that showed higher UV blocking performance.

15. 3. 4 Spreading characteristic of the powders

Of the composite powders prepared, 1) sericite coated with 20% of granular-shaped titanium dioxide, which had relatively large UV blocking performance and a light scattering characteristics similar to that or original sericite, 2) non-treated sericite, and 3) fine powders of titanium dioxide were applied on artificial skin Saplare (Idemitsu Kosan Co., Ltd), and their frictional characteristics were evaluated using TRIBOSTATION Type 32 (Shinto Scientific Co., Ltd., movement speed: 2500 mm/min, measurement distance: 50 mm outward trip, load: $100 g/cm^2$). The inverse of the measurement value was used to show frictional characteristics (figure 6). The larger the number, the higher the extendibility and smoother to spread. The sericite coated with 20% of granular-shaped titanium dioxide spread better than the fine particles of titanium dioxide and spread as smoothly as non-treated sericite. The high extendibility was likely because granular-shaped titanium dioxide uniformly covered the entire surface of sericite (figure 3) and titanium dioxide was little aggregated.

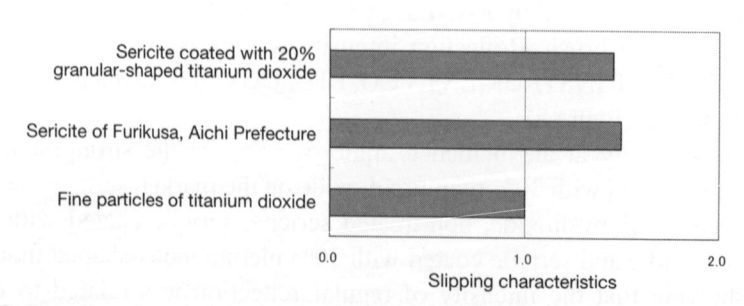

Figure 6 Spreading characteristics of sericite coated with 20% granular-shaped titanium dioxide

15.3.5 Adhesion on the skin

Sericite coated with 20% of granular-shaped titanium dioxide and the powder used as the control were applied on the skin using a makeup puff, and the skin was observed under a microscope (HIROX Co., Ltd, SCOPE KH-2200 MD2). The control powder showed aggregates of titanium dioxide in sulcus cutis. On the other hand, sericite coated with 20% of granular-shaped titanium dioxide spread uniformly on the entire skin (figure 7). Because sericite is flaky and is easy to adhere on the skin in planes and the surface is uniformly covered by titanium dioxide, it was likely that the sericite coated with 20% of granular-shaped titanium dioxide efficiently exposed the titanium dioxide side outwards. Therefore, a high UV blocking performance can be expected.

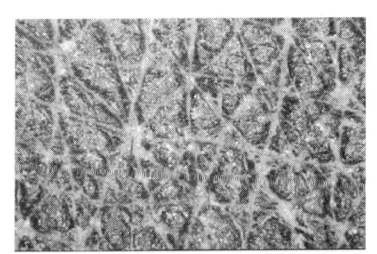

a. Sericite coated with 20% granular-shaped titanium dioxide

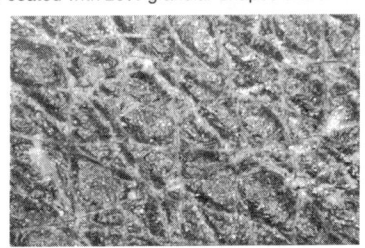

b. Fine particles titanium dioxide + sericite (dry mixture and crushed)

Figure 7 Adhesion on the skin

15.4 Application to powder foundation

15.4.1 Measurement of UV blocking performance

Powder foundations were prepared by mixing the sericite coated with 20% of granular-shaped titanium dioxide and sericite coated with 30% titanium dioxide on the market in proportions shown in table 2, and their UV blocking performances were compared.

A certain amount of a makeup base and the foundation were uniformly applied

Table 2 Prescriptions and UV blocking performances of powdery foundations

Raw materials	Prescription A	Prescription B
Sericite coated with granular-shaped titanium dioxide	10.00	
Sericite coated with titanium dioxide on the market		10.00
Silicone-coated fine particles of titanium dioxide	8.00	8.00
Silicone-coated zinc oxide	2.00	2.00
Silicone-coated sericite	12.00	12.00
Silicone-coated talc	23.00	23.00
Silicone-coated synthetic mica	10.00	10.00
Organic spherical powder	7.00	7.00
Silica	4.00	4.00
Silicone-coated titanium oxide	8.00	8.00
Silicone-coated iron oxide	3.50	3.50
Zinc stearate	1.00	1.00
Methyl paraben	0.50	0.50
Binder	11.00	11.00
Total	100.00	100.00
SPF value	25.8	23.8

on 5×5 cm² pieces of surgical tape (Sumitomo 3M Limited, Blenderm No.1525-5). Ultraviolet rays of 290 to 400 nm were irradiated using a solar simulator (WACOM ELECTRIC CO., LTD. XB-251WI), and the quantity of 305-nm wavelength UV that penetrated through the tape was measured using a UV meter (Tokyo Optical Co., UVR-305/365-D(II)). Foundations whose SPF values were known were similarly measured, and the results were used to calculate the SPF values of the specimens (table 2). The sericite coated with 20% of granular-shaped titanium dioxide contained less titanium dioxide than the titanium dioxide-coated sericite on the market and blocked less UV (figure 4). However, when prepared in foundations, the sericite coated with 20% of granular-shaped titanium dioxide resulted in a larger SPF value. This was likely because, as discussed in Section 15.3.5, sericite is flaky and is easy to adhere on the skin in planes, the surface is uniformly covered by titanium dioxide, and the sericite coated with 20% of granular-shaped titanium dioxide exposed the titanium dioxide side outwards efficiently.

15.5 Conclusions

Composite powders of high UV blocking performance that gives a smooth touch can be prepared by coating the surface of sericite of Furikusa, Aichi Prefecture, with titanium dioxide of various shapes. The coating also changed the optical properties of sericite. Of various shapes tested, needle-shaped titanium dioxide and membranous-shaped titanium dioxide reduced the regular reflection intensity when coated on sericite. On the other hand, a mixture of sericite and fine particles of titanium dioxide, which were dry mixed and crushed, and titanium dioxide-coated sericite on the market were high in regular reflection intensity. In general, the refractive index of titanium dioxide differs depending on crystal system. Even with this point considered, it is interesting that the regular reflection intensity differed by the shape of titanium dioxide.

The UV blocking performances of the composite powders, which were measured by dispersing on silicone resin membranes, were slightly lower than those of the mixture of sericite and fine particles of titanium dioxide, which were dry mixed and crushed, and the titanium dioxide-coated sericite on the market. However, when used to prepare foundations, the composite powders showed larger SPF values *in vitro* than those of the titanium dioxide-coated sericite on the market. This was because the composite powder spread and adhered on the skin uniformly due to the flaky shape of sericite. Thus, it should have a high UV blocking performance when used in practice.

Coating was found to result in various other characteristics. For example, sericite coated with granular-shaped and membranous-shaped titanium dioxide was high in transmissivity of ultraviolet rays but scattered more light than non-coated sericite.

This chapter described an example of controlling the shape of coating particles to control the optical properties of resultant composite powders. Further development in shape control will lead to development of composite powder that has various functions.

References

1. K. Hanada, 26th SCCJ seminar, Society Cosmetic Chemists Japan (2004)
2. K. Hanada, *J. Soc. Cosmet. Chem. Jpn.*, **39** (3), 180 (2005)
3. 2005 Annual Report on Ozone Monitoring, Ministry of the Environment (2006)
4. Report of Ozone Layer Monitoring, 2005, Japan Meteorological Agency (2006)

5. H. Hamaki, *Fragrance Journal*, **30** (7), 33 (2002)
6. K. Ogawa, *et al., J. Soc. Cosmet. Chem. Jpn.*, **39** (3), 33 (2002)
7. F. Sato, *SHIKIZAI KYOKAISHI*, **79** (9), 209 (2005)
8. S. Nishioka, *Ceramics Japan*, **42** (2), 124 (2007)
9. S. Suzuki, *Fragrance Journal*, **22** (6), 51 (1994)
10. I. Asai, Interface Science Practice Lecture 2005, Tokai Branch Office JAPAN Oil Chemists' Society (2005)

Chapter 16
Light Diffusive Inorganic Powder

Naoyuki Enomoto

16.1 Introduction

Functions demanded of cosmetics have diversified, leading to research and development of cosmetics that have unique functions from various perspectives. JGC Catalysts and Chemicals Ltd. has developed various functional materials for cosmetics by applying and deepening inorganic material-based nanotechnologies developed via catalyst preparation. The company specializes particularly in raw materials for makeup. For example, Velvet Veil®, which was developed in 1989, is prepared by coating mica with submicron-size particles of silica and applying the soft focus theory[1]; and SOFT-VISON® with enhanced wrinkle concealing effect has an additional thin film of titanium dioxide on the outmost layer. A more beautiful finish is always being investigated using the optical functions such as light absorption, reflection, refraction, and transmission. A recent trend in development is materials that are translucent, give the illusion of bare skin, and conceal blemishes such as wrinkles, freckles and pores. In this paper, two functional inorganic powders (platelet and spherical) that have light diffusion effects and other unique optical properties are described, which are spherical and plate-shaped powders.

16.2 Light diffusing spherical powder

The characteristics and effects of spherical silica particles with a hollow pore structure, HOLLOWY N-15, are described below.

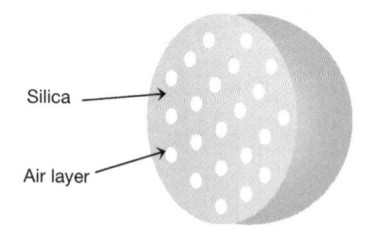

Figure 1 Schematic view of a hollow porous silica particle

16.2.1 Characteristics and effects of HOLLOWY N-15

Ordinary porous spherical silica in the order of microns, which is widely used to improve the feeling onto the skin, has fine pores of several to several hundreds of nanometers opening on the particle surface (so-called "open pore" structure). On the other hand, HOLLOWY N-15 is a silica particle that has a hollow porous structure, a particle diameter of 10μm, and about 16 VOL% of

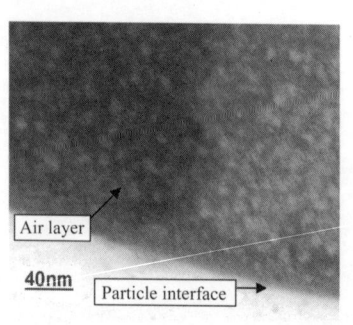

Figure 2 Transmission Electron Microscope image of an ultra thin section of the hollow porous silica (×250,000)

independent pores (closed pores) of a few nanometers which do not open onto the surface. As the surface of the powder is virtually non-porous, the surface area is quite small at about 1 m^2/g, and oil absorption capacity is as low as approx. 20 mL/100g. A schematic view of the cross sectional structure of the particle is shown in figure 1, and a transmission electron microscope image of an ultrathin section of the material is shown in figure 2.

16. 2. 2 Optical properties

As the particles have inner layers of air, the apparent refractive index of HOLLOWY N–15 is approximately 1.38, which is smaller than that of ordinary silica, which is 1.45. To facilitate understanding of its optical properties, the total light transmittance (Tt) and diffusivity (haze) measured by dispersing this and other typical silica particles into solvents (mixtures of water and glycerin) of various refractive indices are shown in figure 3 (diffusivity = (diffuse transmittance / total light transmittance) × 100). In ordinary porous silica (Silica Micro Bead P–1500, mean particle diameter: 5μm, SA: 150 m^2/g), the solvents permeated into the pores, and thus the Tt value increased along with rises in the refractive index of the solvent. When the refractive index was around 1.45, the difference in refractive index between the solvent and the particle was so small that the dispersion liquid looked transparent. At this time, the haze value was also small. Porous silica that contained 20% fine particles (a few dozen nm in particle diameter) of titanium dioxide inside (Silica Micro Bead USB T–20L, mean particle diameter: 5μm, SA: 350 m^2/g) is likely inappropriate for translucent cosmetics although its haze value was large because it contains titanium dioxide (n=2.71), which has a high refractive index, and therefore Tt was low in all regions. On the other hand, the Tt value of the hollow porous silica (HOLLOWY N–15) marked the peak at a refractive index of 1.38 and decreased gradually thereafter. This is because the air layers are much smaller than the wavelength of the light and the inner part of the particle was likely to have acted as a single layer consisting of both silica and air. The haze values of this hollow porous silica were large in all regions because the air layers inside the particle were not affected by the solvents and highly refracted light due to the differences in refraction index between the air layer and silica.

As described above, HOLLOWY N–15 has characteristically high optical properties, such as in light diffusion and transparency.

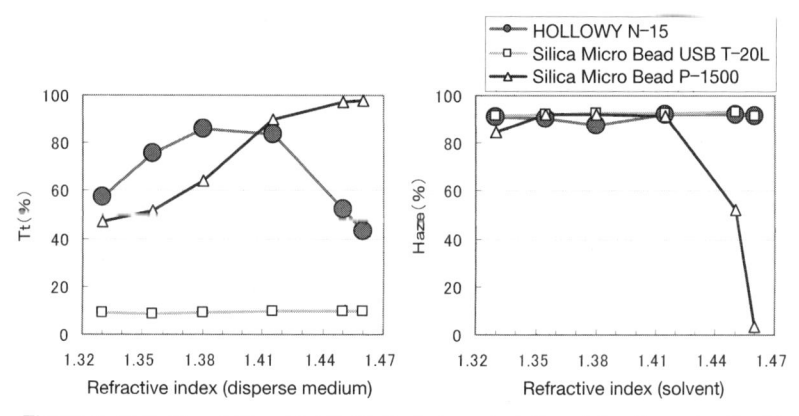

Figure 3 T_t (left) and Haze(right). Optical characteristics of the dispersion liquid

Sample concentration: 1.0% Solvent: water-glycerin Cell: 10mm Measurement: Spectrophotometric Colormeter 300A (NIPPON DENSHOKU INDUSTRIES CO., LTD)

16. 2. 3 Effects of mixing into cosmetics

Cosmetics are required to not only conceal wrinkles immediately after application but also to keep the optical properties unchanged for hours even when they get wet with sweat and sebum. As mentioned earlier, HOLLOWY N–15 shows characteristically shows only small changes in optical properties when it is mixed with solvents of different refractive indices. Unlike ordinary porous silica, it keeps its light diffusing property and does not become transparent even when it gets wet with caprylic triglyceride (refractive index: approx. 1.45), which is believed to be the main constituent of sebum. Therefore, it is expected to maintain moderate transparency and light diffusing properties even when it is wetted with oils and solvents widely used as liquid constituents of cosmetics, which are listed in table 1. Appropriate water-repelling treatment with silicon, etc. of the surface is needed to homogeneously disperse the powder in oil solutions[2]. This powder improves the feeling on the skin of various kinds of cosmetics, including powder foundations and emulsion type skin-care products, such as milky lotion and essences as it has a spherical structure, and can give a light touch and refreshing feeling due to the low oil absorbency. In particular it gives both translucency and wrinkle concealment effects when blended into cosmetics of silicone base as it has a similar refractive index to this powder.

16. 3 Platelet-shaped light diffusive powder

Distinctive features and effects of platelet-shaped light diffusive powder that has a multi-layered coating structure, Cover Leaf® AR–80, are briefly described below.

16. 3. 1 Characteristics and effects of Cover Leaf® AR–80

This powder has high translucency and strong concealment effects of skin blemishes, which enable beautiful makeup finishes and have been achieved using the principle of anti-reflection coats formed on the surface of display panels like LCDs. Anti-reflection coats can be prepared either by canceling out lights by making incident and reflected lights to interfere with each other or restraining the surface reflection by using a film in which the refractive index changes gradually to reduce the difference in refractive index between the substrate and medium (air, solvent, *etc.*). The former method requires the film to be rather thick for strict non-reflecting designs, resulting in drops in uniform extendability, which is demanded of cosmetics. The

Table 1 Refractive index of typical liquid ingredients (Measurement: Abbe refractometer)

INCI NAME	n (25°C)
DIMETHICONE	1.397
ISODODECANE	1.418
ISONONYL ISONONANOATE	1.435
TRIETYLHEXANOIN	1.440
BUTYLENE GLYCOL	1.440
CAPRYLIC / CAPRIC TRIGLYCERIDE	1.450
SQUALANE	1.451
HYDROGENATED POLYISOBUTENE	1.453
MINERAL OIL	1.463

platelet-shaped light diffusive powder used the latter film method and is designed to have coats of titanium dioxide (thickness: about 4 nm), alumina (thickness: about 7 nm) and silica (thickness: about 7 nm) on the surface of talc substrate of about 100-nm thick. The refractive indices of these coats are designed to increase from the outermost layer to the inside. Differences in refractive index between the layers were minimized, and the outermost layer is silica, which has the lowest refractive index among inorganic oxides. In addition, spherical silica particles (80 nm) are adhered to the outermost layer to modify the shape of the powder surface to enhance diffusion of reflected light at the powder surface. A schematic view of a cross section of the platelet-shaped light diffusive powder is shown in figure 4.

16. 3. 2 Optical properties

Total light transmittance (Tt) and diffusivity (haze) measured for each coating step are shown in figure 5. As coating progressed, haze value increased, and the final haze value was approximately double the haze value of talc (substrate). Such a diffusion effect of transmitted light is caused by repeated reflection and refraction of light at the boundary of each layer in the laminate structure. On the other hand, Tt value decreased slightly when coated with titanium dioxide, but increased by coating with alumina and silica to a level higher than that of talc used

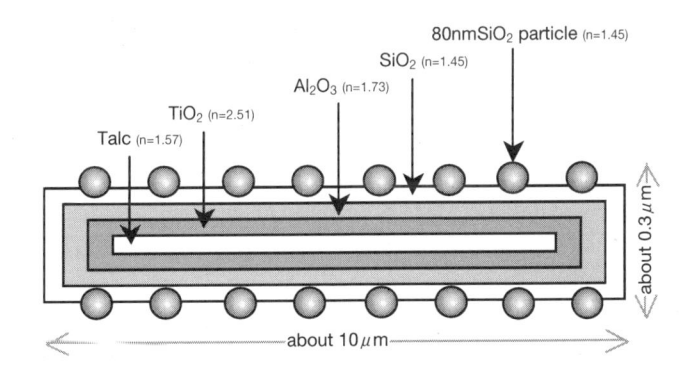

Figure 4 Schematic view of the multilayered coating structure platelet-shaped particle

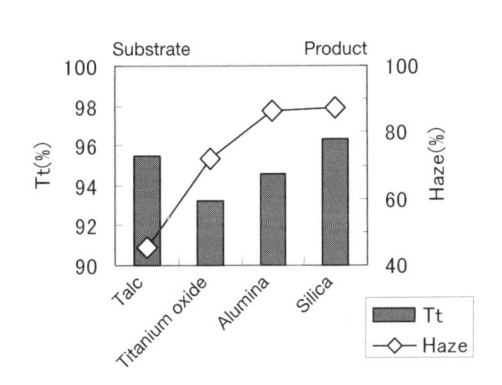

Figure 5 Haze and T_t at each coating stage

Combination ratio: Sample / Nitrocellulose = 20 / 80
Film thicknesses: About 4 μm
Substrate: PET film
Measurement: Haze Computer(SUGA TEST INSTRUMENTS CO., LTD.)

Figure 6 T_t and haze of cream

as the substrate. This is likely attributable to the reductions in reflected light at the surface of the particle, which has a structure of gradually decreasing refraction.

16. 3. 3 Effects of mixing in cosmetics

The expression of the optical properties when blended in cosmetics was examined by preparing creams of the model prescriptions shown in table 2.

Model creams were applied on quartz cells with a slit (width: 30 nm), were dried at 50°C for 10 minutes, and measured for transmittance and diffusivity (figure 6). The creams into which AR–80 was blended showed slightly smaller Tt values and larger haze values than the cream that contained no AR–80 (control). A sensory evaluation showed that Model–1 concealed wrinkles effectively while being highly translucent. Model–2 showed even a higher concealment effect. Photographs of cream application to panelist's eye are shown in figure 7. As shown in the photographs, Model–2 (containing 15% AR–80) concealed crow's feet more effectively than the liquid foundation on the market. As described above, Cover Leaf® AR–80 increases translucency by decreasing light reflection at the surface and increasing light transmission. It also diffuses transmitted light by its refraction-gradient layer structure and thus conceals skin blemishes such as wrinkles and freckles.

16. 4 Conclusions

Some approaches for achieving beauty by utilizing light diffusion effects were described. Not only these but combinations with other microfabrication technologies, such as fine particle preparation, fine pore control, shape control, surface treatment and hybrid inorganic-organic

Table 2 Model prescription of the cream

Ingredient (%)	Control	Model-1	Model-2
Squalane (Olive)	10.0	10.0	15.0
Polyglyceryl-10 Pentast earate (and) Behenyl Alcohol (and) Sodium Stearoyl Lactylate	2.0	2.0	3.0
COVERLEAF AR-80	0.0	5.0	15.0
Butylene Glycol	3.0	3.0	3.0
Methylparaben	0.2	0.2	0.2
Water	69.7	64.7	48.7
Carbomer/1%aq.	10.0	10.0	10.0
Xanthan Gum/2%aq.	5.0	5.0	5.0
Triethanolamine	0.1	0.1	0.1

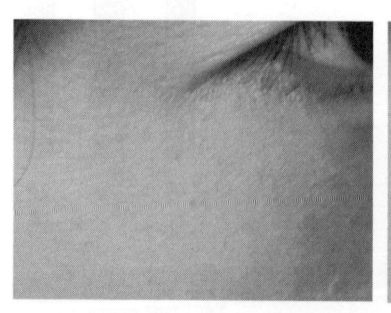

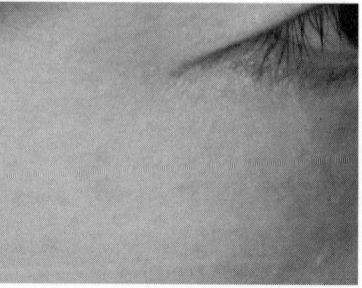

Figure 7 Application of a liquid foundation on the market (left), and application of Model-2 (right)

compounds, are highly potent tools for creating new functional materials. Cosmetics are also demanded to protect the skin from dehydration and ultraviolet rays. A composite powder consisting of zinc oxide coated with a thin film of silica of several nanometers has been reported effective in controlling rough skin[4]. Future studies on skin-care effects of inorganic materials are also greatly expected to produce results.

References

1. N. Nakamura, *J. Soc. Cosmet. Chem.* Jpn., **2**, 2 (1897)
2. Patent No. 3079395
3. M. Matsumoto, *et al.*, *MATERIAL STAGE*, **1**, 1, p.65-71 (2001)
4. E. Kawai, *IFSCC Magazine*, **5**, 4, p.269-275 (2002)

PART 4 Trends in Patents

Chapter 1
Trends of Nanotechnology-related Patents

Yoshihiro Yano

1.1 Introduction

When we think of technical developments, we should not underrate protection of intellectual property rights such as patent rights. Intellectual property rights are a system for protecting the products of creative activities (new technologies and ideas) as properties of the creators. We must seriously think about intellectual property rights also in the fields of cutting-edge nanotechnologies, where competition is severe.

Therefore, this chapter first explains the patent system, which is highly important in intellectual property rights, and then trends of patents in fields where nanotechnologies are involved.

1.2 Intellectual property rights and patent rights

"Intellectual property" or "Intellectual property rights" is defined in the Intellectual Property Basic Act of Japan as the following:
(Reference articles) Intellectual Property Basic Act in Japan
> Article 2 (1) The term "intellectual property" as used in this Act shall mean inventions, devices, new varieties of plants, designs, works and other property that is produced thought creative activities by human beings (including discovered or solved laws of nature or natural phenomena that are industrially applicable), trademarks, trade names and other marks that are used to indicate goods or services in business activities, and trade secrets and other technical or business information that is useful for business activities. (2) The term "intellectual property right" as used in this Act shall mean a patent right, a utility model right, a right that is stipulated by laws and regulations on other intellectual property or right pertaining to an interest that is protected by acts.

Intellectual properties, which are "information with property values", are easy to copy and imitate. Therefore systems for protecting intellectual property rights have developed partly aiming to limit "information" that is freely accessible in order to protect the rights of its creator; and because of this, the systems have been sometimes viewed to impede new creations. However, the patent system, which is one of systems for protecting intellectual property rights, was established to unveil technologies and encourage inventions (Venetian Patent Law, 1474).

Today, patent rights are recognized as exclusive rights (strong exclusive rights enabling excluding imitations) given to inventors in exchange of unveiling the inventions to the world. Article 1 of the Patent Act of Japan[1] clearly states that the purpose of the patent system is to contribute to developments of technologies and industry: "The purpose of this Act is through promoting the protection and the utilization of inventions, to encourage inventions, and thereby to contribute to the development of industry."

1.3 Scope of the Patent Act

What inventions are to be protected by the Patent Act? The Patent Act protects highly advanced creations out of technical ideas that use the law of nature (Patent Act Article 2). Therefore, inventions that do not use the law of nature, such as calculation methods, codes and factitious arrangements, such as financial insurance and tax systems, are out of the scope of the Act. Since they should be technical ideas, discovery itself is also out of the scope. (For example, the Relativity theory discovered by Albert Einstein is not a patent. Dr. Einstein did not make a patent application for the Relativity theory although he had applied patents when he worked as a patent examiner in Bern. He should have understood that theories were out of the scope of the patent system.)

To be patented, creations must meet the aforementioned criteria, be new compared to existing technologies, be not easily derived from exiting technologies, and be stated in a feasible manner. Even state-of-the-art nanotechnologies are not patented unless they satisfy all these criteria.

1.4 Procedure of obtaining patent rights

If you have made an invention, you can apply for a patent by submitting an application to the Patent Office. A flowchart of the procedure of obtaining a patent is shown in figure 1 (prepared based on a flowchart published by the Japan Patent Office).
(1) Application
Any inventions will not be patented unless they are applied for. Prescribed forms should be submitted to the Patent Office. In Japan, the application made first is adopted when two or more applications are made on the same invention. Inventions disclosed before making patent application cannot be applied for.
(2) Formality examination
The application forms are checked for the formality by the Patent Office. Correction is requested when there is a defect.
(3) Publication before examination
After one year and six months from the date of application, the contents of the applied invention are published in a gazette.
(4) Request for examination
The applied invention is examined when the applicant or a third person request for examination and pays the fee. The request should be made in three years from the date of application.
(5) Deeming withdrawal (when no request for examination was made within the term)
Applications for which examination is not requested in three years from the date of application are deemed to be withdrawals.
(6) Substantive examination
Examination is performed by an examiner at the Patent Office. The examiner judges whether the applied invention is to be patented or not. The examination involves examining whether the contents described as invention satisfy the criteria prescribed by the law or not and whether there are causes for rejection or not. Inventions to be patented must meet the following requirements:
They must i) be a technical idea(s) based on the natural laws, ii) be feasible for industrial use, iii) be a technical idea that did not exist before application, iv) be not an invention that is easily

Chapter 1 Trends of Nanotechnology-related Patents 253

Proceedings for getting patent right

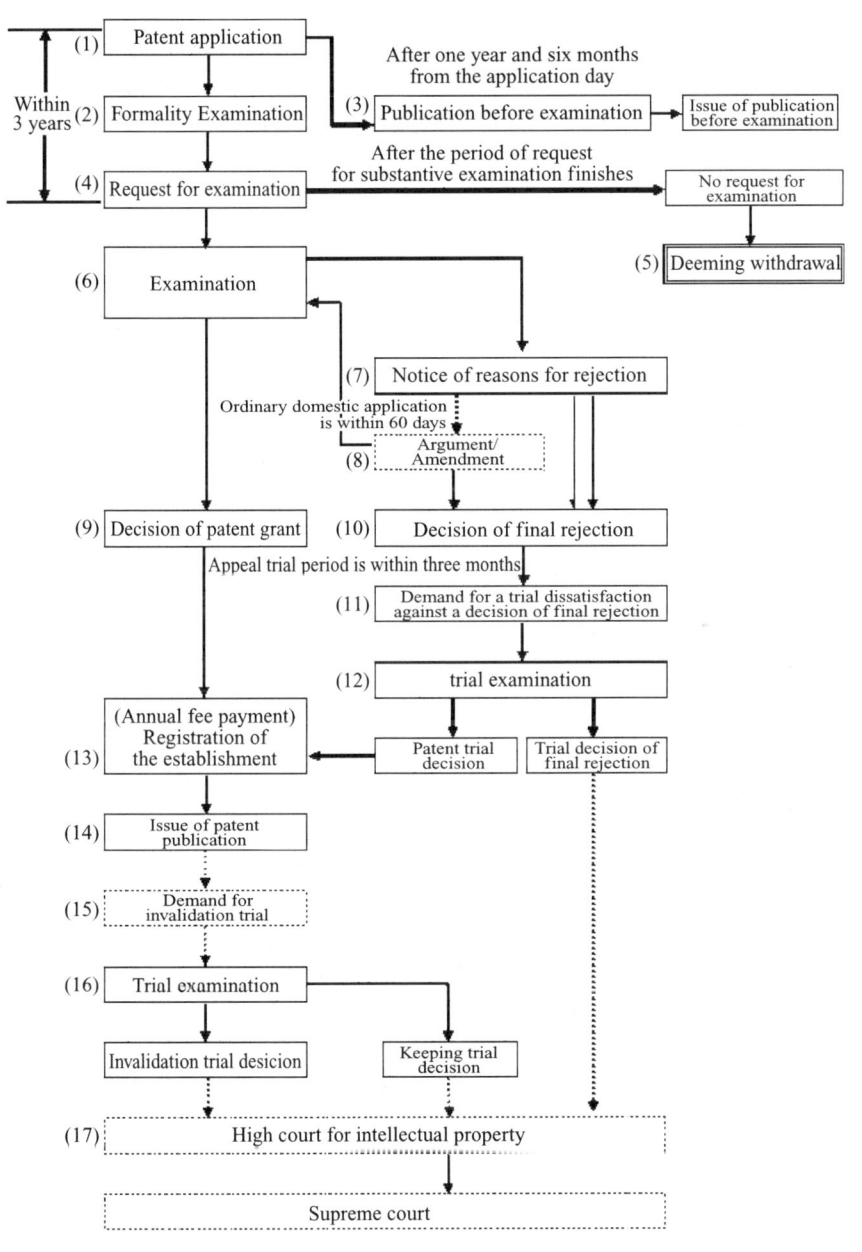

Figure 1 Proceedings for getting patent right

devised by a person who has knowledge of the field, v) be applied before any other person, vi) not offend against public order and decency, and vii) be described according to the rules for writing specifications.

(7) Notice of reasons for rejection
When the examiner finds causes for rejection, its notice is sent to the applicant.
(8) Argument and amendment
The applicant can submit an argument paper describing the differences between the invention applied but rejected and existing technology and/or submit an amendment correcting the scope

of patent application and specifications, etc.

(9) Decision of a patent grant

According to the rule, the invention applied is granted for patent when the examiner finds no cause for rejection. Even when causes for rejection are found, the invention is granted for patent if the causes are resolved by argument or amendment.

(10) Decision of final rejection

When the causes for rejection are not dissolved even by argument or amendment, the invention is decided to be rejected.

(11) Dissatisfied demand for appeal trial

When the applicant has an objection about the causes for rejection mentioned by the examiner, the applicant can request for reexamination (appeal trial).

(12) Appeal trial

In an appeal trial, inventions are examined by a counsel of three or five examiners. Decisions made by the counsel are called "trial decision". When the counsel judges that the causes for rejection have been dissolved, the invention is granted for patent, otherwise the application is rejected.

(13) Registration of the establishment (payment of annual fee)

The application that is granted for patent obtains the patent right only after the applicant pays the annual fee. After payment, the patent registration number is sent to the applicant together with letters patent.

(14) Issue of patent publication

After the registration of the establishment and payment of the annual fee, the contents of the patent right are published in patent publications.

(15) Demand for invalidation trial

When ground for invalidation is found on the patent right even after the registration of establishment, any person can demand for invalidation trial.

(16) Trial examination

Trial examinations responding to a demand for invalidation trial are carried out by a counsel of three or five examiners. When the counsel finds no ground for invalidation to the patent, the patent is decided to be kept. On the other hand, when the counsel judges there is a ground for invalidation, the patent is judged be invalid.

(17) High Court for Intellectual Property

Applicants who want to state dissatisfaction about the trial decisions of rejection or the decision of final rejection or concerned parties who want to state dissatisfaction about the decision of the trial for invalidation of a patent can bring a suit to the High Court for Intellectual Property.

1.5 Trends of nanotechnology patents

A word "nanotechnology" is widely accepted by the public today and frequently appears in newspapers, etc. However, understanding the precise trends of nanotechnology patents it is surprisingly difficult. Nanotechnologies are diverse, are used in a number of different fields, which are not fixed but change rapidly, are relatively new, and undergo rapid development.

The word "nanotechnology" spread when Dr. Kim Eric Drexler wrote a book entitled "Engines of Creation—The Coming Era of Nanotechnology" in 1986. However, Drexler used the term to denote molecular machines, which will be realized the future, and atom control by molecular machines. Today, nanotechnologies include supermolecules, molecular polymers, and minute surface treatment processing etc. These technologies already exited before the word "nanotechnology" was accepted by the public. Therefore, investigations on the trends of nanotechnology-related patents would give different results depending on the definition of the word.

As relatively reliable information, an investigation published by the Technology Trend Group of the Plan Security Research Division, General Affairs Department of the Japanese Patent Office, is described here. The Patent Office has investigated the following fields as fields related to nanotechnology and nano-materials: nano-substances and nano-materials in the electronics, magnetism, and optical applications, etc., nano-substances and nano-materials in structural material application etc., nano-information devices, nano medicine, nano biotechnology, energy and environmental application, surface and interface, measurement technologies and standards, processing and synthesis, basic properties of matter, mathematics, theory and simulation, materials for constructing safe spaces, etc.

The Patent Office carried out extraction and retrieval using the International Patent Classification (IPC) system and original keywords for positioning the aforementioned fields related to nanotechnology and nano-materials. Figure 2–11 are the results of the investigation published by the Patent Office.

Figure 2 shows the monthly number of patent registrations related to nanotechnology, which was about 500–700, in 2005. The total number of registrations was 7135 (See figure 3).

Figure 4 shows the number of patent applications in the fields of nanotechnology and nano-materials, which were published prior to patent examination, in Japan in 2005. There were 2500–4000 applications a month, although the number varied by month. The number of publications prior to patent examination in 2005 increased by about 5 % from that in 2004, which was 38 472 in total (See figure 5).

Figure 6 shows the number of registrations of nanotechnology-and nano-materials-related patents in Japan, the United States (US) and Europe from September 2004 to August 2005. The number of registrations was larger in the US than in Japan and Europe. The number of registrations from September 2004 to August 2005 is shown in figure 7. In Europe, the number of registrations increased by 6 %. It increased by 5 % in Japan, and decreased by 6 % in the US.

Figure 8 shows the number of publications prior to patent examination of inventions related to nanotechnology and nano-materials in Japan, the US and Europe, from September 2004

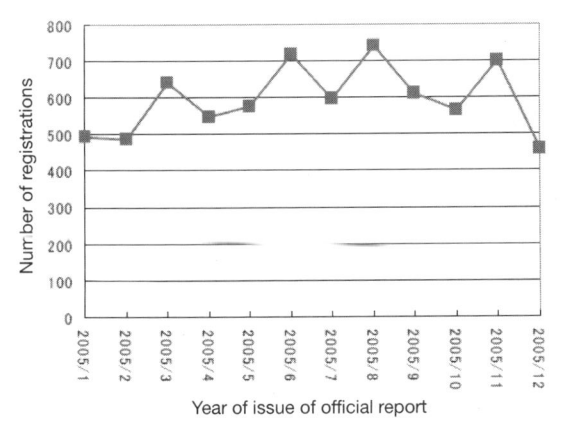

Figure 2 Changes in the number of patent registrations in nanotechnology and nano-materials related fields

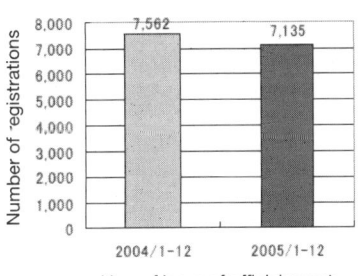

Figure 3 Number of patent registration in one year in nanotechnology and nano-materials related fields

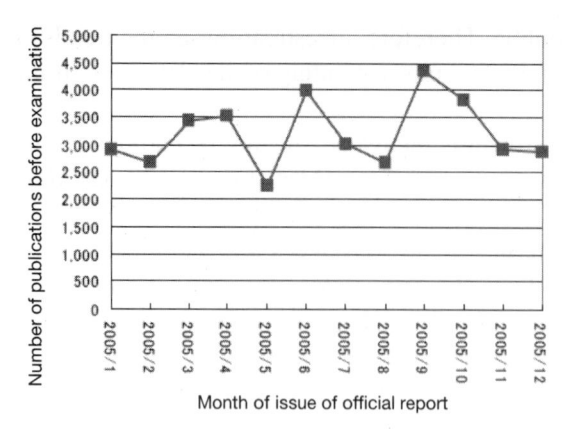

Figure 4 Changes in the number of publications before patent examination in nanotechnology and nano-materials related fields

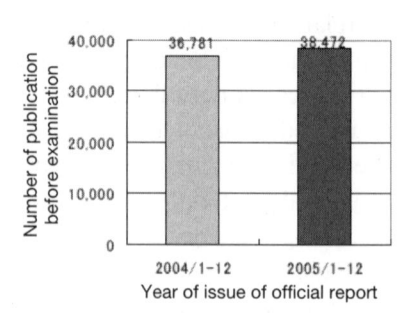

Figure 5 Number of publications before patent examination in one year in nanotechnology and nano-materials related fields

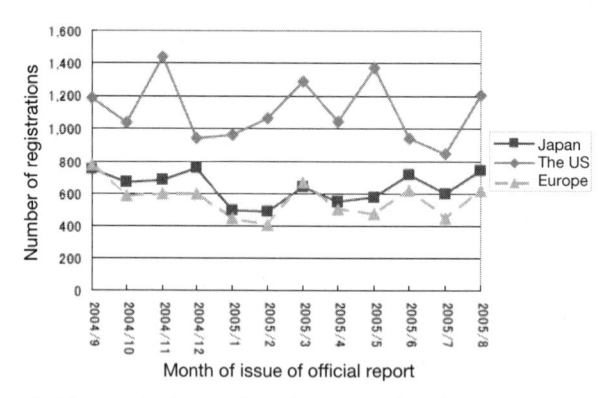

Figure 6 Changes in the number of patent registrations in nanotechnology and nano-materials related fields (Japan, the US, Europe)

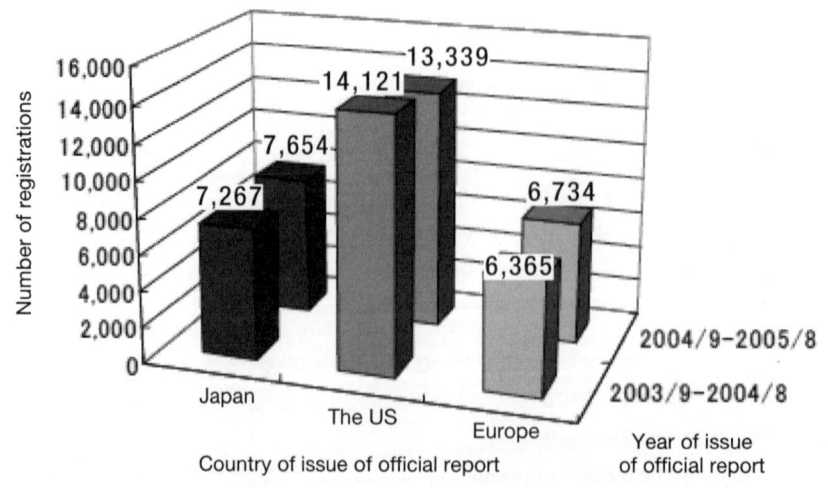

Figure 7 Number of registrations in one year in nanotechnology and nano-materials related fields (Japan, the US, Europe)

to August 2005. We find that the number is comparatively large in Japan. Figure 9 shows the number of publications before examination from September 2004 to August 2005. The number decreased by 1–2 % in Europe from the number in the previous year, and increased by about 10 % in the US and by about 8 % in Japan.

Figure 10 shows the number of registered patents on nanotechnology and nano-materials in Japan, the US and Europe from January 2001 to January 2007. Figure 11 shows the number of publications before examination of nanotechnology and nano-materials related fields in Japan,

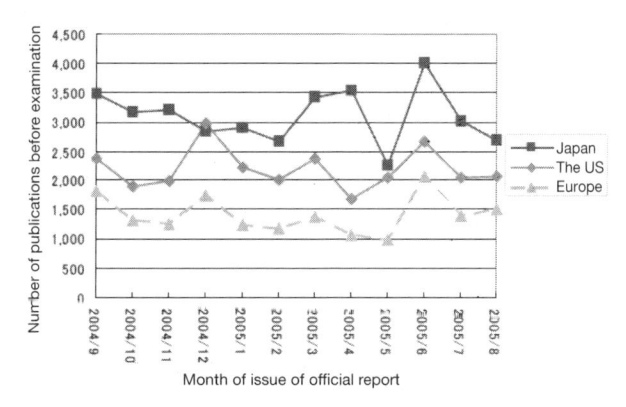

Figure 8 Changes in the number of publications before examination in nanotechnology and nano-materials related fields (Japan, the US, Europe)

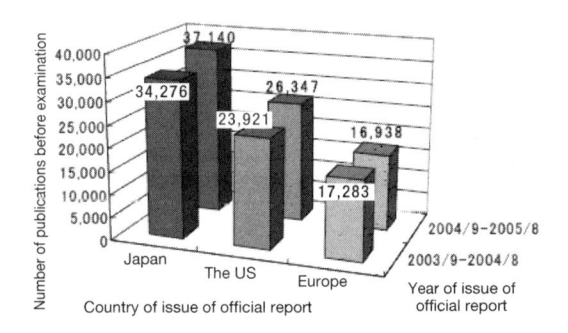

Figure 9 Number of publications before examination in one year in nanotechnology · materials related fields (Japan, the US, Europe)

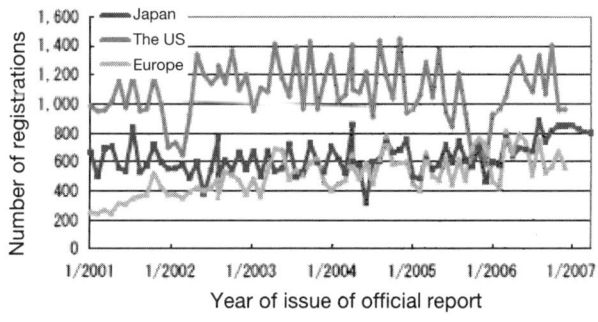

Figure 10 Long-term changes in the number of patent registrations in nanotechnology and nano-materials related fields (Japan, the US, Europe)

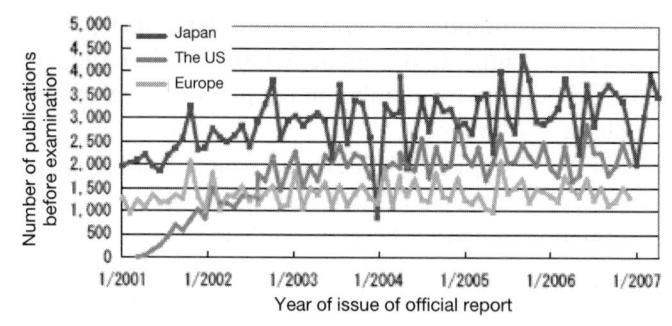

Figure 11 Long-term changes in the number of publications before examination in nanotechnology and nano-materials related fields (Japan, the US, Europe)

the US and Europe from January 2001 to January 2007. From a long span viewpoint, both the number of publications before examination and the number of registrations first increased and then leveled off in all countries.

1. 6 Trends of patents of nanotechnologies in medicine and cosmetics

Most patent applications of the medicine and cosmetic fields are classified under Category A61K of the International Patent Classification (IPC). To understand the overall trends in patent application in the fields of medicine and cosmetics, we searched for applications of IPC Category A61K made in Japan in 1993 to 2006. The number of hits is shown in table 1.

As shown in table 1, 5000–6000 applications of IPC Category A61K have been made and published in Japan every year from 1993 to 2006 although there were several years of exception.

Then, we searched for applications that contained a keyword of "nano" or "minute particle" in their claims from the applications of IPC Category A61K so extracted. "Nano" was used as a keyword because "nano" is widely used in patents related to nanotechnology, such as nano particles, nanosphere, and nanoprocessing, and was judged to be an appropriate keyword. The other keyword, "minute particle" is an essential nanotechnology material in the field of cosmetics. The use of "particle" was estimated to result in too much noise and too many applications that had nothing to do with nanotechnology and was not adequate. Thus, "nano" and "minute particle" were decided to be used to grasp the overall trends of nanotechnologies in the fields of medicine and cosmetics, and noise was not excluded from the result.

As shown in table 1, "nano" appeared in only 14 applications published in 1993. The number jumped to 110 in 2006, showing an eight-fold increase. "Minute particle" appeared in 48 in 1993 and in 100 in 2006, showing an almost double increase. The number of applications that contained either "nano" or "minute particle" in their claims was 62 in 1993 and 205 in 2006, showing a three-fold increase. This number is almost the sum of those that contained "nano" and those that contained "minute particle". Very few applications contained both "nano" and "minute particle" in their claims.

Table 1 shows that the number of applications that contained "nano" and/or "minute particle" in their claims increased year after year although the total number of A61K applications changed very little.

Table 1 Changes in the number of publications before examination of A61K Category

	A61K	"Nano"	"Minute particle"	"Nano" or "Minute particle" ("Nano" and "Minute particle")
1993	5651	14	48	62 (0)
1994	6146	23	70	93 (0)
1995	5804	29	66	92 (3)
1996	5394	22	71	91 (2)
1997	4832	19	79	93 (5)
1998	5228	28	71	98 (1)
1999	5100	20	98	116 (2)
2000	4926	59	97	153 (3)
2001	5097	51	82	133 (0)
2002	5432	68	74	139 (3)
2003	5411	96	73	169 (0)
2004	5964	99	95	191 (3)
2005	5801	86	103	176 (13)
2006	5968	110	100	205 (5)

1.7 Conclusions

This chapter described an intellectual property right system and trends of patents in the nanotechnology and nano-material fields by mainly quoting data published by the Patent Office. Patent has recently attracted attention of the public for large amounts of compensation of patent litigation etc., and is a noticed topic in newspaper etc. For most technologists, the primary interest lies on how to protect their inventions. Before finishing this chapter, I have a few words about points to note while filing for a patent.

In the fields related to nanotechnology, the number of applications has increased. However, in patent disputes, having made a large number of applications is not an effective weapon for winning. Rather, most disputes occur just because of one patent right.

The most important point for obtaining patent rights effectively, in all fields including nanotechnology, is to elaborate the contents of applications. Very few technologists understand this point. Many patent applications are made immediately before congresses, sample works, etc. in a hurry. As a result, applications are prepared in a rather sloppy manner, and many invented technologies are published but fail to obtain patent rights. The tragedy is partially due to patent divisions, but patent application is a cooperative work of technologists and patent divisions, and both are responsible.

In order to win today's patent competitions; methods for applying for patent should be investigated before starting research and development when put in extreme terms. Knowing whether the topic of one's study is within the scope of someone's patent or not is crucial for developing and implementing the technology. Researches should also be conducted by always being aware of the differences between one's and the others' works. It is also important to check what part of the newly developed technology produces results different from existing technologies by conducting experiments before making patent applications.

These works should be performed along with researches. At least the same amount of labor, time and energy as those used for the research should be spent for preparing an application. Information on preceding patents should be collected and experiments should be repetitively conducted to find the characteristics of the invented technology, check the differences in effects with the other patented technologies, and prove that the invented technology is advantageous to the others.

Today, Japan is fighting a technical war of attrition for obtaining international patents. (By the way, it was Erich von Falkenhayn, the Chief of the German Army during World War I, who first discovered the concept of a war of attrition. Falkenhayn discovered that not only temporary concentration of force but also the endurance of force supply would be a key in modern wars.) For Japan to keep its major source of wealth, which is called technologies, technologists involved in nanotechnologies as well as other fields should not underrate patent application. Patent applications of highly elaborated contents are awaited.

References

1. Edited by Patent Office, Industrial Property Right Capitulars, 56 editions [1st book] p. 22, (CJP) Japan Institute of Invention and Innovation (2002)